T0258972

Numerical
Methods
in Photonics

OPTICAL SCIENCES AND APPLICATIONS OF LIGHT
Series Editor
James C. Wyant

*Please visit our website **www.crcpress.com** for a full list of titles*

Numerical
Methods
in Photonics

Andrei V. Lavrinenko
Technical University of Denmark, Kongens Lyngby

Jesper Lægsgaard
Technical University of Denmark, Kongens Lyngby

Niels Gregersen
Technical University of Denmark, Kongens Lyngby

Frank Schmidt
Zuse Institute, Berlin, Germany

Thomas Søndergaard
Aalborg University, Aalborg, Denmark

CRC Press
Taylor & Francis Group
Boca Raton London New York

CRC Press is an imprint of the
Taylor & Francis Group, an **informa** business

MATLAB® and Simulink® are trademarks of the MathWorks, Inc. and are used with permission. The Math-Works does not warrant the accuracy of the text or exercises in this book. This book's use or discussion of MATLAB® and Simulink® software or related products does not constitute endorsement or sponsorship by the MathWorks of a particular pedagogical approach or particular use of the MATLAB® and Simulink® software

CRC Press
Taylor & Francis Group
6000 Broken Sound Parkway NW, Suite 300
Boca Raton, FL 33487-2742

First issued in paperback 2017

© 2015 by Taylor & Francis Group, LLC
CRC Press is an imprint of Taylor & Francis Group, an Informa business

No claim to original U.S. Government works

ISBN-13: 978-1-4665-6388-9 (hbk)
ISBN-13: 978-1-138-07469-9 (pbk)

This book contains information obtained from authentic and highly regarded sources. Reasonable efforts have been made to publish reliable data and information, but the author and publisher cannot assume responsibility for the validity of all materials or the consequences of their use. The authors and publishers have attempted to trace the copyright holders of all material reproduced in this publication and apologize to copyright holders if permission to publish in this form has not been obtained. If any copyright material has not been acknowledged please write and let us know so we may rectify in any future reprint.

Except as permitted under U.S. Copyright Law, no part of this book may be reprinted, reproduced, transmitted, or utilized in any form by any electronic, mechanical, or other means, now known or hereafter invented, including photocopying, microfilming, and recording, or in any information storage or retrieval system, without written permission from the publishers.

For permission to photocopy or use material electronically from this work, please access www.copyright.com (http://www.copyright.com/) or contact the Copyright Clearance Center, Inc. (CCC), 222 Rosewood Drive, Danvers, MA 01923, 978-750-8400. CCC is a not-for-profit organization that provides licenses and registration for a variety of users. For organizations that have been granted a photocopy license by the CCC, a separate system of payment has been arranged.

Trademark Notice: Product or corporate names may be trademarks or registered trademarks, and are used only for identification and explanation without intent to infringe.

Library of Congress Cataloging-in-Publication Data

Lavrinenko, Andrei V.
   Numerical methods in photonics / Andrei V. Lavrinenko, Jesper Lægsgaard, Niels Gregersen, Frank Schmidt, and Thomas Søndergaard.
      pages cm. -- (Optical sciences and applications of light)
   Includes bibliographical references and index.
   ISBN 978-1-4665-6388-9 (hardback)
      1. Photonics--Mathematics. 2. Optoelectronic devices--Mathematical models. 3. Light--Mathematical models. 4. Numerical analysis. I. Title.

TK8304.L335 2014
621.36'501518--dc23
                                                                              2014009744

Visit the Taylor & Francis Web site at
http://www.taylorandfrancis.com

and the CRC Press Web site at
http://www.crcpress.com

Dedication

To our families and parents

Contents

Series Preface

Optical and photonics are enabling technologies in many fields of science and engineering. The purpose of the Optical Sciences and Applications of Light series is to present the state of the art of the basic science of optics, applied optics, and optical engineering, as well as the applications of optics and photonics in a variety of fields, including health care and life sciences, lighting, energy, manufacturing, information technology, telecommunications, sensors, metrology, defence, and education. This new and exciting material will be presented at a level that makes it useful to the practising scientist and engineer working in a variety of fields.

The books in this series cover topics that are a part of the rapid expansion of optics and photonics in various fields all over the world. The technologies discussed impact numerous real-world applications, including new displays in smartphones, computers, and televisions; new imaging systems in cameras; biomedical imaging for disease diagnosis and treatment; and adaptive optics for space systems for defence and scientific exploration. Other applications include optical technology for providing clean and renewable energy, optical sensors for more accurate weather prediction, solutions for more cost-effective manufacturing, and ultra-high-capacity optical fibre communications technologies that will enable the future growth of the Internet. The complete list of areas optics and photonics are involved in is very long and always expanding.

Preface

Simulation and modelling using numerical methods running on computers is today one of the key approaches in scientific work within practically all fields of research. In particular, in the field of photonics, a wide range of numerical methods are needed for studying both fundamental optics and applied branches such as the design, development, and optimization of photonic components. The topic of numerical modelling is thus becoming an increasing priority in photonics education.

This book evolved from research activities and teaching at three different research institutions. Part of the material (the finite-difference time-domain method, the finite-difference frequency-domain method, and the method for nonlinear propagation) was originally developed for a 5 ECTS course entitled 'Numerical Methods in Photonics', which has already been running for 7 years for graduate and undergraduate students at the Department of Photonics Engineering, the Technical University of Denmark (DTU). When it was decided to write a textbook on numerical methods, it was natural to include the popular and powerful modal method also employed at DTU, the finite-element method, which has been used in research and teaching at Zuse Institute Berlin, Germany, for more than 10 years, and the Green's function integral equation methods that have been applied for 10 years in both research and teaching at Aalborg University, Denmark. The book thus covers a wide range of important numerical tools employed in photonics modelling nowadays.

The aim of this textbook is to provide an introduction to a range of numerical methods that are used in photonics for the modelling of propagation and scattering of light in micro- and nano-structures. The propagation and interaction of light at optical wavelengths with such structures is governed by Maxwell's equations, and thus the theoretical methods presented in this book are techniques that either directly solve Maxwell's equations or solve equations deduced from those. While the material in this book can be employed to compute optical modes used, for example, in the quantization of the electromagnetic field or in a semi-classical model of lasing, we will focus only on solving Maxwell's equations. A dedicated treatment of quantum phenomena and branches of photonics which requires going beyond Maxwell's equations and using, for example, quantum mechanics is considered beyond the scope of the book.

This book also aims to provide an introduction to a range of important numerical methods to undergraduate and graduate students with an interest in the photonics and optics of micro- and nano-structures. It assumes that the reader has prior knowledge of optics and Maxwell's equations. A basic knowledge of linear algebra and of programming will also be an advantage. We hope that the book will be useful for engineers and specialists in electromagnetism and optics, as well as for those who prefer to model physical processes using their own tools.

Chapter 2 describes all necessary mathematical operations and notations and provides a basic introduction to optics and Maxwell's equations. Each of the six chapters related to the six numerical methods mentioned earlier have been written under the assumption that the reader is familiar with the material presented in Chapter 2; being familiar with Chapter 2, the following chapters can be read independently of one another. We have also added interlinks between chapters when the same material is referred to or studied in different chapters. For example, implementation of perfectly matched layers for truncation of the numerical domain is discussed for the finite-difference time domain, finite-difference frequency domain, and the Fourier modal methods. Necessary references are provided at the end of each chapter.

Each of the specialized chapters is appended with a set of exercises. We expect that students will be using MATLAB® as a universal platform to write their scripts and execute required computations. All the exercises are strongly recommended for learning and getting a good grasp of the methods. Solutions are available for adopting professors. Furthermore, supplements for both students and teachers will be made available through the publisher's website at http://www.crcpress.com/product/isbn/ 9781466563889.

This book provides a solid foundation in the numerical solutions of Maxwell's equations and gives an introduction to a wide range of methods used for different purposes in photonics. Sometimes, even choosing the optimal numerical method for a particular problem could be a difficult task, and we believe that this book will help make clear decisions with distinct guidelines. The size and format of this textbook does not allow us to make an excursus for all numerical methods. Readers seeking in-depth knowledge on a particular method can refer to the list of references at the end of the chapters, where we point to the relevant literature.

MATLAB® is a registered trademark of The MathWorks, Inc. For product information, please contact:

The MathWorks, Inc.
3 Apple Hill Drive
Natick, MA 01760-2098 USA
Tel: 508-647-7000
Fax: 508-647-7001
E-mail: info@mathworks.com
Web: www.mathworks.com

Authors

Niels Gregersen
Department of Photonics Engineering
Technical University of Denmark
Kongens Lyngby, Denmark

Jesper Lægsgaard
Department of Photonics Engineering
Technical University of Denmark
Kongens Lyngby, Denmark

Andrei V. Lavrinenko
Department of Photonics Engineering
Technical University of Denmark
Kongens Lyngby, Denmark

Frank Schmidt
Numerical Analysis and Modelling
Zuse Institute Berlin
Berlin, Germany

Thomas Søndergaard
Department of Physics and Nanotechnology
Aalborg University
Aalborg, Denmark

Acronyms

1D	One dimensional
2D	Two dimensional
3D	Three dimensional
a-FMM	Aperiodic Fourier modal method
ABC	Absorbing boundary conditions
ADE	Auxiliary differential equation
BC	Boundary conditions
CW	Continuous wave
DBR	Distributed Bragg reflector
DC	Duty cycle
DDA	Discrete dipole approximation
DFT	Discrete Fourier transform
EET	Eigenmode expansion technique
FBG	Fibre Bragg grating
FDTD	Finite-difference time domain
FFT	Fast Fourier transform
FEM	Finite element method
FMM	Fourier modal method
FWHM	Full width at half maximum
GFAIEM	Green's function area integral equation method
GFIEM	Green's function integral equation method
GFSIEM	Green's function surface integral equation method
GFVIEM	Green's function volume integral equation method
(G)NLS	(Generalized) nonlinear Schrödinger equation
LHS	Left-hand side
MM	Modal method
ODE	Ordinary differential equation
PC	Perfect conductor
PEC	Perfect electric conductor
PMC	Perfect magnetic conductor
PML	Perfectly matched layer
RCWA	Rigorously coupled waveguide analysis
RHS	Right-hand side
RK4	Fourth-order Runge–Kutta algorithm
SA	Semi-analytical
SPM	Self-phase modulation
SPP	Surface plasmon polariton
TE	Transverse electric
TEM	Transverse electromagnetic wave
TM	Transverse magnetic
VCSEL	Vertical-cavity surface-emitting laser

1 Introduction

Photonics as a field embraces phenomena that happen on completely different spatial and temporal scales: from nanometres to kilometres and from subfemtoseconds to micro- and milliseconds. Often, such different scales clash in one modelling case. In, for example optical communication systems, the propagation of light through optical fibres with a length from a few metres to hundreds of kilometres is of interest. On the other hand, the size of the electromagnetic modes of such waveguides transverse to the fibre or waveguide axis should be measured in micrometres.

Scattering from micro- and nanostructures is also of interest. Here, near-field distributions of electromagnetic fields can have features in the nanoscale even if, for example visible wavelengths (400–700 nm) are used. Such fine details cannot be seen in the far-field as a consequence of the diffraction limit of light, but they can be seen in the near field using a scanning near-field optical microscope, where, for example a tapered fibre with or without metal coating is inserted into the near optical field and used to pick up light locally. The fibre introduced to pick up light will itself modify the near field. Naturally, theoretical modelling is very important to support and understand such experiments. Many more cases could be mentioned. Nevertheless, the unique property of all these phenomena in a classical description is that they are governed by *Maxwell's equations.*

James Clerk Maxwell presented his well-known set of equations governing electromagnetic fields in 1861. These equations describe the interrelationship between electric and magnetic fields and electric charges and currents. Maxwell's equations are a remarkable set of equations that can explain the behaviour (propagation and scattering) of electromagnetic fields in any kind of a medium, including vacuum, with the help of constitutive equations. The observation of electromagnetic waves was first demonstrated by Heinrich Hertz in 1884.

Maxwell's equations in general have analytical solutions only for very simple cases. When assuming a linear material response, it is possible to some extent to find analytical solutions that describe propagation of light in infinitely long and perfectly straight waveguides with a perfectly circular cross section. However, if the cross section is instead rectangular, or if the waveguide is sharply bent, then this is already no longer possible, and numerical solutions are required. Similarly, scattering of light from a perfect sphere can also be obtained in terms of semi-analytical expressions. However, for other slightly less symmetric structures, then again numerical solutions are required. Such simple structures and related analytical solutions are very useful as test cases and for giving a basic physical insight. However, in modern photonics and nano-optics, one is very often interested in more complex geometries.

Therefore, numerical methods for solving Maxwell's equations are needed. Unlike general numerical techniques for solving differential equations with full or partial derivatives like the Runge–Kutta methods, special routines are applied to solve, namely Maxwell's equations. Such routines are named *Maxwell's solvers*. Following the increasing growth of photonics, more and more computational procedures appear to address various specific needs, and the variety of available Maxwell's solvers has recently been extended a lot. It has become important not only to know how to write a script in a programming language of choice but also to navigate between a variety of routines commercially available or elaborated by research groups regarding any particular concern in simulations. Thus, the *objective* of this book is to present methods for numerically solving Maxwell's equations or equations derived from them such as the wave or Helmholtz equations. This book covers six numerical methods for solving Maxwell's equations. We provide a short summary of each method in the following.

The *finite-difference time-domain* (*FDTD*) method is an example of a universal approach that can be applied to solve Maxwell's equations directly as they are written in textbooks in the time domain without any further approximation or specification of the processes, like plane-wave solution and harmonic time dependence. Maxwell's equations are solved on a grid, which is inserted in the region of interest. The most famous and widespread is the Yee grid (Yee mesh), which arranges six components of electric and magnetic fields in a special order finishing by having only tangential field components on the unit cell interfaces. All derivatives in Maxwell's equations are approximated numerically by finite-difference schemes reducing differential equations to algebraic ones. This leads to an explicit algorithm for updating fields in time that has increased by a small time step. Thus, by applying the updating procedure, we are able to follow all peculiarities of the evolution of fields, which happen during propagation, scattering, absorption, and any other phenomena observed with classical optical waves. The FDTD method can be used to mimic real optical experiments and now very often serves as a reference for checking and validating results of characterization of different optical devices. Moreover, being implemented in the time domain, the FDTD method can be linked with physical phenomena other than the classical electromagnetism nature: atom or molecule kinetics, heat dissipation, diffusion, and so on. A drawback of the FDTD method is the necessity of updating the field components in all points in the numerical space including regions where nothing of interest is studied. However, this drawback is compensated by the linear scaling law of the method with the sizes, very simple meshing of the whole domain, and possibility of effective parallelization of the computation routine.

The *finite-difference frequency-domain method* can be used efficiently in the frequency domain to calculate eigenmodes of straight waveguides or micro- and nanocavities. The main advantage of the method is its generality and in particular its simplicity, which allows for very easy coding of fairly powerful algorithms in high-level languages such as MATLAB®. In fact, such a code is distributed with this book. Implementation of periodic and absorbing boundary conditions is straightforward, although not exposed here. The main disadvantage of the method is that it is less efficient than more sophisticated approaches such as the finite element method (FEM).

The *nonlinear split-step Fourier method* can be used to effectively model nonlinear propagation of short pulses in waveguides using a frequency-domain modal expansion to derive an effective 1+1D propagation equation. In its simplest form, this approach leads to the so-called *nonlinear Schrödinger* (NLS) equation for single-mode problems with second-order dispersion and a constant-mode profile, which lends itself to both analytic and numerical analyses and has been used to describe a wide range of important nonlinear phenomena, especially in fibre-optic waveguides. However, many generalizations of this equation are possible, accounting for effects of higher-order dispersion, delayed nonlinear response, mode profile variations, vectorial effects, and multimode behaviour. The only generalization which seriously affects numerical complexity is the inclusion of several waveguide modes and in some cases mode profile dispersion. In this book, it is shown how a very general propagation equation may be derived, specialized to the case of a single-mode waveguide, and eventually reduced to the NLS equation through various approximations.

The *modal method (MM)* is used in the frequency domain, where a harmonic time dependence at a single frequency is assumed. In the MM, the geometry under study is sliced into layers uniform along a propagation axis, usually the z-axis. Eigenmodes are computed in each layer assuming uniformity along the propagation axis, and the field is expanded on the eigenmodes of each layer. The z dependence of the eigenmodes is described analytically using propagation constants, and the thickness of a layer does not influence the computation time. The scattering at the interfaces between different layers is handled using a mode-matching technique leading to reflection and transmission matrices characterizing each interface. Advantages of the MM include direct access to the propagation constants of the modes as well as intermodal scattering coefficients, allowing for insight into, for example the reflection and transmission of specific optical modes of interest. Furthermore, the method naturally takes advantage not only of uniformity but also periodicity along the propagation axis using the Bloch mode formalism. The method is thus a natural choice for the analysis of gratings and photonic crystal geometries. The MM naturally supports closed and periodic boundary conditions in the plane lateral to the propagation axis; however, to mimic open geometries, the implementation of absorbing boundary conditions is required. Also, light sources such as dipole emitters or current distributions can be implemented in a straightforward way.

Green's function integral equation methods (GFIEM) studied in this book solve Maxwell's equations in the frequency domain. Instead of solving Maxwell's equations on differential form, then equivalent integral equations are constructed, where the total field at any position is directly related to an overlap integral between a Green's function and the field inside or on the surface of a scattering object in a reference geometry. Green's function represents the field emitted by a point source in the reference geometry. The field inside or on the surface of the scattering object is obtained by solving self-consistent equations. The radiation coming from one part of a scatterer is driven by the total field, which includes not only the externally applied field but also contributions to the field due to emission from other parts of the scatterer. One of the strengths of these methods is that the numerical problem can be reduced to either the surface or the inside of the scatterer. Boundary conditions outside the scatterer are automatically taken care of via the choice of Green's function, and there is no requirement for calculating the field in a region outside the scatterer nor to introduce perfectly matched or absorbing layers. Furthermore, if a scatterer placed on a layered structure is considered instead of the same scatterer in free space, then it is sufficient to replace the free-space Green's function with the one for the layered structure, and then repeat the calculation. The size of the numerical problem is not increased, but additional work is needed to obtain Green's function.

The *FEM* studied in this book solves Maxwell's equations in the frequency domain. The realization of the FEM involves two basic steps. First, Maxwell's equations are converted into a so-called variational form that involves integral expressions on the computational domain. Second, the solution space, which should contain a reasonable approximation to the exact solution, has to be constructed. This solution space is obtained by subdividing the computational domain into small geometric patches and by providing a number of polynomials on each patch for the approximation of the solution. The patches together with the local polynomials defined on them are called *finite elements*. The most common examples of finite elements are triangles and rectangles in 2D and tetrahedrons and cuboids in 3D together with constant, linear, quadratic, and cubic polynomials. These locally defined polynomial spaces have to be pieced together to ensure tangential continuity of the electric and magnetic field across the boundaries of neighbouring patches. The strengths of the method are the following: first, complex geometrical shapes can be treated without geometrical approximations, for example curved geometries can be well approximated; second, the finite element mesh can easily be adapted to the behaviour of the solution, for example to singularities at corners; and third, high-order approximations are available and ensure fast convergence of the numerical solution to the exact solution.

The main part of this book is organized in the following way: Chapter 2 introduces Maxwell's equations in the time and frequency domains, the wave equations, equations for obtaining guided modes of waveguides, as well as the notation of the book. Specialized chapters for each of the aforementioned numerical methods then follow. The specialized chapters only refer to Chapter 2 and can thus be read independently from each other.

A very large number of numerical methods and variants of those methods exist for solving Maxwell's equations, and it is not possible to consider all methods in one textbook. We have chosen the methods that we believe are predominantly used in photonics. Methods not covered here include, among others, the multiple multipole method, the beam propagation method, the finite integral and finite volume methods, the method of lines, and the plane-wave expansion method. However, we believe that familiarity with the material in this book will provide a good background for any available Maxwell's solver.

2 Maxwell's Equations

2.1 NOTATION

This book employs the following notations:

Object	Notation	Example
Vector in 3D space	Bold	$\mathbf{r}, \mathbf{E}$
Unit vector	Bold with hat	$\hat{\mathbf{x}}, \hat{\mathbf{y}}, \hat{\mathbf{z}}$
Vector of expansion coefficients	Plain, single bar	$\bar{c}$
Matrix	Plain, double bar	$\bar{\bar{O}}$
Dyadic tensor	Bold	$\mathbf{G}$
Scalar (dot, inner) product	Single dot	$\hat{\mathbf{x}} \cdot \mathbf{E}$
Vector (cross) product	Single cross	$\hat{\mathbf{x}} \times \mathbf{E}$
Tensor (dyadic, outer) product or dyad	No dot	$\mathbf{RR}$

Furthermore, we introduce the shortened designation of partial differential operators:

$$\partial_x \equiv \frac{\partial}{\partial x}, \quad \partial_y \equiv \frac{\partial}{\partial y}, \quad \partial_z \equiv \frac{\partial}{\partial z}, \quad \partial_t \equiv \frac{\partial}{\partial t}. \tag{2.1}$$

2.2 MAXWELL'S EQUATIONS

The basic differential form of Maxwell's equations reads [1,2]

$$\nabla \times \mathbf{E}(\mathbf{r}, t) = -\partial_t \mathbf{B}(\mathbf{r}, t), \tag{2.2}$$

$$\nabla \times \mathbf{H}(\mathbf{r}, t) = \partial_t \mathbf{D}(\mathbf{r}, t) + \mathbf{J}(\mathbf{r}, t), \tag{2.3}$$

$$\nabla \cdot \mathbf{D}(\mathbf{r}, t) = \rho(\mathbf{r}, t), \tag{2.4}$$

$$\nabla \cdot \mathbf{B}(\mathbf{r}, t) = 0, \tag{2.5}$$

where $\mathbf{E}, \mathbf{D}, \mathbf{H}$, and $\mathbf{B}$ are the electric field, the electric displacement, the magnetic field, and the magnetic induction, respectively. The *nabla* operator (known also as *del* operator) $\nabla \equiv \hat{\mathbf{x}}\partial_x + \hat{\mathbf{y}}\partial_y + \hat{\mathbf{z}}\partial_z$ is the vector-differential operator, and $\hat{\mathbf{x}}, \hat{\mathbf{y}}, \hat{\mathbf{z}}$ is a triad of Cartesian orthogonal unit vectors. The functions $\rho(\mathbf{r}, t)$ and $\mathbf{J}(\mathbf{r}, t)$ are free electric charge and current densities.

5

2.3 MATERIAL EQUATIONS

Electric and magnetic field vectors are connected through the constitutive (material) equations:

$$\mathbf{D}(\mathbf{r}, t) = \varepsilon_0 \mathbf{E}(\mathbf{r}, t) + \mathbf{P}(\mathbf{r}, t), \tag{2.6}$$

$$\mathbf{B}(\mathbf{r}, t) = \mu_0 \mathbf{H}(\mathbf{r}, t) + \mathbf{M}(\mathbf{r}, t). \tag{2.7}$$

In these expressions, $\varepsilon_0 \approx 8.854 \times 10^{-12}$ F/m and $\mu_0 \approx 4\pi \times 10^{-7}$ H/m are constants with SI system units. These constants called *vacuum permittivity* and *permeability* are connected through the *speed of light in vacuum* c_0:

$$c_0 = \frac{1}{\sqrt{\varepsilon_0 \mu_0}} = 2.99792458 \times 10^8 \text{ m/s.} \tag{2.8}$$

Their ratio defines the so-called *vacuum impedance*:

$$\eta_0 = \sqrt{\frac{\mu_0}{\varepsilon_0}} = 376.73 \ \Omega. \tag{2.9}$$

The polarization $\mathbf{P}(\mathbf{r}, t)$ and magnetization $\mathbf{M}(\mathbf{r}, t)$ vector functions define the total electric and magnetic dipole moments in a unit volume of material. In vacuum, they are nullified. In a *linear* material, we can write linear-response expressions for these quantities:

$$\mathbf{D}(\mathbf{r}, t) = \varepsilon_0 \mathbf{E}(\mathbf{r}, t) + \varepsilon_0 \int_0^\infty \chi_e(\mathbf{r}, \tau) \mathbf{E}(\mathbf{r}, t - \tau) d\tau, \tag{2.10}$$

$$\mathbf{B}(\mathbf{r}, t) = \mu_0 \mathbf{H}(\mathbf{r}, t) + \mu_0 \int_0^\infty \chi_m(\mathbf{r}, \tau) \mathbf{H}(\mathbf{r}, t - \tau) d\tau, \tag{2.11}$$

where χ_e and χ_m are electric and magnetic *susceptibilities*, respectively. The $\mathbf{D}$ and $\mathbf{B}$ fields depend only on the $\mathbf{E}$ and $\mathbf{H}$ fields at earlier times due to the requirement of causality. In spite of the time-dependent response functions, the constitutive relations in Equations (2.10) and (2.11) have time-translational invariance because χ_e and χ_m do not depend on the time variable t, but only on the integration variable τ. If the properties of the material under consideration are modulated externally, as, for example in an *acousto-optic modulator*, it may be necessary to include an explicit t-dependence into χ_e and/or χ_m.

In the special case of a *nondispersive* medium, the response functions are δ-functions in τ, and the constitutive equations (2.6) and (2.7) are reduced to the form

$$\mathbf{D}(\mathbf{r}, t) = \varepsilon_0 \varepsilon(\mathbf{r}) \mathbf{E}(\mathbf{r}, t), \tag{2.12}$$

$$\mathbf{B}(\mathbf{r}, t) = \mu_0 \mu(\mathbf{r}) \mathbf{H}(\mathbf{r}, t). \tag{2.13}$$

Here, $\varepsilon(\mathbf{r})$ and $\mu(\mathbf{r})$ are dimensionless relative electric *permittivity* (often called *dielectric function*) and magnetic *permeability*, which are typically given in tables of material constants or through some analytical models; see Section 3.5.

Note: ε and μ in the most general linear case are 3×3 matrices. This is important when describing materials with an anisotropic crystal structure. In *isotropic* materials, on the other hand, ε and μ will be scalar functions. Examples of such materials are amorphous materials or high-symmetry (e.g. cubic) crystals.

From the relative permittivity and permeability, one can define the *refractive index n* as

$$n = \sqrt{\mu\varepsilon}, \tag{2.14}$$

which is often also used to characterize materials.

2.4 FREQUENCY DOMAIN

The general form of Maxwell's equations (2.2) through (2.5), containing explicit time derivatives, is the form used in *time-domain methods* such as the FDTD technique to be discussed in Chapter 3. However, in many cases, it can be advantageous to consider the Fourier transform of the time-domain fields:

$$\mathbf{E}(\mathbf{r},t) = \frac{1}{2\pi} \int_{-\infty}^{\infty} \mathbf{E}(\mathbf{r},\omega) e^{\pm i\omega t} d\omega; \quad \mathbf{E}(\mathbf{r},\omega) = \int_{-\infty}^{\infty} \mathbf{E}(\mathbf{r},t) e^{\mp i\omega t} dt, \tag{2.15}$$

$$\mathbf{H}(\mathbf{r},t) = \frac{1}{2\pi} \int_{-\infty}^{\infty} \mathbf{H}(\mathbf{r},\omega) e^{\pm i\omega t} d\omega; \quad \mathbf{H}(\mathbf{r},\omega) = \int_{-\infty}^{\infty} \mathbf{H}(\mathbf{r},t) e^{\mp i\omega t} dt. \tag{2.16}$$

Since the frequency integrations run from $-\infty$ to ∞, the choice of sign in the exponent is purely a matter of convention. We will specify the particular choice in the following material, whenever it leads to some changes in expressions.

We further introduce the Fourier transforms of $\chi_{e,m}$ as

$$\chi_{e,m}(\mathbf{r},\omega) = \int_{-\infty}^{\infty} \chi_{e,m}(\mathbf{r},t) e^{\mp i\omega t} dt \tag{2.17}$$

with the understanding that $\chi_{e,m}(\mathbf{r},t) = 0$ for $t < 0$. The constitutive relation for $\mathbf{D}$ may then be derived as

$$\mathbf{D}(\mathbf{r},\omega) = \int_{-\infty}^{\infty} \mathbf{D}(\mathbf{r},t) e^{\mp i\omega t} dt = \varepsilon_0 \left[\mathbf{E}(\mathbf{r},\omega) + \chi_e(\mathbf{r},\omega) \mathbf{E}(\mathbf{r},\omega) \right]$$

$$= \varepsilon_0 \varepsilon(\mathbf{r},\omega) \mathbf{E}(\mathbf{r},\omega), \tag{2.18}$$

where in the last step, we made use of the *convolution theorem* of Fourier theory, which states that

$$\int_{-\infty}^{\infty} e^{\pm i\omega t} \int_{-\infty}^{\infty} F(\tau)G(t - \tau)d\tau dt = F(\omega)G(\omega). \tag{2.19}$$

Note: Usually, the dielectric constant ε incorporates the effect of any induced currents following Ohm's law. In the case that we wish to study the radiation generated by a given source current density $\mathbf{J}_s$, the resulting fields will induce currents in the materials, especially in metals. The induced currents $\mathbf{J}_c$ can often be described via linear-response expressions, here being Ohm's law, similar to, for example the relation between $\mathbf{E}$ and $\mathbf{D}$. In the time domain, this reads

$$\mathbf{J}_c(\mathbf{r}, t) = \int_0^{\infty} \sigma(\mathbf{r}, \tau)\mathbf{E}(\mathbf{r}, t - \tau)d\tau, \tag{2.20}$$

which then in the frequency domain becomes

$$\mathbf{J}_c(\mathbf{r}, \omega) = \sigma(\mathbf{r}, \omega)\mathbf{E}(\mathbf{r}, \omega). \tag{2.21}$$

Consider the Fourier transform of Equation 2.3 with the previous expression for the induced currents inserted. This results in

$$\nabla \times \mathbf{H}(\mathbf{r}, \omega) = \pm i\omega\varepsilon_0 \left(\varepsilon(\mathbf{r}, \omega) \mp i\frac{\sigma(\mathbf{r}, \omega)}{\omega\varepsilon_0} \right) \mathbf{E}(\mathbf{r}, \omega) + \mathbf{J}_s(\mathbf{r}). \tag{2.22}$$

It is usual to incorporate the effect of any induced currents into the dielectric constant such that the effective dielectric constant becomes

$$\varepsilon_{\text{eff}} = \varepsilon(\mathbf{r}, \omega) \mp i\frac{\sigma(\mathbf{r}, \omega)}{\omega\varepsilon_0}. \tag{2.23}$$

It is common to use ε when referring to the effective dielectric constant.

In a similar way, one finds for the **B–H** relation:

$$\mathbf{B}(\mathbf{r}, \omega) = \mu_0\mu(\mathbf{r}, \omega)\mathbf{H}(\mathbf{r}, \omega). \tag{2.24}$$

So the material constants in the frequency domain are

$$\varepsilon(\omega) = 1 + \chi_e(\omega), \tag{2.25}$$

$$\mu(\omega) = 1 + \chi_m(\omega). \tag{2.26}$$

It can be seen that the frequency dependence of ε and μ, and therefore of the light velocity in the material, comes about from the delayed material response. If χ_e and χ_m are delta functions in time, ε and μ become constants in frequency. This explains why such a material is called *nondispersive*, as in the previous section.

Inserting Equations (2.15) and (2.16) into the dynamic Maxwell's equations (2.2) and (2.3) and using the constitutive relations (Equations 2.18 and 2.24), one obtains

$$\nabla \times \mathbf{E}(\mathbf{r}, \omega) = \mp i\omega \mathbf{B}(\mathbf{r}, \omega) = \mp i\omega \mu_0 \mu(\mathbf{r}, \omega) \mathbf{H}(\mathbf{r}, \omega). \tag{2.27}$$

$$\nabla \times \mathbf{H}(\mathbf{r}, \omega) = \pm i\omega \mathbf{D}(\mathbf{r}, \omega) + \mathbf{J}_s(\mathbf{r}) = \pm i\omega \varepsilon_0 \varepsilon(\mathbf{r}, \omega) \mathbf{E}(\mathbf{r}, \omega) + \mathbf{J}_s(\mathbf{r}), \tag{2.28}$$

where $\mathbf{J}_s$ represents any currents that are not incorporated into the dielectric constant. In the following, we will generally skip the subindex s for simplicity.

Note: These derivations depend on the problem having time-translational invariance (no explicit t-dependence of $\chi_{e,m}$) and the absence of nonlinear terms in the constitutive relations. However, since external modulations and nonlinear effects are usually weak, the frequency-domain equations may still be used as a convenient starting point for perturbative treatments of such effects.

2.5 1D AND 2D MAXWELL'S EQUATIONS

All previous equations are written in vectorial form, which implies a 3D (or at least 2D) character. However, we will also need Maxwell's equations in 1D and 2D representations. We will start out by reducing the 3D equations to the simplest 1D case. An example of a physical situation described by such equations could be plane-wave propagation (excitation field) in the z-direction, with material (optical) properties that are dependent on this coordinate only: $\varepsilon(\mathbf{r}) = \varepsilon(z)$, $\mu(\mathbf{r}) = \mu(z)$. All fields in such 1D problem will depend only on z, t:

$$\mathbf{D}(\mathbf{r}, t) = \mathbf{D}(z, t), \quad \mathbf{E}(\mathbf{r}, t) = \mathbf{E}(z, t), \quad \mathbf{B}(\mathbf{r}, t) = \mathbf{B}(z, t), \quad \mathbf{H}(\mathbf{r}, t) = \mathbf{H}(z, t). \tag{2.29}$$

The *curl* of the electric field then becomes

$$\nabla \times \mathbf{E}(\mathbf{r}, t) \equiv \begin{vmatrix} \hat{\mathbf{x}} & \hat{\mathbf{y}} & \hat{\mathbf{z}} \\ \partial_x & \partial_y & \partial_z \\ E_x(z, t) & E_y(z, t) & E_z(z, t) \end{vmatrix}$$

$$= \hat{\mathbf{x}}(\partial_y E_z(z, t) - \partial_z E_y(z, t)) + \hat{\mathbf{y}}(\partial_z E_x(z, t) - \partial_x E_z(z, t))$$

$$+ \hat{\mathbf{z}}(\partial_x E_y(z, t) - \partial_y E_x(z, t))$$

$$= \hat{\mathbf{x}}(-\partial_z E_y(z, t)) + \hat{\mathbf{y}}(\partial_z E_x(z, t)). \tag{2.30}$$

A similar expression holds for $\nabla \times \mathbf{H}(\mathbf{r}, t)$. Therefore, the vectorial *curl* of $\mathbf{E}$ and $\mathbf{H}$ has vanishing z-components. In the case of an isotropic medium, the right-hand side (RHS) of Equation (2.2) then states that $\partial_t H_z(z, t) = 0$ or $H_z(z, t) = const$, and this constant can be nullified without loss of generality. The same applies for $E_z(z, t)$. It stems from the well-known fact that waves can propagate in (transversely) homogeneous media only as transverse electromagnetic (*TEM*) waves.

For a linearly polarized field, we may choose an x-axis lying parallel to the field vector, which is then given by $\mathbf{E}(z, t) = \hat{\mathbf{x}} E_x(z, t)$, that is it has only one component.

This leads, straightforwardly (see Equation 2.30), to the unidirectional representation of the magnetic field $\mathbf{H}(z,t) = \hat{\mathbf{y}}H_y(z,t)$.

Note: It is possible to show rigorously that a complete set of solutions to the 1D Maxwell's equations may be divided into two independent groups, (E_x, H_y) and (E_y, H_x). Both of the groups are totally identical in their properties. So, it is enough to choose only one group, for example the first as we did here. Note also that circularly, or elliptically, polarized fields may be described as linear combinations of linearly polarized fields. As long as we are considering linear propagation equations, the linearly polarized solutions are therefore sufficient to provide a complete description of the possible fields.

Note: The dimension of a numerical scheme is governed by the dimension of possible spatial variations of material parameters, for example at interfaces. So, in spite of the fact that the electric vector is directed along the x-axis, the magnetic vector along the y-axis, and the direction of propagation is along the z-axis, resembling a 3D case, the problem is actually 1D.

Discarding all zero components of the vectors and spatial derivatives other than over the z coordinate, the Maxwell's equations (2.2) and (2.3) are reduced to scalar forms:

$$\hat{\mathbf{y}} : \partial_z E_x = -\mu_0 \mu \partial_t H_y. \tag{2.31}$$

$$\hat{\mathbf{x}} : -\partial_z H_y = \varepsilon_0 \varepsilon \partial_t E_x + J_x. \tag{2.32}$$

To simplify the appearance of these equations, we omit the spatial and temporal dependencies of fields or material properties and assume that the media are isotropic, so ε and μ are scalar functions of one coordinate.

Now, we proceed to the *2D* case, where (1) the structure we are modelling is uniform in one direction, for example in the y-direction, and (2) an exciting field has no variations in the y-direction. Then all y-derivatives in Maxwell's equations can be omitted. When the current density is absent, the remaining curl equations can be grouped into two separate sets:

$$\partial_z E_x - \partial_x E_z = -\mu_0 \mu \partial_t H_y, \quad -\partial_z H_y = \varepsilon_0 \varepsilon \partial_t E_x, \quad \partial_x H_y = \varepsilon_0 \varepsilon \partial_t E_z; \tag{2.33}$$

$$\partial_z H_x - \partial_x H_z = \varepsilon_0 \varepsilon \partial_t E_y, \quad \partial_z E_y = \mu_0 \mu \partial_t H_x, \quad \partial_x E_y = -\mu_0 \mu \partial_t H_z. \tag{2.34}$$

The former set of equations comprises only H_y, E_x, E_z components, whereas the latter only comprises E_y, H_x, H_z. The two sets are distinguished by the presence or absence of field components along the uniform y-direction. The first set is a *transverse magnetic* (TM) wave, which has no electric field component along y but has a magnetic field along y and is therefore perpendicular to the plane of propagation. The second set is a *transverse electric* (TE) wave, having no magnetic field component along the y-direction, but has an electric field along y and is therefore perpendicular to the plane of propagation.

Note: Be aware that the different definition of polarization can also be found in literature. It depends on which structural properties (or their variations) are considered decisive in a particular case.

2.6   WAVE EQUATIONS

The dynamic frequency-domain Maxwell's equations (2.27) and (2.28) can be used to derive second-order *wave equations* involving only one field vector, either electric or magnetic, and the source current density $\mathbf{J}$. For instance, dividing Equation (2.27) by μ and taking the curl on both sides, one obtains

$$\nabla \times \mu(\mathbf{r}, \omega)^{-1}\nabla \times \mathbf{E}(\mathbf{r}) = \mp i\omega\mu_0\nabla \times \mathbf{H}(\mathbf{r}, \omega). \tag{2.35}$$

Equation (2.28) may now be used on the RHS to obtain

$$\nabla \times \mu(\mathbf{r}, \omega)^{-1}\nabla \times \mathbf{E}(\mathbf{r}, \omega) - \left(\frac{\omega}{c_0}\right)^2 \varepsilon(\mathbf{r}, \omega)\mathbf{E}(\mathbf{r}, \omega) = \mp i\omega\mu_0\mathbf{J}(\mathbf{r}, \omega), \tag{2.36}$$

where Equation (2.8) was also used. In a similar manner, a wave equation for the $\mathbf{H}$ field may be derived:

$$\nabla \times \varepsilon(\mathbf{r}, \omega)^{-1}\nabla \times \mathbf{H}(\mathbf{r}, \omega) - \left(\frac{\omega}{c_0}\right)^2 \mu(\mathbf{r}, \omega)\mathbf{H}(\mathbf{r}, \omega) = \nabla \times \left(\frac{\mathbf{J}(\mathbf{r}, \omega)}{\varepsilon(\mathbf{r}, \omega)}\right). \tag{2.37}$$

Throughout this book, we are mostly concerned with regions located away from source currents. For example in a scattering problem, there will be an incident field, but this field is generated by sources located away from the scattering object. Thus, we now consider the equations without sources.

Note that for a nondispersive medium, the corresponding time-domain equations have equally simple expressions:

$$\mu(\mathbf{r})^{-1}\nabla \times \varepsilon(\mathbf{r})^{-1}\nabla \times \mathbf{H}(\mathbf{r}, t) = -\frac{1}{c_0^2}\frac{\partial^2 \mathbf{H}(\mathbf{r}, t)}{\partial t^2}, \tag{2.38}$$

$$\varepsilon(\mathbf{r})^{-1}\nabla \times \mu(\mathbf{r})^{-1}\nabla \times \mathbf{E}(\mathbf{r}, t) = -\frac{1}{c_0^2}\frac{\partial^2 \mathbf{E}(\mathbf{r}, t)}{\partial t^2}. \tag{2.39}$$

A commonly encountered situation is that of nonmagnetic materials, which have $\mu(\mathbf{r}, \omega) = 1$ everywhere, whereas $\varepsilon(\mathbf{r}, \omega)$ may have both spatial and frequency dependence. In this case, Equation (2.36) becomes

$$\nabla \times \nabla \times \mathbf{E}(\mathbf{r}, \omega) = \left(\frac{\omega}{c_0}\right)^2 \varepsilon(\mathbf{r}, \omega)\mathbf{E}(\mathbf{r}, \omega). \tag{2.40}$$

Using the vector identity

$$\nabla \times \nabla \times \mathbf{F} = -\nabla^2\mathbf{F} + \nabla(\nabla \cdot \mathbf{F}), \tag{2.41}$$

this becomes

$$\nabla^2\mathbf{E}(\mathbf{r}) - \nabla(\nabla \cdot \mathbf{E}(\mathbf{r})) + \varepsilon(\mathbf{r})k_0^2\mathbf{E}(\mathbf{r}) = 0, \tag{2.42}$$

where $k_0 = \omega/c_0$ is the wave number in vacuum. The Gauss law in the absence of free charges (2.4)

$$\nabla \cdot (\varepsilon \mathbf{E}) = 0 \tag{2.43}$$

leads to the expression $\nabla \cdot \mathbf{E} = -\mathbf{E} \cdot (\nabla \ln(\varepsilon(\mathbf{r})))$. Insertion of this result into Equation (2.42) gives

$$\nabla^2 \mathbf{E}(\mathbf{r}, \omega) + \nabla(\mathbf{E}(\mathbf{r}, \omega) \cdot \nabla \ln \varepsilon(\mathbf{r}, \omega)) + \varepsilon(\mathbf{r}, \omega) k_0^2 \mathbf{E}(\mathbf{r}, \omega) = 0. \tag{2.44}$$

A similar approach may be carried through for the $\mathbf{H}$-equation. In a case when light interacts with a nonmagnetic material, it is convenient to take the curl of Equation (2.28) directly, to obtain

$$\nabla \times \nabla \times \mathbf{H}(\mathbf{r}, \omega) = \pm i\omega \varepsilon_0 \nabla \times (\varepsilon(\mathbf{r}, \omega) \mathbf{E}(\mathbf{r}, \omega)), \tag{2.45}$$

which may be rewritten by applying Equations (2.27), (2.28), and (2.41) along with the condition $\nabla \cdot \mathbf{H} = 0$ for a nonmagnetic material and the vector relation $\nabla \times (\varepsilon \mathbf{E}) = \varepsilon \nabla \times \mathbf{E} + \nabla \varepsilon \times \mathbf{E}$:

$$\nabla^2 \mathbf{H}(\mathbf{r}, \omega) + \nabla \ln(\varepsilon(\mathbf{r}, \omega)) \times \nabla \times \mathbf{H}(\mathbf{r}, \omega) + k_0^2 \varepsilon(\mathbf{r}, \omega) \mathbf{H}(\mathbf{r}, \omega) = 0. \tag{2.46}$$

Equations (2.44) and (2.46) are vector equations, with the different vector components being coupled by the spatial derivatives of the ε-function. If the spatial variations in ε are weak, or slow, it may be a permissible approximation to neglect these derivatives, which reduces the wave equations to

$$\nabla^2 \mathbf{E}(\mathbf{r}, \omega) + \varepsilon(\mathbf{r}, \omega) k_0^2 \mathbf{E}(\mathbf{r}, \omega) = 0, \tag{2.47}$$

$$\nabla^2 \mathbf{H}(\mathbf{r}, \omega) + \varepsilon(\mathbf{r}, \omega) k_0^2 \mathbf{H}(\mathbf{r}, \omega) = 0. \tag{2.48}$$

The fact that the individual vector components are decoupled, and all equations are identical, means that the vector wave equation has essentially been reduced to a scalar problem of the form

$$\nabla^2 \Psi + \varepsilon k_0^2 \Psi = 0, \tag{2.49}$$

where Ψ could denote any $\mathbf{E}$ or $\mathbf{H}$ components. Therefore, this approach is commonly denoted the *scalar approximation*. The solutions for all vector components will be identical in the scalar approximation, regardless of the symmetries of the $\varepsilon(\mathbf{r})$ distribution. Note, however, that the solutions should still obey the static Maxwell's equations (2.4) and (2.5), which imposes some constraints on how the different vector components may be combined.

In a homogeneous medium, where material functions do not depend on coordinates, the frequency-domain wave equations can quite generally be reduced to the *Helmholtz* equations:

$$\nabla^2 \mathbf{H}(\mathbf{r}) + k_0^2 \mu \varepsilon \mathbf{H}(\mathbf{r}) = 0, \tag{2.50}$$

$$\nabla^2 \mathbf{E}(\mathbf{r}) + k_0^2 \mu \varepsilon \mathbf{E}(\mathbf{r}) = 0, \tag{2.51}$$

with the corresponding time-domain forms for nondispersive materials:

$$\nabla^2 \mathbf{H}(\mathbf{r}, t) = \frac{\mu\varepsilon}{c_0^2} \frac{\partial^2 \mathbf{H}(\mathbf{r}, t)}{\partial t^2}, \tag{2.52}$$

$$\nabla^2 \mathbf{E}(\mathbf{r}, t) = \frac{\mu\varepsilon}{c_0^2} \frac{\partial^2 \mathbf{E}(\mathbf{r}, t)}{\partial t^2}. \tag{2.53}$$

These equations are important because they are *locally* valid in any region of constants ε and μ. In the commonly encountered situation of piecewise constant ε and μ functions, solutions may therefore be found by matching local solutions to these equations across the material interfaces using appropriate boundary conditions. Since these boundary conditions will be polarization dependent, this does not necessarily break down to a globally scalar problem.

2.7 WAVEGUIDES AND EIGENMODES

An important class of geometries are those featuring uniformity along one axis, which in this section is assumed to be the z-axis. Examples of such geometries are shown in Figure 2.1. Here, the dielectric constant $\varepsilon(\mathbf{r}_\perp)$ has a lateral but no longitudinal dependence, where $\mathbf{r}_\perp$ refers to the coordinates (x, y) of the plane normal to the z-axis. In the following, we will consider nonmagnetic materials without free charges and currents and *monochromatic* fields of the following form:

$$\mathbf{E}(\mathbf{r}, t) = \mathbf{E}(\mathbf{r})e^{-i\omega t}, \quad \mathbf{H}(\mathbf{r}, t) = \mathbf{H}(\mathbf{r})e^{-i\omega t}. \tag{2.54}$$

The frequency-domain Maxwell's equations then read

$$\nabla \times \mathbf{E}(\mathbf{r}) = i\omega\mu_0 \mathbf{H}(\mathbf{r}), \tag{2.55}$$

$$\nabla \times \mathbf{H}(\mathbf{r}) = -i\omega\varepsilon_0\varepsilon \mathbf{E}(\mathbf{r}). \tag{2.56}$$

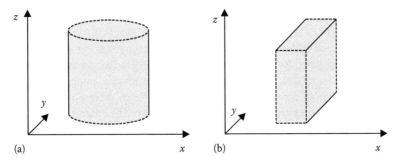

FIGURE 2.1 (a) A circular waveguide and (b) a slab waveguide featuring uniformity along the z-axis.

2.7.1 EIGENVALUE PROBLEM

The z-invariance of ε has two important consequences: Firstly, it allows us to write the electric field as $\mathbf{E}(\mathbf{r}) = \mathbf{e}(\mathbf{r}_\perp)e^{i\beta z}$, with a similar expression for the magnetic field, $\mathbf{H}(\mathbf{r}) = \mathbf{h}(\mathbf{r}_\perp)e^{i\beta z}$. Here, β is denoted the *propagation constant* and is the only parameter controlling the z-dependence of the fields. Secondly, it eliminates the z-derivatives of ε in the wave equations. As we will now show, these simplifications allow us to transform the wave equations into eigenvalue equations involving only the transverse components of the propagating fields.

Considering first the eigenvalue problem (Equation 2.44) for the electric field, the assumption of a straight waveguide allows us to write

$$\nabla_\perp^2 \mathbf{e} + (\nabla_\perp + i\beta\hat{\mathbf{z}})(\mathbf{e} \cdot \nabla\ln\varepsilon(\mathbf{r}_\perp)) + \varepsilon(\mathbf{r}_\perp)k_0^2\mathbf{e} = \beta^2\mathbf{e}, \qquad (2.57)$$

where $\nabla_\perp$ denotes the transverse part of the nabla operator. Equation (2.57) is a second-order eigenvalue problem of the form $\hat{O}_0\mathbf{e} + \hat{O}_1\lambda\mathbf{e} + \hat{O}_2\lambda^2\mathbf{e} = 0$ as we have a β on the left-hand side (LHS). Second-order eigenvalue problems are much more demanding to solve numerically than standard eigenvalue problems. Fortunately, the vanishing z-derivative of ε implies that the scalar equations for the in-plane components $\mathbf{e}_\perp$ in Equation (2.57) are not coupled to e_z, and it thus suffices to solve the equation

$$\nabla_\perp^2 \mathbf{e}_\perp + \nabla_\perp(\mathbf{e}_\perp \cdot \nabla\ln\varepsilon(\mathbf{r}_\perp)) + \varepsilon(\mathbf{r}_\perp)k_0^2\mathbf{e}_\perp = \beta^2\mathbf{e}_\perp, \qquad (2.58)$$

which is a standard eigenvalue problem of the form $\hat{O}\mathbf{e}_\perp = \lambda\mathbf{e}_\perp$. Equation (2.58) is the central equation in the theory of waveguides. By solving it, we obtain eigenmodes $\mathbf{e}_{\perp m}(\mathbf{r}_\perp)$ and eigenvalues $\lambda_m = \beta_m^2$ that are the squares of the propagation constants β_m. In the following, we skip the mode subindex m for simplicity.

For a passive structure, the eigenvalues $\lambda = \beta^2$ are generally real valued; however, the eigenvalues can be either positive or negative leading to complex-valued $\beta = \beta_R + i\beta_I$. Furthermore, solutions with propagation constants of both signs $\pm\beta$ will exist. For $\beta^2 > 0$, we have $\beta = \pm\beta_R$, and the z, t dependence is thus given by $e^{i(\pm\beta_R z - \omega t)}$. We refer to these modes as *propagating modes*. For positive ω, a positive value of β corresponds to *forward propagating* waves, whereas a negative value of β corresponds to *backward propagating* waves. The opposite is true for negative ω-values, or if the sign convention for ω is reversed. It is also noteworthy that Equation (2.57) becomes complex conjugated if the sign of β is reversed, meaning that the eigenmode fields are complex conjugated as well. For $\beta^2 < 0$, we have $\beta = i\beta_I$, a z dependence given by $e^{\pm\beta_I z}$, and these modes will either increase or decrease exponentially along the z-axis. Generally, only the exponentially decreasing fields make physical sense, and we thus refer to all these modes as *evanescent modes*. The evanescent eigenmodes with positive β_I are grouped together with the propagating modes with positive β_R to form the *forward travelling* set of modes, which can be defined in an elegant way using $\beta_R + \beta_I > 0$. Similarly, the evanescent modes with negative β_I are grouped with propagating modes of negative β_R to form the *backward travelling* set, given by $\beta_R + \beta_I < 0$. This classification is summarized in Table 2.1.

TABLE 2.1

Classification Scheme for Forward and Backward Travelling Eigenmodes

Eigenvalue	Propagation Constant	Class	Direction
$\beta^2 > 0$	$+\beta_R$	Propagating	+
$\beta^2 < 0$	$+\beta_I$	Evanescent	+
$\beta^2 > 0$	$-\beta_R$	Propagating	−
$\beta^2 < 0$	$-\beta_I$	Evanescent	−

In structures with a complex permittivity profile, the eigenvalues $\lambda = \beta^2$ become complex. In that case, we simply generalize the classification criterion such that forward travelling set is defined by $\beta_R + \beta_I > 0$ and the backward travelling set by $\beta_R + \beta_I < 0$. Since the propagation constants now feature both a real and an imaginary part, a clear distinction between propagating and evanescent modes is not possible.

Even though Equation (2.58) only involves the in-plane electric field components $\mathbf{e}_\perp$, the complete six-component electromagnetic field description is readily available: when an in-plane profile $\mathbf{e}_\perp$ has been calculated, we can determine the e_z component using the identity

$$e_z = \frac{-\nabla_\perp \cdot (\varepsilon \mathbf{e}_\perp)}{i\beta\varepsilon} \tag{2.59}$$

derived from Equation (2.43). Subsequently, the magnetic field strength $\mathbf{H}(\mathbf{r}_\perp)$ can be calculated from Maxwell's equation (2.55). Using the expression (2.59) for the e_z component, the lateral and longitudinal components of the magnetic field become

$$\mathbf{h}_\perp = \frac{1}{\mu_0\omega}\hat{\mathbf{z}} \times \left(\beta\mathbf{e}_\perp - \nabla_\perp \left[\frac{\nabla_\perp \cdot (\varepsilon \mathbf{e}_\perp)}{\beta\varepsilon} \right] \right), \tag{2.60}$$

$$h_z = \frac{\hat{\mathbf{z}} \cdot \nabla \times \mathbf{e}_\perp}{i\mu_0\omega}. \tag{2.61}$$

Let us now consider the relationship between the fields of a forward propagating $(\mathbf{E}^+, \mathbf{H}^+)$ and a backward propagating $(\mathbf{E}^-, \mathbf{H}^-)$ eigenmodes derived from the same solution $\mathbf{e}_\perp$ to the eigenvalue problem. The forward propagating mode profile $(\mathbf{E}^+, \mathbf{H}^+)$ is given by

$$\begin{bmatrix} \mathbf{E}^+(\mathbf{r}) \\ \mathbf{H}^+(\mathbf{r}) \end{bmatrix} = \begin{bmatrix} \mathbf{e}^+(\mathbf{r}) \\ \mathbf{h}^+(\mathbf{r}) \end{bmatrix} e^{i\beta z} = \begin{bmatrix} \mathbf{e}_\perp(\mathbf{r}_\perp) + e_z(\mathbf{r}_\perp)\hat{\mathbf{z}} \\ \mathbf{h}_\perp(\mathbf{r}_\perp) + h_z(\mathbf{r}_\perp)\hat{\mathbf{z}} \end{bmatrix} e^{i\beta z}. \tag{2.62}$$

Now, inspection of Equations (2.59) through (2.61) reveals that the substitution $\beta \rightarrow -\beta$ results in a backward propagating mode profile given as

$$\begin{bmatrix} \mathbf{E}^-(\mathbf{r}) \\ \mathbf{H}^-(\mathbf{r}) \end{bmatrix} = \begin{bmatrix} \mathbf{e}^-(\mathbf{r}) \\ \mathbf{h}^-(\mathbf{r}) \end{bmatrix} e^{-i\beta z} = \begin{bmatrix} \mathbf{e}_\perp(\mathbf{r}_\perp) - e_z(\mathbf{r}_\perp)\hat{\mathbf{z}} \\ -\mathbf{h}_\perp(\mathbf{r}_\perp) + h_z(\mathbf{r}_\perp)\hat{\mathbf{z}} \end{bmatrix} e^{-i\beta z}, \tag{2.63}$$

and we observe that the forward and backward propagating lateral electric field components are identical $\mathbf{e}_\perp^- = \mathbf{e}_\perp^+ = \mathbf{e}_\perp$, whereas for the relation for the lateral magnetic components, $\mathbf{h}_\perp^- = -\mathbf{h}_\perp^+ = -\mathbf{h}_\perp$ features a minus sign. Furthermore, the electric field z component depends on the propagation direction as $e_z^- = -e_z^+ = -e_z$.

In a similar way, the eigenvalue problem for the magnetic field (Equation 2.46) for the **H** field may be rewritten as

$$\nabla_\perp^2 \mathbf{h} + k_0^2 \varepsilon(\mathbf{r}_\perp)\mathbf{h}(\mathbf{r}) + \nabla\ln(\varepsilon(\mathbf{r}_\perp)) \times \nabla \times \mathbf{h} = \beta^2 \mathbf{h}. \tag{2.64}$$

As for the electric field, the equations for $\mathbf{h}_\perp$ are decoupled from the h_z component, which can be determined unambiguously by the divergence constraint:

$$h_z = \frac{i}{\beta}\nabla_\perp \cdot \mathbf{h}_\perp. \tag{2.65}$$

Finally, the scalar wave equation for a waveguide assumes the form

$$\nabla_\perp^2 \Psi + \varepsilon k_0^2 \Psi = \beta^2 \Psi. \tag{2.66}$$

As before, Ψ could be any transverse field component, but the divergence constraints given by (2.59) and (2.65) should be fulfilled.

In the following sections, we will generally skip subindex $\perp$ for simplicity.

2.7.2　SLAB WAVEGUIDES

For a general waveguide without particular symmetries in the lateral (x, y) plane, the lateral components in the eigenvalue problems (Equations 2.58 through 2.64) are coupled to each other by the $\nabla\ln(\varepsilon(\mathbf{r}_\perp))$ terms. However, in so-called *slab waveguides*, the dielectric constant depends only on one lateral coordinate. A slab waveguide is illustrated in Figure 2.1b, where the dielectric constant $\varepsilon(x)$ depends only on x and not on the y coordinate.

For such a geometry, let us study the special classes of solutions $(\mathbf{e}(x), \mathbf{h}(x))$ independent of the y coordinate. In this case, $\partial_y\mathbf{e} = \partial_y\mathbf{h} = 0$ and the eigenmode problem (Equation 2.58) decouples into two equations for the lateral field components

$$\partial_x^2 e_x + \partial_x\left(e_x\partial_x\ln\varepsilon(x)\right) + \varepsilon_r(x)k_0^2 e_x = \beta^2 e_x, \quad \text{(TM)} \tag{2.67}$$

$$\partial_x^2 e_y + \varepsilon(x)k_0^2 e_y = \beta^2 e_y, \quad \text{(TE)} \tag{2.68}$$

which represent two independent sets of solutions to the eigenmode problem polarized either along the x-axis in Equation (2.67) or along the y-axis in Equation (2.68). The e_z component is computed from Equation (2.59), and since $\partial_y\mathbf{e} = 0$, we observe that only the solution polarized along the x-axis gives a contribution to the e_z field. For solutions e_y to Equation (2.68) polarized along y, we have $e_z = 0$, and for this reason, this set of solutions is referred to as *transverse electric* (TE) modes.

The same decoupling occurs for the magnetic field, and Equation (2.64) becomes

$$\partial_x^2 h_x + \varepsilon(x)k_0^2 h_x + = \beta^2 h_x, \quad \text{(TE)} \tag{2.69}$$

$$\partial_x^2 h_y + \varepsilon(x)k_0^2 h_y - \left(\partial_x\ln\varepsilon(x)\right)\partial_x h_y = \beta^2 h_y, \quad \text{(TM)} \tag{2.70}$$

which again represent two independent solutions for the magnetic field. Here, the h_z component is given by Equation (2.65), and since $\partial_y \mathbf{h} = 0$, the h_z field is zero for solutions h_y polarized along the y-axis. For this reason, solutions to Equation (2.70) are referred to as *transverse magnetic* (TM) modes.

Note: These designations are opposite of the terminology introduced in Section 2.5. In the waveguide community, it is common to define TE/TM modes relative to the axis of propagation, rather than the axis along which the fields and ε-structure are uniform.

The eigenmode problems (Equations 2.68 and 2.69) are equivalent in that they describe electric and magnetic field components of the same optical TE field. This can be seen by applying Maxwell's equation (2.27) to the TE field. For an electric field polarized along the y-axis, we obtain an h_x component given by

$$h_x = -\frac{\beta}{\omega \mu_0} e_y, \tag{2.71}$$

and Equation (2.69) is obtained by inserting Equation (2.71) into (2.68). Similarly, the Equations (2.67) and (2.70) for the TM field are equivalent. From Maxwell's equation (2.28) applied to the TM field, we find that e_x equals

$$e_x = \frac{\beta}{\omega \varepsilon_0 \varepsilon(x)} h_y, \tag{2.72}$$

and insertion of Equation (2.72) into (2.70) gives Equation (2.67).

2.7.3 BOUNDARY CONDITIONS AND EIGENMODE CLASSES

The eigenmode profile describes the mode in the (x, y) plane; however, its z dependence is governed by its propagation constant β alone. It is thus useful to classify different types of eigenmodes according to their eigenvalues. To do this, we consider the propagation constants obtained in a simple waveguide with real-valued relative permittivity profile. Three general classes of geometries illustrated in Figure 2.2 with different boundary conditions and mode classifications exist.

The open geometry sketched in Figure 2.2a extends to $\pm\infty$ along the x-axis. When considering a waveguide in an open geometry, one makes the distinction between guided and radiation modes. Guided modes is the set of discrete modes confined to the waveguide fulfilling $\varepsilon_1 k_0^2 < \beta^2 < \varepsilon_2 k_0^2$, whereas the radiation modes fulfilling $\beta^2 < \varepsilon_1 k_0^2$ form a continuous set and extend to $\pm\infty$. The two solution sets in the β^2 complex plane are illustrated in Figure 2.2d.

While an infinite computational domain can be treated directly using, for example the integral equation method described in Chapter 7, the modal method discussed in Chapter 6 requires that the continuous set of radiation modes are discretized to obtain a finite set of expansion coefficients which can be stored in memory. This can be done by limiting the extent of the computational domain by imposing periodic or closed boundary conditions.

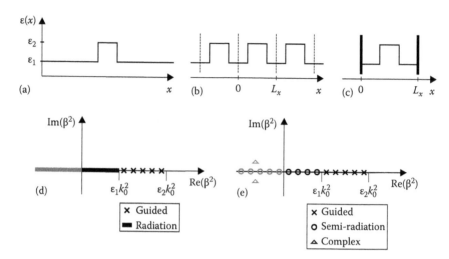

FIGURE 2.2 A waveguide in (a) an open geometry, (b) a periodic geometry, and (c) a closed geometry. (d) Complex β^2 planes illustrating the solution sets for the open geometry and (e) the closed/periodic geometry with guided modes (crosses), the continuous set of radiation modes (bar), semi-radiation modes (circles), and complex modes (triangles). Propagating (black) and evanescent (grey) modes are distinguished.

For the geometry of periodicity L_x in Figure 2.2b, we employ the Bloch theorem and require that the field after one period should reproduce itself multiplied by a phase term $e^{i\alpha}$ such that $\mathbf{E}(x+L_x) = \mathbf{E}(x)e^{i\alpha}$. This condition is known as the *Floquet–Bloch* condition. While the geometry remains infinite, the requirement of periodicity effectively reduces the computational domain to one period, leading to a discretization of the radiation modes as illustrated in Figure 2.2e. While no distinction between the discrete sets of guided and radiation modes is necessary from a computational point of view, the latter set is sometimes referred to as *semi-radiation modes* to distinguish them from the true radiation modes of the open geometry.

A closely related boundary condition is that of the closed geometry shown in Figure 2.2c, where the structure is placed inside a box of length L_x with perfectly conducting walls leading to the requirement $\mathbf{E}(0) = \mathbf{E}(L_x) = 0$. Again, this boundary condition reduces the computational domain to a finite size leading to a discrete set of semi-radiation modes.

As discussed earlier, when β^2 is real, the mode is either propagating or decaying along the z-axis. However, in rare cases, a third class of modes appears: eigenmodes with complex eigenvalues β^2, sometimes referred to as *leaky modes*, can exist even in real-valued permittivity profiles.

2.7.4 ORTHOGONALITY

We will now establish the orthogonality relation for lateral eigenmodes using the Lorentz reciprocity theorem. To derive this theorem, we consider two solutions to Maxwell's equations, where we now include free current sources. The first $(\mathbf{E}_1, \mathbf{H}_1)$

is generated by a current source $\mathbf{J}_1$ and the second $(\mathbf{E}_2, \mathbf{H}_2)$ by a current source $\mathbf{J}_2$. The solutions obey Maxwell's equations:

$$\nabla \times \mathbf{E}_1 = i\omega\mu_0\mu\mathbf{H}_1, \tag{2.73}$$

$$\nabla \times \mathbf{H}_1 = -i\omega\varepsilon_0\varepsilon\mathbf{E}_1 + \mathbf{J}_1, \tag{2.74}$$

$$\nabla \times \mathbf{E}_2 = i\omega\mu_0\mu\mathbf{H}_2, \tag{2.75}$$

$$\nabla \times \mathbf{H}_2 = -i\omega\varepsilon_0\varepsilon\mathbf{E}_2 + \mathbf{J}_2. \tag{2.76}$$

We then consider the expression $\mathbf{H}_2 \cdot (2.73) - \mathbf{E}_1 \cdot (2.76) + \mathbf{E}_2 \cdot (2.74) - \mathbf{H}_1 \cdot (2.75)$ and exploit the identity $\nabla \cdot (\mathbf{A} \times \mathbf{B}) = \mathbf{B} \cdot \nabla \times \mathbf{A} - \mathbf{A} \cdot \nabla \times \mathbf{B}$ to obtain

$$\nabla \cdot (\mathbf{E}_1 \times \mathbf{H}_2 - \mathbf{E}_2 \times \mathbf{H}_1) = \mathbf{J}_1 \cdot \mathbf{E}_2 - \mathbf{J}_2 \cdot \mathbf{E}_1. \tag{2.77}$$

Let us now integrate Equation (2.77) over the volume V surrounded by a surface S as illustrated in Figure 2.3a and use Gauss' theorem to obtain

$$\int_S (\mathbf{E}_1 \times \mathbf{H}_2 - \mathbf{E}_2 \times \mathbf{H}_1) \cdot \hat{\mathbf{n}} dS = \int_V (\mathbf{J}_1 \cdot \mathbf{E}_2 - \mathbf{J}_2 \cdot \mathbf{E}_1) dV, \tag{2.78}$$

where $\hat{\mathbf{n}}$ is the unit vector normal to the surface. Equation (2.78) is known as the Lorentz reciprocity theorem.

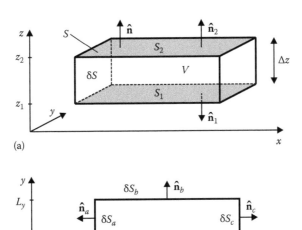

FIGURE 2.3 (a) The bottom S_1, top S_2, and side δS surfaces covering the volume V with normal unit vector $\hat{\mathbf{n}}$. (b) x–y plane of the volume with the side surface subelements δS_a, δS_b, δS_c, and δS_d and corresponding normal unit vectors $\hat{\mathbf{n}}_a$, $\hat{\mathbf{n}}_b$, $\hat{\mathbf{n}}_c$, and $\hat{\mathbf{n}}_d$.

To obtain our orthogonality relation, we now consider a geometry uniform along the z-axis with volume V and thickness $\Delta z = z_2 - z_1$ enclosing our entire lateral domain. We then split our surface into three parts: the bottom surface S_1, the top surface S_2, and the side boundary δS. We first consider the surface integral over the boundary δS by inspecting the x–y plane illustrated in Figure 2.3b. For a rectangular computational domain, the boundary surface integral can be split up into four parts as

$$\int_{\delta S} \mathbf{F} \cdot \hat{\mathbf{n}} dS = \int_{\delta S_a} \mathbf{F} \cdot \hat{\mathbf{n}}_a dS + \int_{\delta S_b} \mathbf{F} \cdot \hat{\mathbf{n}}_b dS + \int_{\delta S_c} \mathbf{F} \cdot \hat{\mathbf{n}}_c dS + \int_{\delta S_d} \mathbf{F} \cdot \hat{\mathbf{n}}_d dS, \qquad (2.79)$$

where we have defined $\mathbf{F} \equiv \mathbf{E}_1 \times \mathbf{H}_2 - \mathbf{E}_2 \times \mathbf{H}_1$ for convenience.

Now, in the particular case of periodic boundary conditions with $e^{i\alpha} = 1$, we have $\mathbf{F}(0, y, z) = \mathbf{F}(L_x, y, z)$ such that

$$\int_{\delta S_a} \mathbf{F}(0, y, z) \cdot \hat{\mathbf{n}}_a dS = \int_{\delta S_c} \mathbf{F}(L_x, y, z) \cdot \hat{\mathbf{n}}_a dS = -\int_{\delta S_c} \mathbf{F}(L_x, y, z) \cdot \hat{\mathbf{n}}_c dS, \qquad (2.80)$$

leading to the cancellation of the first and third terms on the RHS of Equation (2.79). Similarly, the second and fourth terms cancel out.

In the case of a closed structure with perfectly conducting metal walls, the boundary conditions require that the tangential components of the magnetic field be nullified at the boundary leading to $\mathbf{F} \cdot \hat{\mathbf{n}} = 0$ at the walls, and again, the boundary integral disappears.

The remaining contributions from the LHS of Equation (2.78) are

$$\int_{S_1} \mathbf{F} \cdot \hat{\mathbf{n}}_1 dS + \int_{S_2} \mathbf{F} \cdot \hat{\mathbf{n}}_2 dS = -\int F_z(z_1) d\mathbf{r}_\perp + \int F_z(z_2) d\mathbf{r}_\perp \qquad (2.81)$$

$$= \int (F_z(z_2) - F_z(z_1)) d\mathbf{r}_\perp. \qquad (2.82)$$

In a geometry with z invariance, the fields are continuous along z. For small Δz, they can be treated as constants such that the RHS of Equation (2.78) can be converted to a surface integral as

$$\int_V (\mathbf{J}_1 \cdot \mathbf{E}_2 - \mathbf{J}_2 \cdot \mathbf{E}_1) dV = \Delta z \int (\mathbf{J}_1 \cdot \mathbf{E}_2 - \mathbf{J}_2 \cdot \mathbf{E}_1) d\mathbf{r}_\perp. \qquad (2.83)$$

We then equate Equations (2.82) and (2.83), divide by Δz, take the limit $\Delta z \to 0$, and obtain the Lorentz reciprocity theorem for a geometry uniform along the z-axis:

$$\int \frac{\partial}{\partial z} (\mathbf{E}_1 \times \mathbf{H}_2 - \mathbf{E}_2 \times \mathbf{H}_1) \cdot \hat{\mathbf{z}} d\mathbf{r}_\perp = \int (\mathbf{J}_1 \cdot \mathbf{E}_2 - \mathbf{J}_2 \cdot \mathbf{E}_1) d\mathbf{r}_\perp. \qquad (2.84)$$

Let us now assume the absence of free currents such that the RHS of Equation (2.84) is 0. We initially choose forward propagating eigenmodes of the form

$$\begin{bmatrix} \mathbf{E}_1(\mathbf{r}) \\ \mathbf{H}_1(\mathbf{r}) \end{bmatrix} = \begin{bmatrix} \mathbf{e}_{\perp 1}(\mathbf{r}_\perp) + e_{z1}(\mathbf{r}_\perp)\hat{\mathbf{z}} \\ \mathbf{h}_{\perp 1}(\mathbf{r}_\perp) + h_{z1}(\mathbf{r}_\perp)\hat{\mathbf{z}} \end{bmatrix} e^{i\beta_1 z} \tag{2.85}$$

$$\begin{bmatrix} \mathbf{E}_2(\mathbf{r}) \\ \mathbf{H}_2(\mathbf{r}) \end{bmatrix} = \begin{bmatrix} \mathbf{e}_{\perp 2}(\mathbf{r}_\perp) + e_{z2}(\mathbf{r}_\perp)\hat{\mathbf{z}} \\ \mathbf{h}_{\perp 2}(\mathbf{r}_\perp) + h_{z2}(\mathbf{r}_\perp)\hat{\mathbf{z}} \end{bmatrix} e^{i\beta_2 z}, \tag{2.86}$$

and insertion into Equation (2.84) gives

$$(\beta_2 + \beta_1) \int (\mathbf{e}_{\perp 1} \times \mathbf{h}_{\perp 2} - \mathbf{e}_{\perp 2} \times \mathbf{h}_{\perp 1}) \cdot \hat{\mathbf{z}} d\mathbf{r}_\perp = 0. \tag{2.87}$$

We now consider the mode profile $(\mathbf{E}_1, \mathbf{H}_1)$ (2.85) and the mode profile $(\mathbf{E}_2', \mathbf{H}_2')$, which is the backward propagating version of $(\mathbf{E}_2, \mathbf{H}_2)$ given by

$$\begin{bmatrix} \mathbf{E}_2(\mathbf{r}) \\ \mathbf{H}_2(\mathbf{r}) \end{bmatrix} = \begin{bmatrix} \mathbf{e}_{\perp 2}(\mathbf{r}_\perp) - e_{z2}(\mathbf{r}_\perp)\hat{\mathbf{z}} \\ -\mathbf{h}_{\perp 2}(\mathbf{r}_\perp) + h_{z2}(\mathbf{r}_\perp)\hat{\mathbf{z}} \end{bmatrix} e^{-i\beta_2 z}. \tag{2.88}$$

The components of the mode profile are identical to those in Equation (2.86), except for the minus sign in the expressions for the e_z and $\mathbf{h}_\perp$ components as discussed in Section 2.7.1. Insertion of $(\mathbf{E}_1, \mathbf{H}_1)$ and $(\mathbf{E}_2', \mathbf{H}_2')$ into Equation (2.84) results in

$$(\beta_2 - \beta_1) \int (\mathbf{e}_{\perp 1} \times \mathbf{h}_{\perp 2} + \mathbf{e}_{\perp 2} \times \mathbf{h}_{\perp 1}) \cdot \hat{\mathbf{z}} d\mathbf{r}_\perp = 0. \tag{2.89}$$

Now, if $\beta_1 \neq \beta_2$, the integrals in Equations (2.87) and (2.89) must equal zero. Taking their sum, we obtain the orthogonality relation [3,4]:

$$\int (\mathbf{e}_1 \times \mathbf{h}_2) \cdot \hat{\mathbf{z}} d\mathbf{r}_\perp = 0, \tag{2.90}$$

where we have omitted the $\perp$ subindex, which is allowed since we are taking the dot product with $\hat{\mathbf{z}}$.

If $\beta_1 = \beta_2$, the integral in Equation (2.89) may differ from 0. However, in that case, orthogonality of the set of eigenmodes with equal propagation constant β can be obtained by performing a standard Gram–Schmidt orthonormalization process.

The Lorentz reciprocity theorem (Equation (2.78)) and the orthogonality relation (Equation 2.90) are presented here in their unconjugated forms. By making the replacements $(\mathbf{E}_2, \mathbf{H}_2, \mathbf{J}_2) \rightarrow (\mathbf{E}_2^*, \mathbf{H}_2^*, \mathbf{J}_2^*)$ in Equations (2.75) and (2.76), we arrive at their conjugated forms. The corresponding orthogonality relation is given by

$$\int (\mathbf{e}_1 \times \mathbf{h}_2^*) \cdot \hat{\mathbf{z}} d\mathbf{r}_\perp = 0, \tag{2.91}$$

and since one half the real part of the term in parentheses represents the z component of the time-averaged Poynting vector, this relation is also known as the *power orthogonality* relation.

Whereas the power orthogonality relation (Equation 2.91) only holds in geometries with real-valued permittivity profiles, the orthogonality relation (Equation 2.90) is valid in arbitrary geometries, including those with complex permittivity profiles, and is thus more general. The orthogonality relation (Equation 2.90) is used in the modal method in the mode matching taking place at the interfaces between adjacent layers, as described further in Chapter 6.

Note: For periodic boundary conditions with $e^{i\alpha} \neq 1$, the cancellation of the δS boundary terms in Equation (2.79) does not occur, and here, the orthogonality relations do not hold.

REFERENCES

1. J. D. Jackson, *Classical Electrodynamics*, 3rd edn. Hoboken, NJ: Wiley, 1999.
2. J. A. Stratton, *Electromagnetic Theory* (Reissue). Hoboken, NJ: IEEE Press, 2007.
3. H.-G. Unger, *Planar Optical Waveguides and Fibres*. Oxford, U.K.: Clarendon, 1977.
4. W. Snyder and J. Love, *Optical Waveguide Theory*. New York: Chapman & Hall, 1983.

3 Finite-Difference Time-Domain Method

3.1 INTRODUCTION

In this section, we introduce one of the most universal numerical methods in optics and electromagnetism – the *finite-difference time-domain (FDTD) method*. Its universality is a consequence of the method target: the FDTD method is aimed to solve directly Maxwell's equations (2.2) through (2.5) without any further approximations and derivations. Thus, all processes governed by Maxwell's equations can in principle be described numerically by FDTD solvers, which are often referred to as a *brutal force*. Of course, in reality, there is a great deal of particular situations, where FDTD methods can be non-practical due to mesh constrains or processing time limitations, that is, in fibre or waveguide optics, and cavity analysis. But in photonics of micro- and nanostructured systems, this method is a working horse. Very often, results obtained by the FDTD method are used as a reference to validate results of other newly elaborated methods or to support and explain experimental results of characterization. Typically, effectiveness of numerical methods in practical things (processor time, memory costs, etc.) is also mostly weighted against FDTD solutions.

From this perspective, it is reasonable and methodologically correct to start our nomenclature of Maxwell's solvers with the FDTD method. Together with the method, we address here the problems with table (array) functions defined on a mesh, which are not typically encountered while their continuously defined analogues are in use. Thus, we introduce finite-difference schemes and discuss specific properties of optical waves propagating on the mesh, such as numerical dispersion. We derive the FDTD field-evolving mechanism in 1D, 2D, and 3D cases. Further, we briefly outline practicalities in the realization of the FDTD numerical scheme: choice of grid sizes, time steps for stable performance of FDTD, and truncation of a numerical domain with open boundary conditions (BC). For the sake of completeness, we introduce special schemes for description of frequency dispersive materials like metals, which are intensively involved in plasmonics and metamaterials. Extension of the FDTD method to nonlinear materials, materials with gain, and lasing systems is briefly sketched in the end of this chapter. A set of exercises is directed to work on properties of numerical schemes implemented on a grid. As a part of exercises, readers are encouraged to build their own FDTD code in 1D and test it on simple but important examples.

This chapter appears to be rather brief when compared with the *market value* of the FDTD method in photonics. However, there is a good set of original books devoted to the FDTD method [1–5], so we deliberately limited this chapter constraining it to the most crucial aspects in this area. Readers can find extensions and a lot of details of various FDTD realizations in *classical* books listed in the references to this chapter.

Material of this section can be used to solve Exercises 3.1, 3.2, and 3.11.

3.1.1 FINITE-DIFFERENCE APPROXIMATIONS OF DERIVATIVES

Finite differences play the principal role in numerical analysis approximating differential operators. They are intensively used, for example, for solving different types of differential and integro-differential equations. To apply numerical methods, the continuous space-time is approximated by the spatial and temporal grids: discrete sets of points in time and space. Therefore, any differentiable function $u(x)$ defined in space and time must be substituted by a table (array) representing sampling of the function values on the grid. Such discretization of function $u(x)$ on the homogeneous grid defines a table function $u(x_i) \equiv u_i, i = 0, 1, \ldots$, shown graphically in Figure 3.1. Here, function $u(x)$ is sampled at discrete intervals with equal spacing h.

Now, we derive an algorithm of constructing finite-difference approximations for derivative operators at point x. We involve only closest to x neighbour points $x - h$ and $x + h$. Playing with grid values $u(x - h), u(x), u(x + h)$, we can organize three finite-difference schemes to approximate the first-order derivative $u'(x) \equiv du/dx$. First, the so-called *forward difference* scheme D^+u involves function values at x and one step forward $x + h$:

$$D^+u \equiv \frac{u(x + h) - u(x)}{h}. \tag{3.1}$$

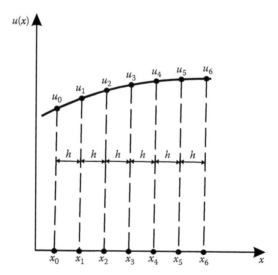

FIGURE 3.1 Finite-difference methods are based on discretization of differentiable function $u = f(x)$ on a grid; samples $u_0 = u(x_0), u_1 = u(x_1)$ are taken at equal intervals h.

Second, stepping backward leads to the *backward difference* scheme D^-u:

$$D^-u \equiv \frac{u(x) - u(x - h)}{h}.$$ (3.2)

And finally, stepping over point x gives the *central difference* scheme $D^c u$:

$$D^c u \equiv \frac{u(x + h) - u(x - h)}{2h}.$$ (3.3)

To evaluate the error, which each of this scheme brings in approximation of the first derivative $u'(x)$, we apply the Taylor expansions of function $u(x)$ at points $x \pm h$ neighbouring to x:

$$u(x \pm h) = u(x) \pm u'(x)h + \frac{u''(x)h^2}{2!} \pm \frac{u'''(x)h^3}{3!} + \frac{u^{IV}(x)h^4}{4!} + O(h^5),$$ (3.4)

where $u''(x) \equiv d^2u/dx^2, u'''(x) \equiv d^3u/dx^3$, and so on, and $O(h^5)$ stands for the estimation of the series truncation error such as comparative to (in other words: on the order of) the 5th power of the magnitude of discretization interval h: $O(h^5) \leq const\ h^5$ for $h \to 0$.

Then, plugging Equation (3.4) in (3.1) for the forward difference scheme, we have

$$D^+u \equiv \frac{u(x + h) - u(x)}{h}$$

$$= \frac{1}{h} \left(u(x) + u'(x)h + \frac{u''(x)h^2}{2!} + \frac{u'''(x)h^3}{3!} + \frac{u^{IV}(x)h^4}{4!} + O(h^5) - u(x) \right)$$

$$= u'(x) + \frac{u''(x)h}{2!} + \frac{u'''(x)h^2}{3!} + \frac{u^{IV}(x)h^3}{4!} + O(h^4) = u'(x) + O(h).$$ (3.5)

By deducting from here derivative $u'(x)$, we obtain the forward difference approximation

$$u'(x) = D^+u + O(h),$$ (3.6)

which tells us that the error of such approximation is of the first order of magnitude with respect to the grid step h. Consequently, this scheme serves as the first-order approximation to the derivative $u'(x)$.

In the same way, we can prove that the backward difference scheme works pretty much with the same precision:

$$u'(x) = D^-u + O(h).$$ (3.7)

The accuracy of the derivative approximation can be improved if the central difference scheme is employed:

$$D^c u \equiv \frac{u(x+h) - u(x-h)}{2h}$$

$$= \frac{1}{2h}\left(u(x) + u'(x)h + \frac{u''(x)h^2}{2!} + \frac{u'''(x)h^3}{3!} + \frac{u^{IV}(x)h^4}{4!} + O(h^5)\right)$$

$$- \frac{1}{2h}\left(u(x) - u'(x)h + \frac{u''(x)h^2}{2!} - \frac{u'''(x)h^3}{3!} + \frac{u^{IV}(x)h^4}{4!} + O(h^5)\right)$$

$$= u'(x) + \frac{u'''(x)h^2}{3!} + O(h^4), \tag{3.8}$$

which yields

$$u'(x) = D^c u - \frac{u'''(x)h^2}{3!} + O(h^4) = D^c u + O(h^2). \tag{3.9}$$

If we compare the truncation errors of derivative approximations by the finite-difference schemes, then the second-order central difference scheme is definitely more accurate providing quadratic dependence on the grid spacing. It is illustrated graphically in Figure 3.2. The exact tangent in the point P (dashed line) is approximated more accurately by the slope of the central difference line (dotted line AB) then by the slopes of backward (AP) and forward difference (PB) lines. It is evident that the central difference scheme (3.9) requires the existence of the $u'''(x)$, while for the backward and forward differences, only the second-order differentiability of the function $u(x)$ is needed.

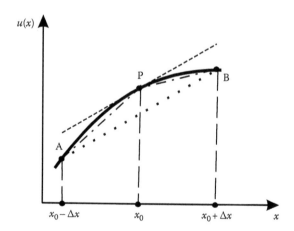

FIGURE 3.2 The central difference scheme applied here to calculate the approximate derivative of function $u(x)$ is more precise than backward and forward difference schemes. The accurate value of $u'(x_0)$ is given by the slope of the dashed line. The approximate values are given by the slopes of the dotted (AB) and dashed-dotted (AP and PB) lines.

Note: Remember that the derivative of function $u(x)$ at point x_0 is equal to the tangent of angle α, which is defined by the slope of the tangent line at point x_0: $\left. \dfrac{du}{dx} \right|_{x_0} = \tan \alpha$.

The set of points, where the values of a function are used for approximation of a derivative of this function, is called the *stencil* (of the corresponding differential operator). So, forward, backward, and central finite-difference schemes of the first-order derivative have two-point stencils. Among them, the D^c operator possesses the best (second-order) accuracy.

Note: Everything has its price. The price for the second-order accuracy of the central difference scheme is that we involve in calculations of the derivatives in point x function values at two other points $x \pm h$. You can extend this rule by the analogy to the third-, fourth-, or higher-order approximation, where at least 3, 4, or 5 others than x points will be involved in the calculation scheme.

The second-order derivative can be approximated by the finite differences with the same second-order accuracy in the following way:

$$u''(x) \equiv \frac{d^2u}{dx^2} \approx \frac{u(x+h) - 2u(x) + u(x-h)}{h^2}. \tag{3.10}$$

More about discretizations and finite-difference approximations for differential operators can be read in Section 3.3.

3.1.2 FINITE-DIFFERENCE APPROXIMATION OF 1D MAXWELL'S EQUATIONS

In the following sections, we implement finite-difference schemes for solving time-domain Maxwell's equation. Finite-difference solutions of Maxwell's equations in the frequency domain are considered in Chapter 4.

Electric and magnetic fields in a 1D problem are functions of two variables: spatial coordinate z and time t. For sampling continuous field functions, we introduce a homogeneous grid along the z coordinate with step h. It is represented by a discrete set of equally spaced points:

$$Z : \{z_i|_{i=0,1,2,\dots,N}, z_i = z_0 + ih\}. \tag{3.11}$$

The similar uniform grid is applied for the time variable. This is why the method is called the FDTD method:

$$T : \{t_n|_{n=0,1,2,\dots}, t_n = t_0 + n\delta t\}, \tag{3.12}$$

where δt is a step in time. Acronym FDTD was coined by Allen Taflove in 1980 [6]. We already know that the best result in approximation of the first derivatives with a two-point stencil provides the second-order central difference scheme D^c (3.9). We exploit it to approximate Equation (2.32) at the point (z_i, t_n) keeping current $j = 0$ for simplicity. We will refer to grid points in a shorthand manner as (i, n) and to finite differences in time and space as D_t^c and D_z^c, correspondingly.

So we have

$$\left.\frac{\partial E_x}{\partial t}\right|_i^n = -\frac{1}{\varepsilon_0 \varepsilon_i} \left.\frac{\partial H_y}{\partial z}\right|_i^n \Rightarrow \left.D_t^c E_x\right|_i^n = -\frac{1}{\varepsilon_0 \varepsilon_i} \left.D_z^c H_y\right|_i^n + O(h^2) + O(\delta t^2), \quad (3.13)$$

where $\varepsilon_i \equiv \varepsilon(z_i)$ is the grid-defined permittivity. Direct application of formula (3.8) gives

$$\left.D_t^c E_x\right|_i^n = \frac{E_x(z_i, t_n + \delta t) - E_x(z_i, t_n - \delta t)}{2\delta t} \equiv \frac{E_i^{n+1} - E_i^{n-1}}{2\delta t}, \quad (3.14)$$

$$\left.D_z^c H_y\right|_i^n = \frac{H_y(z_i + h, t_n) - H_y(z_i - h, t_n)}{2h} \equiv \frac{H_{i+1}^n - H_{i-1}^n}{2h}. \quad (3.15)$$

Here, we adopt the shortening designation: $E_x(z_i, t_n) \equiv E_i^n$, $H_y(z_i, t_n) \equiv H_i^n$, which will be used throughout the chapter.

It is possible to increase the accuracy of the finite-difference schemes by four times by using finer half-step grids $z_i = z_0 + ih/2$, $t_n = t_0 + n\delta t/2$; however, the number of grid points also grows by four times. Interestingly, it is possible to reduce the number of array points for storing electric and magnetic fields by a factor of two each, keeping the accuracy of the scheme as $O(h^2)$. Such trick is justified by the fact that electric field E_x is defined only by magnetic fields in the nearest neighbour points and vice versa. So E_x is positioned in every second site of the fine grid $z_i = z_0 + ih/2$, say with even i. So does H_y, but with odd i points. The same is applied for the time dependencies. One can consider the case as if there are two separate systems of grids: one only for electric field E_x and another only for magnetic field H_y. The grids are shifted by half steps in time and space relatively to each other.

We apply the half-step offsets for the t-grid of the electric field and the z-grid of the magnetic field. This yields

$$\left.D_t^c E_x\right|_i^n = \frac{E_i^{n+1/2} - E_i^{n-1/2}}{\delta t}, \quad (3.16)$$

$$\left.D_z^c H_y\right|_i^n = \frac{H_{i+1/2}^n - H_{i-1/2}^n}{h}, \quad (3.17)$$

where we introduce now the half-step offsets directly in the sub- and superscripts notation: $E_i^{n+1/2} \equiv E(z_0 + ih, t_0 + n\delta t + 1/2)$ and similar for other fields.

Spatial arrangement of the electric and magnetic fields at shifted points is known as the *staggered* grid in opposite to the collocated grid, where electric and magnetic fields are defined in the same points. The procedure of fields evolution implemented on the staggered grid with time offsets is called the *leapfrog* scheme in analogy of *electric* and *magnetic* frogs leaps.

So, instead of (3.13) through (3.15), we have

$$\frac{E_i^{n+1/2} - E_i^{n-1/2}}{\delta t} = -\frac{1}{\varepsilon_0 \varepsilon_i} \frac{H_{i+1/2}^n - H_{i-1/2}^n}{h}. \quad (3.18)$$

This equality is fulfilled with an error proportional to the squares of time and space increments, which is omitted. The expression can be reshuffled to explicitly express the latest-in-time field component, that is, $E_i^{n+1/2}$ through the previous ones. Deriving the formula for the latest electric field gives us a tool to predict (update) electric field in the next time moment if we know the inputs to this formula: the previous (old) electric field in this point $E_i^{n-1/2}$ and magnetic fields in the two closest neighbour points a half of the time step earlier $H_{i\pm1/2}^n$. It reads

$$E_i^{n+1/2} = E_i^{n-1/2} - \frac{\delta t}{h\varepsilon_0\varepsilon_i}(H_{i+1/2}^n - H_{i-1/2}^n). \tag{3.19}$$

The leapfrog scheme of the FDTD method is illustrated in Figure 3.3. Equation (3.19) realizes connections shown by blue arrows.

Now, we approximate the second Maxwell's equation (2.31). It will serve for updating the magnetic vector, so we centre the finite-difference stencils at mesh point $(z_{i+1/2}, t_{n+1/2})$.

$$\frac{\partial H_y}{\partial t}\bigg|_{i+1/2}^{n+1/2} = -\frac{1}{\mu_0\mu_{i+1/2}}\frac{\partial E_x}{\partial z}\bigg|_{i+1/2}^{n+1/2} \Rightarrow$$

$$D_t^c H_y\big|_{i+1/2}^{n+1/2} = -\frac{1}{\mu_0\mu_{i+1/2}}D_z^c E_x\big|_{i+1/2}^{n+1/2}, \tag{3.20}$$

$$\mu_{i+1/2} \equiv \mu(z_{i+1/2}). \tag{3.21}$$

Then, finite differences on the staggered grid with leapfrogging in time give

$$\frac{H_{i+1/2}^{n+1} - H_{i+1/2}^n}{\delta t} = -\frac{1}{\mu_0\mu_{i+1/2}}\frac{E_{i+1}^{n+1/2} - E_i^{n+1/2}}{h}, \tag{3.22}$$

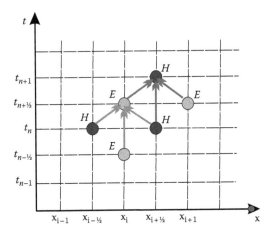

FIGURE 3.3 (See colour insert.) Leapfrog time updates of electric (green) and magnetic (red) fields are shown by arrows: blue for the electric fields update (3.19) and brown for the magnetic (3.23).

with straightforward transformation into the second updating equation for the magnetic field

$$H_{i+1/2}^{n+1} = H_{i+1/2}^n - \frac{\delta t}{h\mu_0\mu_{i+1/2}}\left(E_{i+1}^{n+1/2} - E_i^{n+1/2}\right). \tag{3.23}$$

Equation (3.23) is represented in the FDTD scheme (Figure 3.3) by a set of brown arrows.

Equations (3.19) and (3.23) are the core expressions of the FDTD method in 1D. They reflect the essential property of the electromagnetic field, which can exist and propagate in vacuum – generating the electric field by altering magnetic one and vice versa.

3.1.3 FORTRAN, C, MATLAB®, ETC., ADAPTATION OF THE FDTD METHOD

To implement economically the FDTD method in Fortran, C++, or any other pro-gramming language, the E and H array indexes must be integers. Therefore, we imply virtual half-step shifts on staggered grids to restore their integer numbering. It really doesn't matter to which fields to apply positive or negative shifts. It is our conven-tion, but once implying, we should keep it in mind all the time during the program utilization.

Our optional choice here is to push all time points for the electric field by $\delta t/2$ forward and all z points for the magnetic field by $-h/2$ backward. Then Equations (3.19) and (3.23) read

$$E_i^{n+1} = E_i^n - \frac{\delta t}{h\varepsilon_i}(H_i^n - H_{i-1}^n), \tag{3.24}$$

$$H_i^{n+1} = H_i^n - \frac{\delta t}{h\mu_i}(E_{i+1}^{n+1} - E_i^{n+1}). \tag{3.25}$$

Note: A careful analysis of these equations allows to stress that in the formal way, it brings us to the forward difference scheme for the electric field but to the backward difference scheme for magnetic field. Such scheme of disctretization greatly simplifies the application of finite differences providing that we keep in mind only one collocated grid with *integer* points for all fields.

The second adaptation rule is connected with numerical values of the fields. Initial Maxwell's equations (2.2) through (2.5) are written in the SI units, where the electric field is measured in V/m and magnetic in A/m. However, in many of photonic or electromagnetic applications such as calculation of transmission and reflection coefficients and mapping fields, the exact values of the fields are redundant. Moreover, the magnetic field of an electromagnetic wave measured in SI units is several orders of magnitudes less than its electric field. This is why it might be useful to redefine fields in the FDTD algorithm making it simpler and more symmetric. We already discussed the units for the permittivity and permeability and some fundamental constants associated with them (see Equations (2.8) and (2.9)). These constants can be used to symmetrize the appearance of 1D Maxwell's equations and to equalize the numerical values of electric and magnetic fields.

If we use the vacuum impedance (2.9) and define new electric field in such way

$$\hat{E} = \frac{E}{\eta_0}, \tag{3.26}$$

then inserting it into Equations (3.24) and (3.25) leads to the symmetrical updating scheme:

$$\hat{E}_i^{n+1} = \hat{E}_i^n - \frac{Q}{\varepsilon_i}(H_i^n - H_{i-1}^n), \tag{3.27}$$

$$H_i^{n+1} = H_i^n - \frac{Q}{\mu_i}(\hat{E}_{i+1}^{n+1} - \hat{E}_i^{n+1}). \tag{3.28}$$

As we see further, dimensionless parameter

$$Q = \frac{c_0 \delta t}{h} \tag{3.29}$$

plays an important role in analysis of the stability of the FDTD method and choice of the time increment size.

Equations (3.27) and (3.28) with the redefined electric field or their SI analogues (3.24) and (3.25) are the basic formulas for the realization of the 1D FDTD evolution scheme.

3.1.4 FDTD METHOD IN 3D

Extension of the FDTD method to 3D problems is straightforward. We should start from vectorial Maxwell's equations (2.2) and (2.3) and consistently discretize them on the homogeneous Yee grid (or Yee mesh); see a unit Yee cell in Figure 3.4. This grid was originally proposed by Kane Yee in 1966 [7], the author of the FDTD method. The Yee grid realizes leapfrog time update of the fields on a staggered 3D mesh and thus represents the core of the FDTD method. Components of the electric fields are defined in the middle of the edges of the cube and magnetic field components in the centres of cube faces. Such disposition was done on purpose to arrange continuity of relevant field components (tangential components of **E** and **H**) at the interfaces. The 2D Yee grid is considered in Section 4.5.

Applying the finite-difference scheme on the Yee grid (with leapfrogging), we get with some rearranging the set of explicit fields updating expressions:

$$E_x|_{i+1/2,j,k}^{n+1/2} = E_x|_{i+1/2,j,k}^{n-1/2} + \frac{\delta t}{\varepsilon_0 \varepsilon_{i+1/2,j,k}} \times \left(\frac{H_z|_{i+1/2,j+1/2,k}^n - H_z|_{i+1/2,j-1/2,k}^n}{\delta y} \right.$$

$$\left. - \frac{H_y|_{i+1/2,j,k+1/2}^n - H_y|_{i+1/2,j,k-1/2}^n}{\delta z} \right),$$

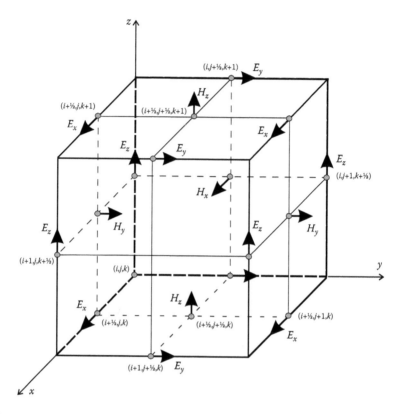

FIGURE 3.4 The Yee grid in 3D. Not all of the fields components in the Yee cell are shown for the sake of clarity of presentation.

$$
E_y|_{i,j+1/2,k}^{n+1/2} = E_y|_{i,j+1/2,k}^{n-1/2} + \frac{\delta t}{\varepsilon_0 \varepsilon_{i,j+1/2,k}} \times \left(\frac{H_x|_{i,j+1/2,k+1/2}^n - H_x|_{i,j+1/2,k-1/2}^n}{\delta z} \right.
$$

$$
\left. - \frac{H_z|_{i+1/2,j+1/2,k}^n - H_z|_{i-1/2,j+1/2,k}^n}{\delta x} \right),
$$

$$
E_z|_{i,j,k+1/2}^{n+1/2} = E_z|_{i,j,k+1/2}^{n-1/2} + \frac{\delta t}{\varepsilon_0 \varepsilon_{i,j,k+1/2}} \times \left(\frac{H_y|_{i+1/2,j,k+1/2}^n - H_y|_{i-1/2,j,k+1/2}^n}{\delta x} \right.
$$

$$
\left. - \frac{H_x|_{i,j+1/2,k+1/2}^n - H_x|_{i,j-1/2,k+1/2}^n}{\delta y} \right), \tag{3.30}
$$

$$
H_x|_{i,j+1/2,k+1/2}^{n+1} = H_x|_{i,j+1/2,k+1/2}^n + \frac{\delta t}{\mu_0 \mu_{i,j+1/2,k+1/2}}
$$

$$
\times \left(\frac{E_y|_{i,j+1/2,k+1}^{n+1/2} - E_y|_{i,j+1/2,k}^{n+1/2}}{\delta z} - \frac{E_z|_{i,j+1,k+1/2}^{n+1/2} - E_z|_{i,j,k+1/2}^{n+1/2}}{\delta y} \right),
$$

$$H_y\big|_{i+1/2,j,k+1/2}^{n+1} = H_y\big|_{i+1/2,j,k+1/2}^{n} + \frac{\delta t}{\mu_0 \mu_{i+1/2,j,k+1/2}}$$

$$\times \left(\frac{E_z\big|_{i+1,j,k+1/2}^{n+1/2} - E_z\big|_{i,j,k+1/2}^{n+1/2}}{\delta x} - \frac{E_x\big|_{i+1/2,j,k+1}^{n+1/2} - E_x\big|_{i+1/2,j,k}^{n+1/2}}{\delta z} \right),$$

$$H_z\big|_{i+1/2,j+1/2,k}^{n+1} = H_z\big|_{i+1/2,j+1/2,k}^{n} + \frac{\delta t}{\mu_0 \mu_{i+1/2,j+1/2,k}}$$

$$\times \left(\frac{E_x\big|_{i+1/2,j+1,k}^{n+1/2} - E_x\big|_{i+1/2,j,k}^{n+1/2}}{\delta y} - \frac{E_y\big|_{i+1,j+1/2,k}^{n+1/2} - E_y\big|_{i,j+1/2,k}^{n+1/2}}{\delta x} \right).$$

In analogy with the 1D case (the Note in the previous section), we can get the 3D FDTD equations on a pseudo-collocated cubic mesh ($\delta x = \delta y = \delta z = h$) just following the rule that **E**-vector derivatives are approximated by forward differences and **H**-vector derivatives by backward differences [8]. Here, they are

---

Fields Update Equations in 3D

$$E_{x;i,j,k}^{n+1} = E_{x;i,j,k}^{n} + \frac{Q}{\varepsilon_{i,j,k}} \left(H_{z;i,j,k}^{n} - H_{z;i,j-1,k}^{n} - H_{y;i,j,k}^{n} + H_{y;i,j,k-1}^{n} \right),$$

$$E_{y;i,j,k}^{n+1} = E_{y;i,j,k}^{n} + \frac{Q}{\varepsilon_{i,j,k}} \left(H_{x;i,j,k}^{n} - H_{x;i,j,k-1}^{n} - H_{z;i,j,k}^{n} + H_{z;i-1,j,k}^{n} \right),$$

$$E_{z;i,j,k}^{n+1} = E_{z;i,j,k}^{n} + \frac{Q}{\varepsilon_{i,j,k}} \left(H_{y;i,j,k}^{n} - H_{y;i-1,j,k}^{n} - H_{x;i,j,k}^{n} + H_{x;i,j-1,k}^{n} \right),$$

$$H_{x;i,j,k}^{n+1} = H_{x;i,j,k}^{n} + \frac{Q}{\mu_{i,j,k}} \left(E_{y;i,j,k+1}^{n+1} - E_{y;i,j,k}^{n+1} - E_{z;i,j+1,k}^{n+1} + E_{z;i,j,k}^{n+1} \right),$$

$$H_{y;i,j,k}^{n+1} = H_{y;i,j,k}^{n} + \frac{Q}{\mu_{i,j,k}} \left(E_{z;i+1,j,k}^{n+1} - E_{z;i,j,k}^{n+1} - E_{x;i,j,k+1}^{n+1} + E_{x;i,j,k}^{n+1} \right),$$

$$H_{z;i,j,k}^{n+1} = H_{z;i,j,k}^{n} + \frac{Q}{\mu_{i,j,k}} \left(E_{x;i,j+1,k}^{n+1} - E_{x;i,j,k}^{n+1} - E_{y;i+1,j,k}^{n+1} + E_{y;i,j,k}^{n+1} \right),$$

(3.31)

Note: It is easy to reproduce all equations starting with any of them by taking into account the cycling procedure of changing components $x \to y \to z \to x$ and relevant grid indices $i \to j \to k \to i$.

Note: Equations should be strictly used in the *field-update* order, because all electric and magnetic field components are interweaved in the right-hand parts of expressions (3.31). For instance, first, all electric vector components are updated, then all magnetic vector components.

3.1.5 FDTD METHOD IN 2D

In a 2D case due to the absence of any dependencies of fields on the x-coordinate, six scalar Maxwell's equations split into two groups called TE and TM waves (see Section 2.5). In the finite-difference adaptation, it means that there are no i indices and all finite-difference operators D_x must be nullified. Then from (3.30), we deduce

$$H_x|_{j+1/2,k+1/2}^{n+1} = H_x|_{j+1/2,k+1/2}^{n} + \frac{\delta t}{\mu_0 \mu_{j+1/2,k+1/2}}$$

$$\times \left(\frac{E_y|_{j+1/2,k+1}^{n+1/2} - E_y|_{j+1/2,k}^{n+1/2}}{\delta z} - \frac{E_z|_{j+1,k+1/2}^{n+1/2} - E_z|_{j,k+1/2}^{n+1/2}}{\delta y} \right),$$

$$E_y|_{j+1/2,k}^{n+1/2} = E_y|_{j+1/2,k}^{n-1/2} + \frac{\delta t}{\varepsilon_0 \varepsilon_{j+1/2,k}} \left(\frac{H_x|_{j+1/2,k+1/2}^{n} - H_x|_{j+1/2,k-1/2}^{n}}{\delta z} \right),$$

$$E_z|_{j,k+1/2}^{n+1/2} = E_z|_{j,k+1/2}^{n-1/2} + \frac{\delta t}{\varepsilon_0 \varepsilon_{j,k+1/2}} \left(-\frac{H_x|_{j+1/2,k+1/2}^{n} - H_x|_{j-1/2,k+1/2}^{n}}{\delta y} \right),$$

for the TE mode and

$$E_x|_{j,k}^{n+1/2} = E_x|_{j,k}^{n-1/2} + \frac{\delta t}{\varepsilon_0 \varepsilon_{j,k}} \left(\frac{H_z|_{j+1/2,k}^{n} - H_z|_{j-1/2,k}^{n}}{\delta y} - \frac{H_y|_{j,k+1/2}^{n} - H_y|_{j,k-1/2}^{n}}{\delta z} \right),$$

$$H_y|_{j,k+1/2}^{n+1} = H_y|_{j,k+1/2}^{n} + \frac{\delta t}{\mu_0 \mu_{j,k+1/2}} \left(-\frac{E_x|_{j,k+1}^{n+1/2} - E_x|_{j,k}^{n+1/2}}{\delta z} \right),$$

$$H_z|_{j+1/2,k}^{n+1} = H_z|_{j+1/2,k}^{n} + \frac{\delta t}{\mu_0 \mu_{i+1/2,j+1/2,k}} \left(\frac{E_x|_{i+1/2,j+1,k}^{n+1/2} - E_x|_{i+1/2,j,k}^{n+1/2}}{\delta y} \right),$$

for the TM mode.

Thorough analysis of different multidimensional grids and various aspects of the FDTD implementation can be found in books [1,2,5].

3.2 NUMERICAL DISPERSION AND STABILITY ANALYSIS OF THE FDTD METHOD

Numerical methods, in particular the FDTD method, deal with discrete rather than continuous functions. Substitution of continuous functions by their table approximates defined on a grid causes many analytically derived properties to fail.

For example, limits, derivatives, and integral operations become approximated by finite differences of the table function values, which, in turn, are determined on a grid. Thus, results in such approximations are grid dependent. As a direct consequence, even a plane wave propagation in vacuum described on a grid acquires dispersion and anisotropy. Such spurious properties exhibited by wave propagation due to the tabulation of functions are named numerical dispersion, numerical anisotropy, etc. Therefore, we can anticipate that some of the well-known properties of the trial solutions of Maxwell's equations should be carefully revised.

In this section, we analyze what implications in numerical electrodynamics are brought by discrete functions defined on a grid. We are taking a careful look on these properties and their influence on the FDTD algorithm. Such learning is the essential part of researchers' adaptation to the numerical realm ruling in modelling. The section contains basic knowledge on stability analysis, numerical dispersion, and check of the FDTD scheme for the divergence-free behaviour of the field-updating engine. Regarding the topics described here, Exercises 3.3 through 3.10 can be recommended for practicing with the material.

3.2.1 DISPERSION EQUATION IN 3D

The classical dispersion equation, which links frequency ω and wave vector $\mathbf{k}$ for a monochromatic plane wave in vacuum, is

$$\omega^2 = \mathbf{k}^2 c_0^2. \tag{3.32}$$

This equation acts in the continuous space, where fields are represented by continuous functions. When fields are defined on a grid, it has to be substituted by another equation, which we derive later.

We start with *curl* Maxwell's equations (2.2) and (2.3) in vacuum:

$$\frac{\partial \mathbf{E}}{\partial t} = \frac{1}{\varepsilon_0} \nabla \times \mathbf{H}, \tag{3.33}$$

$$\frac{\partial \mathbf{H}}{\partial t} = -\frac{1}{\mu_0} \nabla \times \mathbf{E}, \tag{3.34}$$

rewritten regarding the time derivative terms. Switching to scalar equations with the field components and substituting derivatives by the corresponding central difference operators (3.14), (3.15) gives

$$D_t^c E_x = \frac{1}{\varepsilon_0}(D_y^c H_z - D_z^c H_y), \quad D_t^c E_y = \frac{1}{\varepsilon_0}(D_z^c H_x - D_x^c H_z),$$

$$D_t^c E_z = \frac{1}{\varepsilon_0}(D_x^c H_y - D_y^c H_x), \quad D_t^c H_x = -\frac{1}{\mu_0}(D_y^c E_z - D_z^c E_y), \tag{3.35}$$

$$D_t^c H_x = -\frac{1}{\mu_0}(D_z^c E_x - D_x^c E_z), \quad D_t^c H_x = -\frac{1}{\mu_0}(D_x^c E_y - D_y^c E_x).$$

These equations must have a solution like a travelling grid-based plane wave:

$$\mathbf{E}(x, y, z, t) \equiv \mathbf{E}(a, b, c, n) = \mathbf{E}_0 \exp\left[i\left(\omega n \delta t - k_x a \delta x - k_y b \delta y - k_z c \delta z\right)\right], \quad (3.36)$$

and similar for the magnetic vector. Note that we follow $e^{i\omega t}$ convention. Here, by applying different grid steps in x-, y-, z-directions, we acknowledge an arbitrary rectangular 3D grid. A central difference time operator applied on the field $\mathbf{E}$ in point (a, b, c, n) of such grid gives

$$D_t^c \mathbf{E}(x, y, z, t)\big|_{abc}^n = \frac{\mathbf{E}_{abc}^{n+1/2} - \mathbf{E}_{abc}^{n-1/2}}{\delta t}$$

$$= \mathbf{E}(a, b, c, n) \frac{e^{i\omega\delta t/2} - e^{-i\omega\delta t/2}}{\delta t}$$

$$= \mathbf{E}(a, b, c, n) \frac{2i\sin(\omega\delta t/2)}{\delta t}, \quad (3.37)$$

where definition of the sine function through a complex exponential function is applied. In analogy, a spatial central difference gives

$$D_x^c \mathbf{E}(x, y, z, t)\big|_{abc}^n = \mathbf{E}(a, b, c, n) \frac{-2i\sin(k_x\delta x/2)}{\delta x}, \quad (3.38)$$

and corresponding expressions for other coordinates.

Eventually, six scalar equations (Equation (3.35)) represent a homogeneous system of linear equations and can be written in the matrix form:

$$\begin{pmatrix} T & 0 & 0 & 0 & -Z/\varepsilon_0 & Y/\varepsilon_0 \\ 0 & T & 0 & Z/\varepsilon_0 & 0 & -X/\varepsilon_0 \\ 0 & 0 & T & -Y/\varepsilon_0 & X/\varepsilon_0 & 0 \\ 0 & Z/\mu_0 & -Y/\mu_0 & T & 0 & 0 \\ -Z/\mu_0 & 0 & X/\mu_0 & 0 & T & 0 \\ Y/\mu_0 & -X/\mu_0 & 0 & 0 & 0 & T \end{pmatrix} \begin{pmatrix} E_x \\ E_y \\ E_z \\ H_x \\ H_y \\ H_z \end{pmatrix} = 0. \quad (3.39)$$

where $T = \sin(\omega\delta t/2)/\delta t$, $X = \sin(k_x\delta x/2)/\delta x$, $Y = \sin(k_y\delta y/2)/\delta y$, and $Z = \sin(k_z\delta z/2)/\delta z$.

This system has a non-trivial solution if and only if the determinant of the matrix with coefficients

$$\begin{vmatrix} T & 0 & 0 & 0 & -Z/\varepsilon_0 & Y/\varepsilon_0 \\ 0 & T & 0 & Z/\varepsilon_0 & 0 & -X/\varepsilon_0 \\ 0 & 0 & T & -Y/\varepsilon_0 & X/\varepsilon_0 & 0 \\ 0 & Z/\mu_0 & -Y/\mu_0 & T & 0 & 0 \\ -Z/\mu_0 & 0 & X/\mu_0 & 0 & T & 0 \\ Y/\mu_0 & -X/\mu_0 & 0 & 0 & 0 & T \end{vmatrix} = 0. \quad (3.40)$$

Considering it as a 2×2 block matrix $\overline{\overline{M}} = \begin{pmatrix} \overline{\overline{A}} & \overline{\overline{B}} \\ \overline{\overline{C}} & \overline{\overline{D}} \end{pmatrix}$, the determinant $|\overline{\overline{M}}|$ can be

computed via $|\overline{\overline{M}}| = |\overline{\overline{AD}} - \overline{\overline{AC}}(\overline{\overline{A}})^{-1} \overline{\overline{B}}| = |(\overline{\overline{A}})^2 - \overline{\overline{CB}}|$ [9], where we take into account that $\overline{\overline{A}} = \overline{\overline{D}}$ and matrices $\overline{\overline{A}}$ (diagonal) and $\overline{\overline{C}}$ commute $\overline{\overline{AC}} = \overline{\overline{CA}}$. After cumbersome calculations with expansion of the determinant of a 3×3 matrix, the 3D equation describing numerical dispersion is derived:

$$\frac{1}{c_0^2 \delta t^2} \sin^2 \frac{\omega \delta t}{2} = \left(\frac{1}{\delta x} \sin \frac{k_x \delta x}{2} \right)^2 + \left(\frac{1}{\delta y} \sin \frac{k_y \delta y}{2} \right)^2 + \left(\frac{1}{\delta z} \sin \frac{k_z \delta z}{2} \right)^2. \quad (3.41)$$

This equation noticeably differs from (3.32) certifying a new property acquired by plane (and all other) waves propagating on a grid – *numerical dispersion*. Numerical dispersion means dependence of a phase velocity on the wavelength (material dispersion), direction of propagation (anisotropy), and lattice discretization (numerical artefact); see Exercises 3.4, 3.5, and 3.10. It can lead to pulse distortion, artificial anisotropy, and spurious refraction at the boundaries of different grids even solving problem of wave propagation in vacuum or uniform dielectric.

It is straightforward to derive equations that describe numerical dispersion on 2D and 1D grids. Instead of repeating all steps, it is enough to reduce (3.41) to two dimensions by omitting k_z term and to 1D grid by omitting the terms k_y and k_z:

$$\frac{1}{c_0^2 t^2} \sin^2 \frac{\omega \delta t}{2} = \left(\frac{1}{\delta x} \sin \frac{k_x \delta x}{2} \right)^2 + \left(\frac{1}{\delta y} \sin \frac{k_y \delta y}{2} \right)^2, \quad (3.42)$$

$$\frac{1}{c_0^2 t^2} \sin^2 \frac{\omega \delta t}{2} = \left(\frac{1}{\delta x} \sin \frac{k_x \delta x}{2} \right)^2. \quad (3.43)$$

Using trigonometric expression $\sin^2 \alpha = \cos 2\alpha - 1$, the last equation can be transformed into

$$\cos \omega \delta t - 1 = Q^2 (\cos k_x \delta x - 1). \quad (3.44)$$

3.2.2 NUMERICAL STABILITY CRITERIA

It is clear that the choice of time and coordinate grid spacings influences the accurateness and robustness of a numerical scheme a lot. We are especially interested in any constrains on time step δt, because it defines the scenery and dynamical resolution of fields evolution in time-domain methods. A δt too small will make the total computation routine extensive; a δt too big will reasonably violate the causal connection between fields updates in different grid points. To estimate the restrains on the acceptable time step range, we need a numerical scheme stability analysis.

Assuming that the frequency can be extended to complex values $\omega = \omega' + i\omega''$, Equation (3.41) can be solved in a very formal way. We introduce variable ξ:

$$\xi = c_0 \delta t \sqrt{\left(\frac{1}{\delta x} \sin\left(\frac{k_x \delta x}{2}\right)\right)^2 + \left(\frac{1}{\delta y} \sin\left(\frac{k_y \delta y}{2}\right)\right)^2 + \left(\frac{1}{\delta z} \sin\left(\frac{k_z \delta z}{2}\right)\right)^2}. \quad (3.45)$$

Then dispersion equation (3.41)

$$\sin^2\left(\frac{\omega \delta t}{2}\right) = \xi^2 \quad (3.46)$$

can be straightforwardly solved for the frequency

$$\omega = \frac{2}{\delta t} \sin^{-1} \xi \equiv \frac{2}{\delta t} \arcsin \xi. \quad (3.47)$$

In order to have a travelling wave solution (like the one described by Equation (3.36)), which can propagate without losses for infinite time (the ideal case), frequency ω must be constrained to real values. So, condition $\omega'' \equiv Im(\omega) = 0$ claims that $|\xi| \leq 1$. Assuming wave vector $\mathbf{k}$ being real (lossless propagation) and limiting all real sin-functions by their supreme values, we arrive at the inequality [7]

$$\delta t \leq \frac{1}{c_0} \left\{ \left(\frac{1}{\delta x}\right)^2 + \left(\frac{1}{\delta y}\right)^2 + \left(\frac{1}{\delta z}\right)^2 \right\}^{-1/2}. \quad (3.48)$$

This condition is called the *Courant stability criterion* or *Courant–Friedrichs–Lewy condition* (CFL) in mathematical literature. It connects the sizes of grid intervals in space and time increment, effectively limiting δt in the FDTD method with the fixed grid. For the cubic 3D mesh,

$$\delta t \leq \frac{\delta x}{c_0 \sqrt{3}}, \quad (3.49)$$

and for the square 2D mesh,

$$\delta t \leq \frac{\delta x}{c_0 \sqrt{2}}. \quad (3.50)$$

The easiest connection between δt and δx is in the 1D case:

$$\delta t \leq \frac{\delta x}{c_0}. \quad (3.51)$$

It is convenient to operate with the unitless ratio $c_0 \delta t / \delta x$ often called the *Courant number*, which has been already labelled by Q (3.29). It is clear that in a 1D space, the

Courant criteria (3.51) dictates that $Q \le 1$, where the equality defines the so-called magic step.

It is worth to illustrate in 1D what can happen if the Courant stability criteria (3.48) is violated and $Q > 1$. Then, certainly, $\xi > 1$. It means that frequency ω is a complex number, which stems from Equation (3.47). To handle complex numbers and inverse trigonometric functions, the following mathematical equality is useful:

$$\arcsin \alpha = -i \ln(i\alpha + \sqrt{1 - \alpha^2}). \tag{3.52}$$

Note: The proof of this equality can be done with the help of the exponential representation of sin function (the Euler formula):

$$\sin \beta = \frac{e^{i\beta} - e^{-i\beta}}{2i}, \tag{3.53}$$

which can be rewritten and solved as a quadratic equation with respect to $e^{i\beta}$, with subsequent transformation first to β and eventually to $\alpha = \sin \beta$.

Transforming arcsin function in Equation (3.47) accordingly to (3.52), we obtain

$$\omega \delta t = -2i \ln(i\xi + \sqrt{1 - \xi^2}) = -2i \ln\left(e^{i\pi/2}(\xi + \sqrt{\xi^2 - 1})\right)$$

$$= \pi - 2i \ln(\xi + \sqrt{\xi^2 - 1}). \tag{3.54}$$

Substituting the derived complex product in the exponential phase function, it is easy to get

$$e^{i\omega t} = e^{in\omega \delta t} = e^{i\pi n}\left(\xi + \sqrt{\xi^2 - 1}\right)^{2n}, \tag{3.55}$$

which indicates an unlimited growth of electric and magnetic fields with time ($t = n\delta t$) taking into account that $\xi > 1$.

To illustrate the stability criteria, we performed 1D simulations of propagation of a pulse in a homogeneous medium with different parameters Q (Figure 3.5). Three values are used: $Q = 0.99$, $Q = 1.00$, and $Q = 1.01$, which in the terms of a time increment equal to 99%, 100% (magic step), and 101% of the upper limit of δt. The whole time scale of 0.3 ps is equivalent to about 3000 time steps. Results for the time steps obeying the stability criteria (3.51) nearly overlap and show very stable field-updating process. However, even small violation of the criteria leads to the constant unlimited exponential growth of the computed fields; see the exponential growth of the largest field component in the whole domain with time in Figure 3.5. Note that changing parameter Q in Equation (3.27) actually means the change in the ratio $\delta t / \delta x$. For the same time, scaling implies that we vary the grid sized accordingly to alterations in parameter Q.

3.2.3 DIVERGENCE-FREE CHARACTER OF THE FDTD METHOD

So far, we have been dealing with two *curl* Maxwell's equations (2.2) and (2.3). Two other equations (2.4) and (2.5) are automatically obeyed by continuous

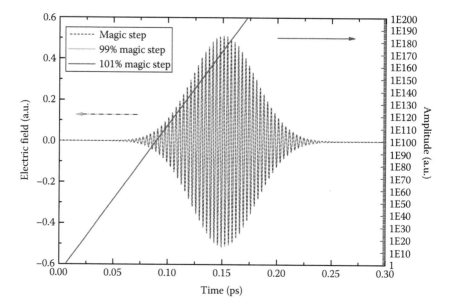

FIGURE 3.5 (**See colour insert.**) Fields in the FDTD scheme with different choice of the time step. Left scale: the electric fields in the pulse for $Q = 0.99$ and $Q = 1.00$ are unseparated. Right scale: the amplitude of the biggest electric field in the numerical domain for $Q = 1.01$.

electromagnetic fields in isotropic homogeneous dielectrics due to the *curl* properties of the Maxwellian electromagnetic field: closed field lines equalize the number of lines entering and leaving any closed volume. However, this property is not automatically guaranteed for grid-defined functions. So, a robust numerical method for solving Maxwell's equations (so-called Maxwell's solver) must satisfy discrete *div* equations too.

Note: Failure of nullifying the divergence operator $\nabla \cdot \mathbf{E}$ in vacuum means physically existence of free electric charges with density $\rho \neq 0$. If we consider properties of electromagnetic fields, then generation of such spurious charges may lead to instabilities in numerical algorithms and nonphysical results.

The FDTD method proposed by K. Yee with the leapfrog updating scheme on the staggered grid is divergence-free [2]. The proof for the case of electric field follows.

Operator ∇ applied through the scalar product on electric and magnetic vectors corresponds to the differential form of Maxwell's equations. The Gauss theorem links the divergence operator integrated over volume V with the flux of the vectorial field through closed surface S of this volume:

$$\int_V \nabla \cdot \mathbf{E} dV = \oint_S \mathbf{E} \cdot \mathbf{n} dS. \tag{3.56}$$

Therefore, it is enough to show that the FDTD algorithm does not generate any spurious electric charges in a closed volume while evolving electric fields in time.

We choose one unit cell of the Yee grid. What we are going to show is that no false charges appear in the fields evolution according to the finite-difference algorithm for the solution of Maxwell's equations or simply

$$\frac{\partial}{\partial t} \oint_S \mathbf{E} \cdot \mathbf{n} dS = 0. \tag{3.57}$$

For a cubic-shaped Yee cell centred at $i + 1/2, j + 1/2, k + 1/2$ (Figure 3.4), it reads that fluxes of the electric field through all six faces of the unit cell cube are not changing with time. Then

$$
\begin{aligned}
\frac{\partial}{\partial t} \oint_S \mathbf{E} \cdot \mathbf{n} dS = & \left(\frac{\partial E_x}{\partial t} \Big|_{i+1,j+1/2,k+1/2} - \frac{\partial E_x}{\partial t} \Big|_{i,j+1/2,k+1/2} \right) \delta y \delta z \\
& + \left(\frac{\partial E_y}{\partial t} \Big|_{i+1/2,j+1,k+1/2} - \frac{\partial E_y}{\partial t} \Big|_{i+1/2,j,k+1/2} \right) \delta x \delta z \\
& + \left(\frac{\partial E_z}{\partial t} \Big|_{i+1/2,j+1/2,k+1} - \frac{\partial E_z}{\partial t} \Big|_{i+1/2,j+1/2,k} \right) \delta x \delta y. \tag{3.58}
\end{aligned}
$$

Partial time derivatives of field $\mathbf{E}$ can be approximated by the central difference schemes with field $\mathbf{H}$, accordingly to the component disposition on the Yee grid (see Equation (2.2)). For example,

$$
\varepsilon_0 \frac{\partial E_x}{\partial t} \Big|_{i+1,j+1/2,k+1/2} = \frac{H_{z;i+1,j+1,k+1/2} - H_{z;i+1,j,k+1/2}}{\delta y}
$$

$$
- \frac{H_{y;i+1,j+1/2,k+1} - H_{y;i+1,j+1/2,k}}{\delta z},
$$

$$
\varepsilon_0 \frac{\partial E_x}{\partial t} \Big|_{i,j+1/2,k+1/2} = \frac{H_{z;i,j+1,k+1/2} - H_{z;i,j,k+1/2}}{\delta y}
$$

$$
- \frac{H_{y;i,j+1/2,k+1} - H_{y;i,j+1/2,k}}{\delta z},
$$

$$
\varepsilon_0 \frac{\partial E_y}{\partial t} \Big|_{i+1/2,j+1,k+1/2} = \frac{H_{x;i+1/2,j+1,k+1} - H_{x;i+1/2,j+1,k}}{\delta z}
$$

$$
- \frac{H_{z;i+1,j+1,k+1/2} - H_{z;i,j+1,k+1/2}}{\delta x},
$$

$$
\varepsilon_0 \frac{\partial E_y}{\partial t} \Big|_{i+1/2,j,k+1/2} = \frac{H_{x;i+1/2,j,k+1} - H_{x;i+1/2,j,k}}{\delta z}
$$

$$
- \frac{H_{z;i+1,j,k+1/2} - H_{z;i,j,k+1/2}}{\delta x},
$$

$$\varepsilon_0 \frac{\partial E_z}{\partial t}\bigg|_{i+1/2,j+1/2,k+1} = \frac{H_{y;i+1,j+1/2,k+1} - H_{y;i,j+1/2,k+1}}{\delta x}$$

$$- \frac{H_{z;i+1/2,j+1,k+1} - H_{z;i+1/2,j,k+1}}{\delta y},$$

$$\varepsilon_0 \frac{\partial E_z}{\partial t}\bigg|_{i+1/2,j+1/2,k} = \frac{H_{y;i+1,j+1/2,k} - H_{y;i,j+1/2,k}}{\delta x}$$

$$- \frac{H_{z;i+1/2,j+1,k} - H_{z;i+1/2,j,k}}{\delta y}.$$

Here we exploited the cycle substitution $x \to y \to z \to x$ in subindexes and coordinates with the relevant changes in grid numbers. After substituting all finite differences in expression (3.58), the time derivatives are transformed to summation over 24 magnetic field components. Same field components can be grouped. For example, there are eight H_z terms, all multiplied by δz. Combining all such terms, it is easy to notice that they appear in pairs with joint cancelation, the first and the forth, the second and the seventh, etc.:

$$H_{z;i+1,j+1,k+1/2} - H_{z;i+1,j,k+1/2} - H_{z;i,j+1,k+1/2} + H_{z;i,j,k+1/2}$$

$$- H_{z;i+1,j+1,k+1/2} + H_{z;i,j+1,k+1/2} + H_{z;i+1,j,k+1/2} - H_{z;i,j,k+1/2} = 0. \qquad (3.59)$$

With some patience, the same result can be obtained for other x- and y-components. So we have just confirmed that the total flux of the electric field is not changing in time. It can be easily equalized to 0, meaning the absence of electric charges in vacuum at the zero time. Thus, we conclude that the unit Yee cell does not generate any fictitious charges and divergence of the electric field on the grid is, therefore, equal to zero.

3.3 MAKING YOUR OWN 1D FDTD

Now, having derived updating equations for electric and magnetic fields (3.24) and (3.25) or in the normalized form (3.27) and (3.28), we can actually proceed with time evolution of the fields.

The task of this section is to pave a smooth and easy way of making your own code for numerical simulations. When one gets a problem to be solved numerically (especially for the first time), typical questions are as follows: how to start, what to arrange first, how to describe the geometry of the system under study, how to assess the number of time steps and the size of simulation domain, etc. All these possible questions are considered later. The material of this section is not a set of imperative rules, but rather guidelines how and what to do. We choose as the target the realization of a 1D FDTD scheme, keeping in mind that all essential features of FDTD simulations can be revealed in such a simple layout. Exercises 3.12 through 3.18 are the important part of learning the FDTD simulations.

Roughly, the computation process can be divided into four major steps (Figure 3.6), which are studied later in successive sections.

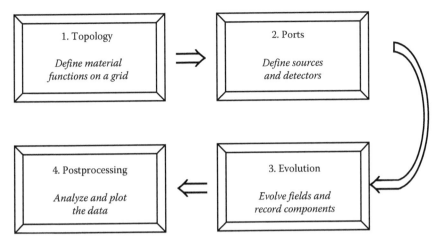

FIGURE 3.6 Major steps in the organization of the computational process with the FDTD method.

3.3.1 STEP 1: SETTING MATERIAL PROPERTIES ON A GRID

At this step, we have to describe the physical space, that is, to define material table functions $\varepsilon(z_i) = \varepsilon_i$, $\mu(z_i) = \mu_i$, etc. We agree upon using uniform meshes with space interval $\delta x = h$ and time interval δt (3.11) and (3.12). This is convenient especially at the beginning and still can provide results with sufficient accuracy. The question is how to choose space interval h?

Typically (see Exercises 3.2 and 3.3), the finer the mesh, the less the numerical dispersion error. On the other hand, the memory allocation might be a problem for a very fine mesh and extended sizes of the physical domain. So, reasonable numbers of the grid size in concerns with the wavelength are [1]

$$N_\lambda = \frac{\lambda_0}{h} \geq 10n \tag{3.60}$$

where
  λ_0 is some characteristic vacuum wavelength in our problem
  n is the refractive index of the medium

If the physical domain consists of several media, the highest refractive index is taken into account. For a broad-spectrum source of electromagnetic waves, the shortest λ of interest is used.

However, another concern is connected with the smallest characteristic size or length observed in the layout of the physical domain. At least two mesh points should be allocated in representation of each of material pieces or inclusions.

If the computational problem is formulated for a periodic structure, then there is another natural parameter for specification of the mesh interval, namely, the lattice constant Λ. We shall take care of keeping

$$N_\Lambda = \Lambda/h \geq 10n. \tag{3.61}$$

Objects of high attention are interfaces between different media. If the nearest grid points from both sides of the interface are deep enough in both media, then functions like ε_i are unambiguously defined:

$$\varepsilon(x_i) = \varepsilon_1, \varepsilon(x_i + h) = \varepsilon_2. \tag{3.62}$$

However, sometimes, it happens that grid point $x_i + h$ is lying exactly or in the nearest proximity to the interface between two media. The simplest way is to avoid such situation just by changing the grid step h to a bit bigger or smaller values. However, it is not very convenient to check the rule especially if we have complex patterning in three dimensions. So the mean value by simple scalar averaging,

$$\varepsilon(x_i + h) = \frac{\varepsilon(x_i) + \varepsilon(x_i + 2h)}{2} = \frac{\varepsilon_1 + \varepsilon_2}{2}, \tag{3.63}$$

can be recommended. The next step can be the weighted averaging, when the weight coefficients are associated with the distances from the grid points to the interface. And finally, vectorial nature of fields and slanted boundaries requires tensorial averaging performed separately for different field components; see, for example, [10]. More detailed study of the dielectric-function averaging is given in Section 4.5.2.

Now, we consider that by these operations, material parameter functions ε_i and μ_i have been defined on the 1D grid occupying cells from 1 to N.

Having defined spatial step h automatically guides us to the upper limit for the time stepping through the stability criteria in 1D (3.51):

$$\delta t \leq \frac{h}{c_0}, \tag{3.64}$$

or by using parameter Q (3.29)

$$Q \leq 1.0. \tag{3.65}$$

Utilizing unitless parameter Q in formulas instead of δt has some advantages in formalizing calculational stuff, because it substitutes combinations of femtoseconds $= 10^{-15}$ s, nanometers $= 10^{-9}$ m, and $c_0 = 3 \cdot 10^8$ m/s (typical values in photonics calculations) with more simple numbers like 0.5, 0.9, and 1.0.

The lower boundary for the time stepping is not regulated by the Courant condition. Certainly, the less the δt, the more steps needed to *transport* a laser pulse through the region of interest. Meanwhile, numerical dispersion also depends on the Q number. In particular, for the 1D case, it can be suppressed if one takes the magic step $Q = 1.0$, which we also highly recommend to do.

The physical domain ends in space. How to terminate it in terms of an accurate and robust calculation procedure is a very important question. The algorithm of termination of a physical domain defines BC. The stencils of the finite-difference schemes $D^c, D^{\pm}$ (3.1), (3.2), and (3.3) comprise neighbouring grid points. Thus, to fulfill consistency of the FDTD calculational scheme, two auxiliary points $i = 0$ and $i = N+1$ must be added to the grid. Fields in these points are involved in the updating schemes for electromagnetic fields at the periphery of the physical domain in points $i = 1$ and $i = N$. However, they are not updated accordingly to the FDTD algorithm. The way, how the fields in these two points are defined, actually specifies the type of the BCs. According to the virtual collocation of the grids for electric and magnetic fields, we need only one auxiliary point for each of the field, for example, we need E_{N+1} at the very right end of the domain (forward difference scheme for field E) to update H_N and H_0 at the very left (backward difference scheme for field H) to update E_1.

Following formulations of the BCs are possible in the FDTD.

1. Periodic BCs: Applicable if the modelled structure has a periodical arrangement (lattice), however, the wave vector or (in 1D) wave number has to be known. There are

$$E_{N+1} = E_1 \exp(-ikhN), \quad H_0 = H_N \exp(ikhN), \quad (3.66)$$

where k is a wave number.

2. Perfect electric conductor (PEC): The so-called closed simulated domain. A model for a metallic wall with infinite conductivity, which nullifies the continuous tangential component of electric field at the boundary $\mathbf{E}_t = 0$. In 1D, it is simply

$$E_{N+1} = 0. \quad (3.67)$$

3. Perfect magnetic conductor (PMC): The so-called closed simulated domain. A model for a metallic wall with infinite magnetic conductivity, which acts pretty much the same as PEC, but on the magnetic field $\mathbf{H}_t = 0$. In 1D, it is simply

$$H_0 = 0. \quad (3.68)$$

4. Perfect electromagnetic conductor (PC): The so-called closed simulated domain. A model for a metallic wall with infinite electric and magnetic conductivities, for example, superconductor.

$$E_{N+1} = 0, \quad H_0 = 0. \quad (3.69)$$

In general case, both continuous electric and magnetic field tangential components should go to zero $\mathbf{E}_t = \mathbf{H}_t = 0$.

5. Absorbing boundary conditions (ABC): Very important class of BC in photonics. They are mimicking boundaries with open space (or simply absence of any boundaries). We will consider them in details in the next section.
6. Hard-source BCs: An option to introduce a non-penetrative source, which generates a one-way travelling wave: from the right side (big i) to the left. For example, a harmonic (monochromatic) electric source

$$E^n_{N+1} = A\sin(\omega t_n) = A\sin(\omega \delta t n) = A\sin\left(\frac{2\pi Q}{N_\lambda}n\right), \qquad (3.70)$$

where A is the amplitude of the wave. We will consider sources in the next section. A hard magnetic source can be formulated as the left domain boundary.

Note: The physical domain together with the BCs, which terminate the physical space and can be rather extended, forms the *numerical or computational domain*, where computations are performed.

3.3.2 STEP 2: SETTING SOURCES AND DETECTORS

Initially, we have nullified arrays of electric and magnetic fields. If there are not any electromagnetic sources, an updating procedure will only add numerical noise to the values of the fields. Hence, the source of the electromagnetic field should be specified. It might be not even a source (an antenna) in the common sense, generating alternating (in time) electric and magnetic fields, but simply a set of points with non-zero E and H. The options are the following:

1. Initial fields distribution: As it has been just mentioned any set of grid points with non-zero fields defined before the first time step update can serve as a source. The remarkable property of the FDTD method is that it describes Maxwell's fields only. The distribution we arranged most-probably happens to be non-Maxwellian; nevertheless, only Maxwellian-driven fields will propagate in the region with zero-defined fields. Possible choice is a plane wave Gaussian pulse:

$$E^0_i = A\exp\left[-\left(\frac{i-x_c}{\alpha}\right)^2\right], \quad H^0_i = \frac{1}{\eta_i}E^0_i, \qquad (3.71)$$

where
  A is the amplitude of the pulse
  x_c is the coordinate of its centre (may be defined not only as an integer, but also as a real number)
  α is the numerical parameter determining the strength of the pulse descending
  $\eta_i = \sqrt{\mu_i/\varepsilon_i}$ is the relative impedance of the medium (remember that the vacuum impedance $\eta_0 = \sqrt{\mu_0/\varepsilon_0}$ we have already included in redefined $\hat{E}$)!

Such approach can be understood as making a snapshot of the travelling pulse and setting this time moment as the initial one.

Another choice is simply constant fields in the interval of grid points from a to b

$$E_i^0 = A \neq 0, \quad H_i^0 = \frac{1}{\eta} E_i^0, \quad \{i = a, \dots, b\}. \tag{3.72}$$

It is even possible to forget about the magnetic field of the source and keep it initially nullified. In spite that $H = 0$ everywhere is definitely a non-Maxwellian field, the *proper* Maxwellian values for both electric and magnetic fields will be restored undergoing the time updating scheme (3.27) and (3.28).

Pros and cons of the initial field distribution scheme are lying together in its simplicity. It is very easy to implement, but it can be easily created as non-Maxwellian, so with caution it is better to keep it away from regions containing any structures of interest or detectors.

2. Hard source: It was introduced as a BC (see formula (3.70)); however, it can be placed in any grid point of the computational domain. The shortcoming of such approach is that the point, where the hard source is placed, is impenetrable for electromagnetic waves from both sides. So all fields scattered in the domain to the left of the hard source have no influence on the fields in the right part of the domain and vice versa. This is why it is preferable to apply it mostly as a BC.

3. Additive or current source: It is the most convenient and flexible way of introducing a source in the FDTD method. The idea of this approach is a straightforward consequence of Ampere's law (2.32), where $J_x = J_i$ is an arbitrary current density. This current can be specified to be whatever is desired. For example, a *harmonic source* or in other words a *continuous wave* (CW) source

$$J_i(t) = J_i \sin(\omega_0 t), \quad J_i^n = J_i \sin(\omega_0 n \delta t), \tag{3.73}$$

where
J_i is the amplitude of the current
ω_0 is its frequency

The amplitude is independent of time, but can be a function of position. If it is defined in one grid point, it will serve as a point source, like a point dipole. Certainly, we can make a bar of dipoles arranging a set of points for the amplitude function and profiling them with, for example, the Gaussian-like shape. However, such profile has a limited value in 1D. More important is a temporal profile of the signal. The CW source is a monochromatic model. It generates fields on one frequency. All real signals have some time envelopes. So, instead of simple harmonic function with a constant amplitude (3.73), more useful are pulses, where bandwidth can be regulated by special parameters:

$$J_i(t) = J_i \sin(w_0 t) e^{-((t-t_0)/\alpha_t)^2}, \quad J_i^n = J_i \sin(w_0 n \delta t) e^{-((n\delta t - t_0)/\alpha_t)^2},$$

$$J_i(t) = J_i \sin(w_0 t) \sin(\Omega t), \quad J_i^n = J_i \sin(w_0 n \delta t) \sin(\Omega n \delta t),$$

$$0 \le n\delta t \le \pi/\Omega, \tag{3.74}$$

$$J_i(t) = J_i \sin(w_0 t) \cos(\Omega t), \quad J_i^n = J_i(w_0 n \delta t) \cos(\Omega n \delta t),$$

$$0 \le n\delta t \le \pi/(2\Omega).$$

The rule of thumb is that the shorter is the pulse in time, the broader is its spectral/frequency band and vice versa. Therefore, the bandwidth of the signal generated by the source can be adjusted by varying parameters like α_t or Ω. The sign and the amplitude of the source functions are not a concern except the cases when the current source is specified by the problem itself.

Note: In some programming languages, for example, Fortran, exponentially unlimited decay of the function (3.74) should be terminated in order to avoid stepping out of the permitted range of values. Typically, we limit the number of steps when a current source exists (when $J_i(t) \ne 0$).

At the same step, we should think about *detectors* (sensors or monitors), a special set of grid points in which fields will be stored in prescribed time steps, for example, in all time steps from 1 to N_t. Basically, these points can be placed where it is desired, however, with precautions to be kept away (at least several grid points) from the sources.

3.3.3 STEP 3: EVOLVING FIELDS

This is the most obvious part, where the explicit field-updating scheme (3.27) and (3.28) or its equivalent in SI units (3.24) and (3.25) is applied. During the time marching, new electric and magnetic fields are updated consequently with the help of the *old* components and stored for the designated detector points.

There are two special comments about the total number of time steps specified for execution.

1. Typically, fields are assumed to be either in the time domain or in the frequency domain. Time-domain $E(r, t)$ and frequency-domain $E(r, w)$ vector fields are linked through the integral Fourier transform like Equations (2.15) and (2.16):

$$\mathbf{E}(\mathbf{r}, w) = \int_{-\infty}^{+\infty} \mathbf{E}(\mathbf{r}, t) e^{-iwt} dt, \tag{3.75}$$

$$\mathbf{E}(\mathbf{r}, t) = \frac{1}{2\pi} \int_{-\infty}^{+\infty} \mathbf{E}(\mathbf{r}, w) e^{iwt} dw. \tag{3.76}$$

The time-domain fields result from the FDTD method, but often spectral knowledge about structure behaviour is needed, for instance, transmission and reflection spectra. Conversion of the time-domain fields into the frequency-domain ones is straightforward (see the recipe (3.75)), but implementation of the integration on a grid has different algorithms. The simplest is the *discrete Fourier transform* (DFT) – just an approximation of the integration procedure in Equation (3.75) by summation. If big arrays of data stored for long time intervals have to be converted, optimized procedures are required to shorten the execution time. One of such options is the so-called *fast Fourier transform* method (FFT) [11]. The FFT is a special algorithm, based on the binary decomposition of decimal numbers, which helps to save some time by the special arrangement of conversion operations. Correspondingly, it is applied when the number of time steps equals to the power of 2. All time steps exceeding the nearest power of two value will be ignored at the postprocessing part, for example, if a program makes 50,000 time steps, then only $32,768 = 2^{15}$ will be employed in the FFT procedure. So planning simulations for analyzing transmission spectra or other information in the frequency domain with the employment of the FFT conversion, this fact should be taken into account.

2. The resolution in the frequency domain is directly connected with the total evolution time $T = N_t \delta t$, where N_t is the total number of time steps: the bigger T, the finer presentation of fields in frequency domain we have. While δt cannot be taken arbitrary large due to the numerical stability reasons, the only way to refine the spectral resolution is to increase parameter N_t.

For crude spectral calculations, we recommend that the number of time steps will be at least 5 times the biggest number of grid points in the numerical domain. It is enough to resolve the basic features of spectra and is not very long from the processing time point of view. For better resolution, the total number of time steps must be increased at least by one order of magnitude. Certainly, calculations with cavities with very long transient periods for feeding a resonator and its relaxation dynamics require individual evaluation of needed time steps for every structure.

Note: In some cases, the conventional DFT can be superior to the FFT, though slower, if we need to address any specific frequency or use particular number of time steps not correlated with the power of 2. The requirement for the frequency resolution in a certain range is the same as for the FFT and is defined through N_t.

More information about implementation of frequency-domain to time-domain conversions with Fourier transforms, in particular, with the FFT procedure can be found in Sections 5.5 and 7.9.

3.3.4 STEP 4: POSTPROCESSING OF INFORMATION

This is actually the loosest and most informal part. The FDTD method is one of the most universal methods in electrodynamics and optics, so it produces various valuable information. We are going to reveal several output options typically encountered in research in photonics.

1. By default, after termination of the program, we have the last field arrays in all grid points. It will give us the snapshot picture of fields. Therefore, by dumping of global information from field arrays, we can plot the data showing field distribution in the region of interest by the end of the time interval $T = N_t \delta t$. This in turn can show confinement of the fields in cavities and waveguides, symmetry properties of the modes, and enhancement or inhibition of wave amplitudes in special regions. By Fourier transformation of fields from the real space into the k-space (wave vector space), information about spatial harmonics is available. It is important, for example, with diffraction on gratings, where diffracted beams are characterized by the wave vector.

2. Data stored in detectors show dependencies of fields on time in designated points. We can trace the shape and duration of signals, restore the phase and group velocities of pulses, and analyze dynamics of electromagnetic processes.

3. After processing the data stored by the FFT, we obtain all frequency-relevant (wavelength relevant) information about fields, including amplitudes, vector components, and phases of the complex field vectors. We can identify transmission or reflection spectral coefficients by normalizing transmitted and reflected fields (intensities or fluxes) with the incident ones. Scattering on bodies of different shapes is easily traced.

4. It is possible to combine two approaches by utilizing a CW source. Then information provided in time will also refer to the frequency response of a system due to the monochromatic waves emitted by the source. However, due to the abrupt start of the process, the transient period characterized by presence of the multifrequency input has to be excluded from observations. After some case-dependent time, perturbations caused by the transient period will die out, and electromagnetic processes in the system will reach the so-called *steady state*. The steady state in this case is a periodical (time-dependent) field distribution in the system specific for its harmonic excitation on a certain frequency. It can characterize, for example, the cavity/waveguide mode profile.

Note: In principle, the FDTD evolution scheme can work with both real and complex fields. It is the question of convenience, because even with metallic inclusions, the fields update can be formulated in a pure real format. However, fields become complex after the DFT or FFT. If the aim is to keep fields real throughout the whole computational routine, then a special (real) formulation of the Fourier transform should be employed; see, for example, [11].

3.4 ABSORBING BOUNDARY CONDITIONS

The basic idea behind BCs is to provide a *correct* field in the auxiliary point (points) included in stencils for the numerical scheme. In the previous section, we have already talked about some of them. They help to reduce the numerical domain taking into account symmetry or periodicity of the system. However, they are not universal.

Moreover, they do not possess an important property: to terminate the numerical domain, but leave the physical domain unconditionally *opened*.

Let us assume that we are going to model a photonic wire with a bend, that is, a ridge dielectric waveguide with typical cross section around 300 nm × 500 nm. It is not periodic and has no symmetry; it reflects uncoupled light and scatters light on surface imperfections, etc. It is *open* in the sense that light being disturbed by the bend or shape imperfections can easily leave it. The same behaviour has to be revealed in simulations with any code based on a suitable numerical method.

That is why the ABCs are of great importance in photonics. In the perfect case, any ABC has to work only for exit: waves are allowed only to leave the simulation domain, but not to return back after *numerical* reflection. This is a significant problem in 2D and 3D case. Hereafter, you can find some descriptions of basic recipes implemented in modern computational schemes.

Note: Sometimes the term *radiation/radiating or open BCs* is used instead of ABC to stress the openness (unidirectional propagation) provided by the rules of termination of the physical space. Read more about radiating BCs in Chapter 7.

3.4.1 Analytical Absorbing Boundary Conditions

1. Outgoing waves in 1D: A 1D modelling is somehow privileged to have a *magic step*, when numerical equations (3.19) and (3.23) are equivalent to the 1D wave equation approximated by finite differences. So our wish to have only exiting fields on the both outer sides of the numerical grid can be easily implemented in the formal way:

$$E_{N+1}^{n+1} = E_N^n, \quad H_0^{n+1} = H_1^n. \tag{3.77}$$

   Such termination means nothing but a naive attempt to realize the concept of the outward-going waves: an updated field in the certain point is the field in the nearest neighbouring point in the previous time step.

2. Engquist–Majda one-way wave equations: The approach, which was named after Engquist and Majda [12], advances the previous concept of unidirectional propagation. We take a reduced wave equation that describes a wave travelling only in one specific direction. Then fields at the outer grid points made to obey this equation satisfying conditions for the absence of inward-going waves.

Let's illustrate the concept with a 1D scheme. The 1D wave equation in vacuum

$$\frac{\partial^2 E_x}{\partial z^2} - \frac{1}{c_0^2} \frac{\partial^2 E_x}{\partial t^2} = 0, \tag{3.78}$$

or in operator notation,

$$\hat{G}_{zt} E_x \equiv \left(\partial_{zz}^2 - \frac{1}{c_0^2} \partial_{tt}^2 \right) E_x = 0, \tag{3.79}$$

can be factorized by representing operator $\hat{G}_{zt}$ as a product of two operators

$$\hat{G}_{zt} = \hat{G}_{zt}^{+}\hat{G}_{zt}^{-} = \left(\partial_z + \frac{1}{c_0}\partial_t\right)\left(\partial_z - \frac{1}{c_0}\partial_t\right). \tag{3.80}$$

Each of this operators describes one-way z-propagation, either in positive direction as $\hat{G}_{zt}^{+}$ does or in negative as for $\hat{G}_{zt}^{-}$. Now, the ABCs are readily formulated:

$$\text{At } z = z_1, \quad \hat{G}_{zt}^{-}H_y = 0, \tag{3.81}$$

$$\text{At } z = z_N, \quad \hat{G}_{zt}^{+}E_x = 0. \tag{3.82}$$

Implying the D^{-} scheme for the first equation and D^{+} for the second, it is possible to derive an explicit-updating procedure for H_0^n and E_{N+1}^{n+1} for further stepping in time.

It is more complicated in 2D, because it is hard to factorize the 2D wave equation, for example, in the (x, y) plane:

$$\hat{G}_{xyt}\mathbf{E} \equiv \left(\partial_{xx}^2 + \partial_{yy}^2 - \frac{1}{c_0^2}\partial_{tt}^2\right)\mathbf{E} = 0, \tag{3.83}$$

because partial derivatives become interweaved through the factorization. Moreover, different field components are involved with different continuity conditions at the interfaces. However, quite efficient approach was proposed in [13], which is now called Mur's ABC. Read more about analytical ABC in book [2].

3.4.2 PERFECTLY MATCHED LAYER: BASIC IDEA

Another ABC approach was proposed in 1994 by J. P. Berenger [14]. He coined the name *perfectly matched layer* (PML). PML ABCs are now the most powerful and popular method of organization non-reflecting (open) boundaries in both frequency and time domains.

The basic idea can be illustrated on a simple example of a plane interface between two media, characterized by pairs of material parameters ε_1, μ_1 and ε_2, μ_2. A plane wave is normally incident onto the interface (Figure 3.7). By employing the concept of impedance, Fresnel formula for the amplitude of the reflection wave stands

$$r = \frac{\eta_1 - \eta_2}{\eta_1 + \eta_2}, \tag{3.84}$$

where

$$\eta_1 = \sqrt{\frac{\mu_1}{\varepsilon_1}}, \quad \eta_2 = \sqrt{\frac{\mu_2}{\varepsilon_2}}. \tag{3.85}$$

Claiming reflectionless behaviour, that is, $r = 0$, the impedance matching condition must be obeyed:

$$\eta_1 = \eta_2. \tag{3.86}$$

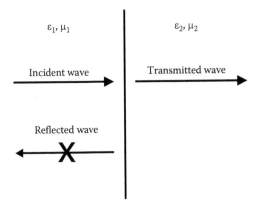

FIGURE 3.7 Reflection and transmission on the plane interface between two different media characterized by parameters ε and μ.

From the point of view of energy fluxes, the absence of reflection means that the impinging energy is completely transported in the reflectionless regime into the region 2. There are only two possibilities here: either energy leaks away as it happens in a transparent infinite medium or it is absorbed. The former option cannot be realized, because the numerical domain must be terminated. Hence, the latter option is the only one, which can be effectively implemented in multidimensional problems with complex material topology. So, a PML, which terminates the physical domain in one particular direction, permits energy to enter freely (without reflections) into it. This energy is dissipated afterwards due to absorbing properties exhibited by the PML.

To remind about the electromagnetic properties of media, we note that nonabsorbing or transparent dielectric has real constants ε and μ, while absorbing one is described by complex parameters, which in the simplest model with constant conductivities are

$$\varepsilon(\omega) = \varepsilon\left(1 + \frac{\sigma}{i\omega\varepsilon}\right), \quad \mu(\omega) = \mu\left(1 + \frac{\sigma^*}{i\omega\mu}\right), \tag{3.87}$$

where now ε and μ are real constants and σ and σ* are conventional electric conductivity and (virtual!, just for convenience) magnetic conductivity, also some real constants.

Assuming the first medium nonabsorbing Equation (3.86) gives

$$\sqrt{\frac{\mu_1}{\varepsilon_1}} = \sqrt{\frac{\mu_2\left(1 + \frac{\sigma^*}{i\omega\mu_2}\right)}{\varepsilon_2\left(1 + \frac{\sigma}{i\omega\varepsilon_2}\right)}}. \tag{3.88}$$

Further assuming natural match of parameters for reflectionless behaviour at the interface

$$\varepsilon_1 = \varepsilon_2 = \varepsilon; \quad \mu_1 = \mu_2 = \mu, \tag{3.89}$$

we infer that complex factors in the right-hand side of (3.86) have to compensate each other:

$$\frac{1 + \frac{\sigma^*}{i\omega\mu}}{1 + \frac{\sigma}{i\omega\varepsilon}} = 1,$$

or

$$\frac{\sigma^*}{\mu} = \frac{\sigma}{\varepsilon}. \tag{3.90}$$

So far, we see that a plane wave experiences no reflections and transports 100% of its energy across the boundary between the physical domain and PML. What will be the further propagation of a plane wave in the PML? Propagation properties are described by the wave numbers. In the first medium

$$k_1 = \frac{\omega}{c_0} n_1 = \frac{\omega}{c_0} \sqrt{\varepsilon_1 \mu_1} \tag{3.91}$$

is real, so the wave propagates without decay. In the second medium

$$k_2 = \frac{\omega}{c_0} n_2 = \frac{\omega}{c_0} \sqrt{\varepsilon_2 \mu_2 \left(1 + \frac{\sigma^*}{i\omega\mu_2}\right)\left(1 + \frac{\sigma}{i\omega\varepsilon_2}\right)} \tag{3.92}$$

is complex. Due to the matching properties (3.89) and (3.90), it is

$$k_2 = k_1 \left(1 + \frac{\sigma}{i\omega\varepsilon_1}\right). \tag{3.93}$$

Real parts of the wave numbers in the first and the second media, as it stems from the equation earlier, are equal. This means that waves at the interface are really matched temporally (by the same ω) and spatially (by the same $Re(k)$). The non-zero imaginary part of k_2 manifests the exponential decay of the wave amplitude in the PML, that is, absorption.

$$\exp(-ik_2'' z) = \exp\left(-i\frac{-i\sigma k_1}{\omega\varepsilon_1} z\right) = \exp\left(-\frac{\sigma k_1}{\omega\varepsilon_1} z\right). \tag{3.94}$$

3.4.3   PERFECTLY MATCHED LAYER: GENERALIZATION AND REALIZATION

In 2D and 3D cases, the picture of perfect matching is not so simple, especially for the oblique incidence. As it was shown by Gedney [15], it requires the fulfillment of specific tensorial conditions, which depend on orientation of the interfaces. For a 1D PML terminating a computational domain in the z-direction, it reads

$$\overline{\overline{\varepsilon}} = \varepsilon \overline{\overline{\Psi}}, \quad \overline{\overline{\mu}} = \mu \overline{\overline{\Psi}}, \tag{3.95}$$

where $\overline{\overline{\Psi}}$ is a diagonal matrix

$$\begin{pmatrix} \psi_z & 0 & 0 \\ 0 & \psi_z & 0 \\ 0 & 0 & 1/\psi_z \end{pmatrix}. \tag{3.96}$$

By the form of matrix $\overline{\overline{\Psi}}$, it is obvious that the PML is a anisotropic (uniaxial) medium described by proportional tensorial dielectric and magnetic functions. Parameter ψ_z has general form [15]:

$$\psi_z = \kappa_z + \frac{\sigma_z}{i\omega\varepsilon}, \tag{3.97}$$

but in most of ordinary cases, $\kappa_z = 1$, thus providing from (3.87) that $\varepsilon(\omega) = \varepsilon\psi_z$ and $\mu(\omega) = \mu\psi_z$.

Expressions for dielectric and magnetic functions (3.87) belong to the frequency domain; however, the FDTD method is a time-domain method, and the field-updating schemes like (3.31) contain time-domain permittivities and permeabilities. Therefore, it is necessary to transform material functions of the PMLs into the time-domain presentation.

It is a straightforward procedure in a 1D case. First, we will perform some auxiliary derivation. By multiplying both sides of Equation (3.75) with $i\omega$ and direct application of integration rule $\int u dv = uv - \int v du$, it is possible to prove that the Fourier transform of the frequency-domain vector field $i\omega \mathbf{E}(\mathbf{r}, \omega)$ in the time domain is vector field $\partial \mathbf{E}(\mathbf{r}, t)/\partial t$. In formal way, this rule tells that operation of multiplication by factor $i\omega$ in the frequency domain corresponds to the time derivative operation in the time domain

$$i\omega \times f(\omega) \, \{\text{in frequency domain}\} \leftrightarrow \partial f(t)/\partial t \, \{\text{in time domain}\}. \tag{3.98}$$

Note: You can check this rule on a simple example by taking the time derivative of monochromatic vector field $\mathbf{E}(\mathbf{r}, t) = \mathbf{E}(\mathbf{r}, \omega)e^{i\omega t}$, which immediately confirms (3.98).

Further, we start with the *curl* magnetic field equation (Ampere's law) in the time domain without currents (2.3):

$$\nabla \times \mathbf{H}(\mathbf{r}, t) = \frac{\partial \varepsilon_0 \varepsilon \mathbf{E}(\mathbf{r}, t)}{\partial t}, \tag{3.99}$$

which can be easily converted into the frequency domain by use of the rule (3.98):

$$\nabla \times \mathbf{H}(\mathbf{r}, \omega) = i\omega\varepsilon_0\varepsilon(\omega)\mathbf{E}(\mathbf{r}, \omega). \tag{3.100}$$

Substituting frequency-domain dielectric constant (3.87) in the x and y components of a z-PML (see Equations (3.95) and (3.97)) gives

$$(\nabla \times \mathbf{H}(\mathbf{r}, \omega))_x = i\omega\varepsilon_0\varepsilon E_x(\mathbf{r}, \omega) + \varepsilon_0\sigma E_x(\mathbf{r}, \omega). \tag{3.101}$$

Inverse transformation of this equation into the time domain is trivial due to the presence of constants ε and σ. It reads

$$(\nabla \times \mathbf{H}(\mathbf{r}, t))_x = \frac{\partial \varepsilon_0\varepsilon E_x(\mathbf{r}, t)}{\partial t} + \varepsilon_0\sigma E_x(\mathbf{r}, t). \tag{3.102}$$

It is, actually, nothing else but a usual expression of Ampere's law (2.3), with a substitution of Ohm's law $\mathbf{J} = \sigma\mathbf{E}$ instead of the current density.

Derivatives in 1D presentation of formula (3.102) have to be substituted by time and coordinate operators $D^c|_i^n$ (3.8). Instead of Equation (3.18), it appears:

$$\frac{E_i^{n+1/2} - E_i^{n-1/2}}{\delta t} = -\frac{1}{\varepsilon_i}\left(\frac{H_{i+1/2}^n - H_{i-1/2}^n}{\varepsilon_0 h} + \sigma_i E_i^n\right). \tag{3.103}$$

The last term in brackets is taken exactly at point (i, n), because it is not connected to any finite difference. However, by default, the electric field components are defined in half integer time points. To proceed further, we need to approximate E_i^n, for example, as the mean value between two subsequent time steps

$$E_i^n = \frac{E_i^{n+1/2} + E_i^{n-1/2}}{2}. \tag{3.104}$$

Substitution of approximation (3.104) in Equation (3.103) provides a formal arithmetic equation on $E_i^{n+1/2}$. It gives

$$E_i^{n+1/2} = \frac{2\varepsilon_i - \sigma_i\delta t}{2\varepsilon_i + \sigma_i\delta t}E_i^{n-1/2} - \frac{2\delta t}{h\varepsilon_0(2\varepsilon_i + \sigma_i\delta t)}\left(H_{i+1/2}^n - H_{i-1/2}^n\right). \tag{3.105}$$

The similar derivation applied to Faraday's law (2.2) leads to

$$H_{i+1/2}^{n+1} = \frac{2\mu_{i+1/2} - \sigma_{i+1/2}^*\delta t}{2\mu_{i+1/2} + \sigma_{i+1/2}^*\delta t}H_{i+1/2}^n - \frac{2\delta t}{h\mu_0\left(2\mu_{i+1/2} + \sigma_{i+1/2}^*\delta t\right)}\left(E_{i+1}^{n+1/2} - E_i^{n+1/2}\right). \tag{3.106}$$

Note: The first updating scheme (3.105) can be also used in the case of an absorbing medium, when σ is the frequency-independent conductivity of the medium responsible for losses. It suits perfectly the microwave analysis and actually all low-frequency (non-optical) electromagnetism as soon as conductivity can be kept as constant (frequency independent). There are no any straightforward material analogue for σ^*; it is a constant required purely for the PML's formulation. So, for an ordinary absorbing medium, σ^* should be nullified.

Both updating schemes can be processed in the same way as we did for Fortran, C++, etc., adaptation; see, for example, (3.24) through (3.28). We should also take into account that parameters σ and σ^* in PMLs are connected through Equation (3.90). This allows us to write down the end-user formulas for the realization of PMLs in the unitless 1D FDTD computational scheme:

$$\hat{E}_i^{n+1} = \frac{2\varepsilon_i - \sigma_i\delta t}{2\varepsilon_i + \sigma_i\delta t}\hat{E}_i^n - \frac{2Q}{2\varepsilon_i + \sigma_i\delta t}\left(H_i^n - H_{i-1}^n\right). \qquad (3.107)$$

$$H_i^{n+1} = \frac{2\mu_i - \sigma_i^*\delta t}{2\mu_i + \sigma_i^*\delta tt}H_i^n - \frac{2Q}{2\mu_i + \sigma_i^*\delta t}\left(\hat{E}_{i+1}^{n+1} - \hat{E}_i^{n+1}\right). \qquad (3.108)$$

Fortunately, for a transverse electromagnetic wave, we don't need an updating scheme for the normal components of the fields in PMLs, that is, for E_z in PML-z. The problem here is in the $1/\psi_z$ dependency on frequency, which instead of a *good* equation (3.100) gives a *bad* one:

$$(\nabla \times \mathbf{H}(\mathbf{r}, \omega))_z = i\omega\varepsilon_0\varepsilon\frac{1}{\psi_z}E_z(\mathbf{r}, \omega) = i\omega\varepsilon_0\varepsilon\frac{i\omega\varepsilon}{i\omega\varepsilon + \sigma}E_z(\mathbf{r}, \omega)$$

$$= i\omega\varepsilon_0\varepsilon\left(1 - \frac{\sigma}{i\omega\varepsilon + \sigma}\right)E_z(\mathbf{r}, \omega)$$

$$= i\omega\varepsilon_0\varepsilon E_z(\mathbf{r}, \omega) - \frac{i\omega\varepsilon_0\varepsilon\sigma}{i\omega\varepsilon + \sigma}E_z(\mathbf{r}, \omega) \qquad (3.109)$$

Now the frequency-domain to time-domain conversion is not as trivial as for other components, at least the simple substitution rule (3.98) cannot be applied to the right-hand side.

Note: In a 3D case, expression (3.96) represents a PML layer at one face of the physical domain. The most challenging are the edges and corners of the domain, where PMLs formulated in different directions (PML-x, PML-y, PML-z) interweave. The explicit schemes for the time-domain implementation of unsplit 2D and 3D PMLs are presented in [16–18].

From the practical point of view, the PML's strength, defined by parameter σ_z (3.97), is chosen coordinate dependent with the gradual increase away from the interface in a polynomial fashion. For example, for the outer right PML-z, we have

$$\psi_z = 1 + \frac{\sigma_z}{i\omega\varepsilon}, \qquad \sigma_z = \sigma_{max}\left(\frac{z - z_{PML}}{d}\right)^m, \qquad (3.110)$$

where
$2 \leq m \leq 4$
d is the thickness of PML
z_{PML} is the first coordinate of the PML (or the last of the physical domain before the PML)

σ_{max} is the maximal value of the conductivity in the PML. By a pragmatic approach, it is better not to stick to the value of σ_{max} itself, but rather look at the coefficient $\zeta_i \equiv \sigma_i \delta t / 2\varepsilon_i$ in Equation (3.107) or (3.108). We can impose polynomial grading on the coefficient ζ directly:

$$\zeta_z = \zeta_{max} \left(\frac{z - z_{PML}}{d} \right)^m = \zeta_{max} \left(\frac{i - i_{PML}}{d} \right)^m, \qquad (3.111)$$

where

     ζ_{max} can be chosen in the range $0.1 \div 10$

     i_{PML} stands for the entrance coordinate of the PML-z

For the PML-z termination of the computational space from the left side, the z-dependence should be reversed:

$$\zeta_z = \zeta_{max} \left(\frac{d - i}{d} \right)^m. \qquad (3.112)$$

Here now, d is also the exit coordinate of the left PML (or the first grid point in the physical space).

3.5   FDTD METHOD FOR MATERIALS WITH FREQUENCY DISPERSION

In Chapter 2, we considered constitutive relations for media with frequency dispersion. In general case, frequency dispersion leads to complex time-domain expression of the constitutive equations through the convolution theorem (2.10), when, for example, displacement vector $\mathbf{D}(\mathbf{r}, t)$ is determined by electric fields in all previous time moments, indicating somehow that the medium traces *the* history of electric field values.

In some particular cases, it is possible to simplify significantly the convolution theorem and avoid backward time integration. Instead, an additional differential equation is formulated, which is also numerically integrated through an explicit finite-difference scheme linked with the main FDTD algorithm. These cases are described by models, which were developed to describe different physical processes. We consider three of such models, which are widely spread in photonics and plasmonics problems: Debye material and Drude and Lorentz dispersion terms.

3.5.1   FREQUENCY DISPERSION MODELS

3.5.1.1   Debye Material

The Debye material is a material whose dielectric properties satisfy the so-called Debye relaxation model [19]. This model describes response of the material with polar molecules considered as a set of dipoles undergoing perturbation by an external alternative electrical field. The susceptibility of the Debye material depends on frequency in the following way [19]:

$$\chi_D(\omega) = \frac{\chi_0}{1 + i\omega\tau}, \tag{3.113}$$

where

τ is the relaxation time, which defines the time scale of relaxation properties in the material

χ_0 is the constant connected with the static orientational polarizability of molecules

Note: We accept the exp($i\omega t$) time dependence for a harmonic process: $\mathbf{E}(t) = \mathbf{E}_0 \exp(i\omega t)$. Adopting the reverse exp($-i\omega t$) dependence will lead to formal change of the sign associated with frequency ω everywhere in related formulas. For example, Debye susceptibility will be

$$\chi_D(\omega) = \frac{\chi_0}{1 - i\omega\tau}. \tag{3.114}$$

Debye susceptibility (3.113) can be formally converted in the time domain by the Fourier transformation. However, we are mostly interested in the time-domain relation between fields $\mathbf{E}$ and $\mathbf{D}$; thus, we are further stuck with the convolution theorem (2.10). The hardships of the time-domain conversion can be circumvented by the use of an auxiliary differential equation (ADE) as it will be shown later.

3.5.1.2 Drude Model

This model describes free-electron gas formed by conductive electrons in metal. The electrons are driven by an external electric field $\mathbf{E}(t) = \mathbf{E}_0 \exp(i\omega t)$. The simple equation of motion for electrically driven charge $-e$ with some typical resistance force proportional to the velocity reads

$$m\ddot{\mathbf{r}} = -e\mathbf{E} - \gamma m\dot{\mathbf{r}}, \quad \dot{\mathbf{r}} \equiv \frac{d\mathbf{r}}{dt}, \quad \ddot{\mathbf{r}} \equiv \frac{d^2\mathbf{r}}{dt^2}. \tag{3.115}$$

Converting it into the frequency domain according to the rule (3.98) leads to

$$m(i\omega)^2 \mathbf{r}(\omega) = -e\mathbf{E}(\omega) - i\omega\alpha m\mathbf{r}(\omega). \tag{3.116}$$

The radius vector as a function of frequency is then extracted as

$$\mathbf{r}(\omega) = \frac{e}{m(\omega^2 - i\gamma\omega)}\mathbf{E}(\omega), \tag{3.117}$$

and particle dipole moment $\mathbf{p}(\omega) = -e\mathbf{r}(\omega)$. Polarization vector is defined as the total dipole moment in a unit volume: $\mathbf{P}(\omega) = n\mathbf{p}$, where n is the concentration of free electrons. We derive

$$\mathbf{P}(\omega) = -\frac{ne^2}{m}\frac{\mathbf{E}(\omega)}{\omega^2 - i\gamma\omega}. \tag{3.118}$$

So, susceptibility

$$\chi_{Dr}(\omega) = -\frac{\omega_p^2}{\omega^2 - i\gamma\omega}, \tag{3.119}$$

where ω_p is the plasma frequency $\omega_p^2 = ne^2/m\varepsilon_0$. The dielectric function in the Drude model is

$$\varepsilon(\omega) = \varepsilon_\infty - \frac{\omega_p^2}{\omega^2 - i\gamma\omega}. \tag{3.120}$$

Constant $\gamma = 1/\tau$ is an inverse relaxation time or collision frequency, $\varepsilon_\infty = \varepsilon(\infty)$ – the corrected dielectric permittivity in the limit of extremely high frequencies.

3.5.1.3 Lorentz Model

Let us assume the resonance behaviour of a bound particle (electron) with charge Q and mass m. It can be accounted by including a linear returning force $-k\mathbf{r}$ in the equation of motion for electrically driven charge Q with the standard resistance force:

$$m\ddot{\mathbf{r}} = Q\mathbf{E} - \gamma_l m\dot{\mathbf{r}} - k\mathbf{r}. \tag{3.121}$$

In the frequency domain, it reads

$$m(i\omega)^2\mathbf{r}(\omega) = Q\mathbf{E}(\omega) - i\omega\gamma_l m\mathbf{r}(\omega) - k\mathbf{r}(\omega). \tag{3.122}$$

Following the same path as for the Drude model through the radius vector, dipole moment, and polarization, the susceptibility in the Lorentz model becomes

$$\chi_L(\omega) = \frac{\omega_l^2}{\omega_0^2 + i\gamma_l\omega - \omega^2}, \tag{3.123}$$

where $\omega_l^2 = nQ^2/m\varepsilon_0$, $\omega_0 = \sqrt{k/m}$ is the resonant oscillation frequency and γ_l is the damping coefficient. Correspondingly, permittivity function in the Lorentz model is

$$\varepsilon(\omega) = \varepsilon_\infty + \frac{\omega_l^2}{\omega_0^2 + i\gamma_l\omega - \omega^2}. \tag{3.124}$$

Note: Here, we exploit the single-pole Lorentz model. In some problems, several Lorentzial poles are required to adequately describe the dielectric response of bounded electrons in intraband transitions. Such models are successfully applied for simulating metals in the visible range [20].

3.5.2 NUMERICAL IMPLEMENTATION OF FREQUENCY DISPERSION IN FDTD THROUGH AUXILIARY EQUATION

This approach is the working horse of all metal-involved FDTD simulations in photonics, plasmonics, and metamaterials. It has insignificant variations in terms of what characteristic is taken into consideration – polarization vector $\mathbf{P}(t)$ or polarization current $\mathbf{J}_p(t) = \partial \mathbf{P}(t)/\partial t$. Here, we focus on the polarization current approach.

Ampere's law in the frequency domain (3.101),

$$\nabla \times \mathbf{H}(\omega) = i\omega\varepsilon_0\varepsilon(\omega)\mathbf{E}(\omega) + \mathbf{J}(\omega), \tag{3.125}$$

is written with the use the frequency-dependent electric current $\mathbf{J}(\omega) = \sigma\mathbf{E}(\omega)$ with constant conductivity. Rewriting this expression with the assistance of Equation (2.25) results in

$$\nabla \times \mathbf{H}(\omega) = i\omega\varepsilon_0\mathbf{E}(\omega) + i\omega\varepsilon_0\chi_e(\omega)\mathbf{E}(\omega) + \mathbf{J}(\omega). \tag{3.126}$$

The polarization current in the frequency domain is

$$\mathbf{J}_p(\omega) = i\omega\varepsilon_0\chi_e(\omega)\mathbf{E}(\omega), \tag{3.127}$$

where we applied the formal rules of the time- to frequency-domain conversion (multiplication with $i\omega$ instead of time derivative). Transformation of this equation into the time domain is quite straightforward with the reverse formal rules of conversion, apart from the second term with a product of two frequency-dependent functions. By denoting the result of the convolution theorem in the second term as simply $\mathbf{J}_p(t)$, we can finish the conversion to the time domain

$$\nabla \times \mathbf{H}(t) = \varepsilon_0\frac{\partial \mathbf{E}(t)}{\partial t} + \mathbf{J}_p(t) + \sigma\mathbf{E}(t). \tag{3.128}$$

The central question is how does this polarization current look like in the time domain. Explicit expressions for the time-dependent polarization current can be obtained applying particular models of frequency dispersion of materials.

3.5.2.1 Debye Material

In the case of the Debye material with susceptibility given by expression (3.113), the polarization current in the frequency domain (3.127) is

$$\mathbf{J}_p(\omega) = i\omega\varepsilon_0\frac{\chi_0}{1 + i\omega\tau}\mathbf{E}(\omega). \tag{3.129}$$

By getting rid of the denominator, we obtain

$$\mathbf{J}_p(\omega) + i\omega\tau\mathbf{J}_p(\omega) = i\omega\varepsilon_0\chi_0\mathbf{E}(\omega). \tag{3.130}$$

Conversion of this equation into the time domain is straightforward:

$$\mathbf{J}_p(t) + \tau\frac{\partial \mathbf{J}_p(t)}{\partial t} = \varepsilon_0\chi_0\frac{\partial \mathbf{E}(t)}{\partial t}. \tag{3.131}$$

The wanted time-dependent polarization current is a solution of the first-order differential equation (3.131) assuming that the electric field derivative is known. It is called the ADE for the FDTD method in dispersive media [21] and [22].

A solution of the ADE (3.131) must be sought in the same way as for Maxwell's equations: the derivatives can be expressed through finite differences and linked with the field-updating scheme. Applying the second-order central differences to the time derivatives centering in the time moment $n + 1/2$, and the mean value of the polarization current gives

$$\frac{\mathbf{J}_p^{n+1} + \mathbf{J}_p^n}{2} + \tau \frac{\mathbf{J}_p^{n+1} - \mathbf{J}_p^n}{\delta t} = \varepsilon_0 \chi_0 \frac{\mathbf{E}^{n+1} - \mathbf{E}^n}{\delta t}. \tag{3.132}$$

It is straightforward now to express the updated polarization current:

$$\mathbf{J}_p^{n+1} = \frac{2\tau - \delta t}{2\tau + \delta t} \mathbf{J}_p^n + \frac{2\varepsilon_0 \varepsilon_d}{2\tau + \delta t} \left(\mathbf{E}^{n+1} - \mathbf{E}^n \right). \tag{3.133}$$

The polarization current is presented in Ampere's law (3.128). Applying the finite-difference scheme of Ampere's law in the conventional time moment $n + 1/2$

$$\nabla \times \mathbf{H}^{n+1/2} = \varepsilon_0 \frac{\left(\mathbf{E}^{n+1} - \mathbf{E}^n \right)}{\delta t} + \mathbf{J}_p^{n+1/2} + \sigma \frac{\mathbf{E}^{n+1} + \mathbf{E}^n}{2}, \tag{3.134}$$

it is easy to see that it contains $\mathbf{J}_p^{n+1/2}$, but not $\mathbf{J}_p^{n+1}$ and $\mathbf{J}_p^n$, which we get through the updating scheme (3.133). The easiest way to proceed further it is to use again the averaging formula

$$\mathbf{J}_p^{n+1/2} = 1/2 \left(\mathbf{J}_p^{n+1} + \mathbf{J}_p^n \right) \tag{3.135}$$

and insert expression (3.133) here. Then (3.135) is ready for implementation in Ampere's law (3.134) providing an arithmetic equation for updated $\mathbf{E}^{n+1}$. By this, we keep the scheme for consequent updating the fields and polarization current explicit.

Fields Update Algorithm for Debye Dispersive Material

The scheme of updating fields in the FDTD method applied for materials with Debye model of frequency dispersion looks like this:

1. Knowing $\mathbf{H}^{n-1/2}, \mathbf{E}^n$ update magnetic field to $\mathbf{H}^{n+1/2}$ through Faraday's law

$$\nabla \times \mathbf{E}(\mathbf{r}, t) = -\frac{\partial \mathbf{B}(\mathbf{r}, t)}{\partial t};$$

2. Using $\mathbf{H}^{n+1/2}, \mathbf{E}^n, \mathbf{J}^n$ find new electric field $\mathbf{E}^{n+1}$ with (3.134)
3. With the help of $\mathbf{E}^{n+1}, \mathbf{E}^n$, and $\mathbf{J}^n$ update polarization current to $\mathbf{J}^{n+1}$ according to (3.133)

Note: The updating equation for the polarization current (3.133) includes not only the updated electric field but also the *old* electric field in the previous time moment. So, both $\mathbf{E}^{n+1}$ and $\mathbf{E}^n$ arrays have to be stored for the method processing.

3.5.2.2   Drude Model of Dispersion

The Drude dispersion model is described through the Drude susceptibility (3.119), which can be rewritten as follows:

$$\chi_{Dr}(\omega) = -\frac{\omega_p^2}{\omega^2 - i\gamma\omega} = \frac{\omega_p^2}{i\omega\,(i\omega + \gamma)}. \tag{3.136}$$

It leads to the polarization current:

$$\mathbf{J}_p(\omega) = \frac{i\omega\varepsilon_0\omega_p^2}{i\omega(i\omega + \gamma)}\mathbf{E}(\omega), \tag{3.137}$$

which brings us to the frequency-domain equation

$$i\omega\mathbf{J}_p(\omega) + \gamma\mathbf{J}_p(\omega) = \varepsilon_0\omega_p^2\mathbf{E}(\omega). \tag{3.138}$$

Corresponding ADE equation in the time domain,

$$\frac{\partial \mathbf{J}_p(t)}{\partial t} + \gamma\mathbf{J}_p(t) = \varepsilon_0\omega_p^2\mathbf{E}(t), \tag{3.139}$$

can be discretized on a grid

$$\frac{\mathbf{J}_p^{n+1} - \mathbf{J}_p^n}{\delta t} + \frac{\gamma}{2}\left(\mathbf{J}_p^{n+1} + \mathbf{J}_p^n\right) = \varepsilon_0\omega_p^2\frac{\mathbf{E}^{n+1} + \mathbf{E}^n}{2}, \tag{3.140}$$

providing the updating scheme for the Drude polarization current

$$\mathbf{J}_p^{n+1} = \frac{2 - \gamma\delta t}{2 + \gamma\delta t}\mathbf{J}_p^n + \frac{\varepsilon_0\omega_p^2\delta t}{2 + \gamma\delta t}\left(\mathbf{E}^{n+1} + \mathbf{E}^n\right). \tag{3.141}$$

As usual, we then follow expression (3.135) to find $\mathbf{J}_p^{n+1/2}$. The complete updating scheme is pretty much the same as for the Debye case.

3.5.2.3   Lorentz Model of Dispersion

In case of the Lorentz model of dispersion, the polarization current in the frequency domain is

$$\mathbf{J}_p(\omega) = i\omega\varepsilon_0\frac{\omega_l^2}{\omega_0^2 + i\gamma_l\omega - \omega^2}\mathbf{E}(\omega). \tag{3.142}$$

Walking the same line as for the Debye medium, we obtain the frequency-domain equation:

$$\left(\omega_0^2 + i\omega\gamma_l - \omega^2\right)\mathbf{J}_p(\omega) = i\omega\varepsilon_0\omega_l^2\mathbf{E}(\omega). \tag{3.143}$$

After its transformation into the time domain, the ADE for the Lorentz model is derived:

$$\omega_0^2 \mathbf{J}_p(t) + \gamma_l \frac{\partial \mathbf{J}_p(t)}{\partial t} + \frac{\partial^2 \mathbf{J}_p(t)}{\partial t^2} = \varepsilon_0 \omega_l^2 \frac{\partial \mathbf{E}(t)}{\partial t}. \tag{3.144}$$

Now, the ADE comprises the second-order derivative; therefore, the centering point of the finite-difference approximation will be n. So central finite differences applied around time step n accordingly to expression (3.10) lead to

$$\omega_0^2 \mathbf{J}_p^n + \gamma_l \frac{\mathbf{J}_p^{n+1} - \mathbf{J}_p^{n-1}}{2\delta t} + \frac{\mathbf{J}_p^{n+1} - 2\mathbf{J}_p^n + \mathbf{J}_p^{n-1}}{\delta t^2} = \varepsilon_0 \omega_l^2 \frac{\mathbf{E}^{n+1} - \mathbf{E}^{n-1}}{2\delta t}. \tag{3.145}$$

Then, updated current is

$$\mathbf{J}_p^{n+1} = 2\frac{2 - \omega_0^2 \delta t^2}{2 + \gamma_l \delta t}\mathbf{J}_p^n + \frac{\gamma_l \delta t - 2}{2 + \gamma_l \delta t}\mathbf{J}_p^{n-1} + \frac{\varepsilon_0 \omega_l^2 \delta t}{2 + \gamma_l \delta t}\left(\mathbf{E}^{n+1} - \mathbf{E}^{n-1}\right). \tag{3.146}$$

This value is used to find $\mathbf{J}_p^{n+1/2}$ according to Equation (3.135). The updating scheme for the Debye medium is applied here as well. However, approximation of second-order derivatives requires four additional arrays to be stored in the fields updating procedure: $\mathbf{E}^n, \mathbf{E}^{n-1}, \mathbf{J}_p^n, \mathbf{J}_p^{n-1}$.

3.5.3 LINEAR POLARIZATION MODEL FOR DISPERSIVE MATERIALS IN FDTD

Instead of manipulating with the polarization current $\mathbf{J}_p = \partial \mathbf{P}/\partial t$, it is possible to account for the dispersive properties of materials directly through the polarization vector $\mathbf{P}$. Such formulation, however, claims employing of vector $\mathbf{D}$

$$\mathbf{D} = \varepsilon_0 \varepsilon_\infty \mathbf{E} + \mathbf{P}. \tag{3.147}$$

Let us consider, for example, contribution from the Debye and Lorentz polarization models $\mathbf{P} = \mathbf{P}_D + \mathbf{P}_L$. Corresponding polarization vectors in the frequency domain are easily restored from susceptibilities χ_D (3.113) and χ_L (3.123).

In the case of the Debye material,

$$\mathbf{P}_D(\omega) = \varepsilon_0 \chi_D(\omega)\mathbf{E}(\omega) = \frac{\varepsilon_0 \chi_0}{1 + i\omega\tau}\mathbf{E}(\omega). \tag{3.148}$$

Following the well-known path (remove the denominator and transform the equation into the time domain), we obtain the differential equation for the polarization vector:

$$\tau_d \frac{\partial \mathbf{P}_D(t)}{\partial t} + \mathbf{P}_D(t) = \varepsilon_0 \chi_d \mathbf{E}(t). \tag{3.149}$$

Conventionally, first-order derivatives are approximated by finite differences at time point $n + 1/2$. The algorithm, analogous to what we did for the polarization current within the Debye model, gives

$$\mathbf{P}_D^{n+1} = \frac{2\tau - \delta t}{2\tau + \delta t}\mathbf{P}_D^n + \frac{\varepsilon_0 \chi_0 \delta t}{2\tau + \delta t}\left(\mathbf{E}^{n+1} + \mathbf{E}^n\right). \tag{3.150}$$

For the Lorentz polarization,

$$\mathbf{P}_L(\omega) = \varepsilon_0 \chi_L(\omega)\mathbf{E}(\omega) = \frac{\varepsilon_0 \omega_l^2}{\omega_0^2 + i\omega\gamma_l - \omega^2}\mathbf{E}(\omega). \tag{3.151}$$

The time-domain differential equation is then

$$\frac{\partial^2 \mathbf{P}_L(t)}{\partial t^2} + \gamma_l \frac{\partial \mathbf{P}_L(t)}{\partial t} + \omega_0^2 \mathbf{P}_L(t) = \varepsilon_0 \omega_l^2 \mathbf{E}(t). \tag{3.152}$$

Due to the presence of the second-order derivative, the finite-difference scheme is applied at point n. Eventually, the updating scheme reads

$$\mathbf{P}_L^{n+1} = 2\frac{2 - \omega_0^2 \delta t^2}{2 + \delta t \gamma_l}\mathbf{P}_L^n + \frac{\delta t \gamma_l - 2}{2 + \delta t \gamma_l}\mathbf{P}_L^{n-1} + \frac{2\varepsilon_0 \omega_l^2 \delta t^2}{2 + \delta t \gamma_l}\mathbf{E}^n. \tag{3.153}$$

Updated polarization terms (3.150) and (3.153) are gathered in (3.147):

$$\mathbf{D}^{n+1} = \varepsilon_0 \varepsilon_\infty \mathbf{E}^{n+1} + \mathbf{P}_D^{n+1} + \mathbf{P}_L^{n+1}. \tag{3.154}$$

Fields Update Algorithm for Linear Dispersive Material

The FDTD cycle now is performed in the following sequence.

1. First, **D** is updated from Ampere's law:

$$\nabla \times \mathbf{H}(\mathbf{r}, t) = \frac{\partial \mathbf{D}(\mathbf{r}, t)}{\partial t} + \mathbf{J}(\mathbf{r}, t).$$

2. Then, by substituting it into (3.154), the linear equation regarding $\mathbf{E}^{n+1}$ is solved.
3. Finally, $\mathbf{E}^{n+1}$ is used to update the magnetic field through the Faraday equation:

$$\nabla \times \mathbf{E}(\mathbf{r}, t) = -\frac{\partial \mathbf{B}(\mathbf{r}, t)}{\partial t}.$$

Note: There are no clear advantages which scheme – with polarization vector or polarization current – to apply. Both of them lead to introduction and solution of ADEs. The polarization vector is more transparent in the physical sense, for instance, in the case of extending formalism to the nonlinear materials. However, it involves operations with additional electric field – displacement vector **D**, what makes more expensive algorithm in terms of processor resources.

3.5.4 PIECEWISE LINEAR RECURSIVE CONVOLUTION SCHEME

There is an alternative method to incorporate a frequency dispersive medium in the FDTD algorithm without formulating and solving an ADE. Schematically, it employs electric displacement vector $\mathbf{D}$, which is used in Ampere's law as follows.

The frequency-domain material equation

$$\mathbf{D}(\omega) = \varepsilon_0 \varepsilon_\infty \mathbf{E}(\omega) + \varepsilon_0 \chi_e(\omega)\mathbf{E}(\omega) \tag{3.155}$$

is transformed in the time domain with the help of the convolution theorem (2.10):

$$\mathbf{D}(t) = \varepsilon_0 \varepsilon_\infty \mathbf{E}(t) + \varepsilon_0 \int_0^t \mathbf{E}(t - \tau)\chi_e(\tau)d\tau. \tag{3.156}$$

To apply the last formula directly, we have to store electric field values in all previous time steps, what is basically very inefficient. To circumvent such storage while calculating the convolution integral $\mathbf{I}_n = \int_0^{n\delta t} \mathbf{E}(n\delta t - \tau)\chi_e(\tau)d\tau$, it was proposed to utilize a special type of an electric field function, which helps to estimate this integral by a progressively accumulating scheme. This method is called the *piecewise linear recursive convolution*.

We will give just a brief overview of the calculating principle. Assume the electric field being constant at each step δt, then

$$n = 0; \quad \mathbf{I}_0 = 0;$$

$$n = 1; \quad \mathbf{I}_1 = \int_0^{\delta t} \mathbf{E}(\delta t - \tau)\chi_e(\tau)d\tau = \mathbf{E}^1 \chi_e^0 + \mathbf{E}^0 \chi_e^1;$$

$$n = 2; \quad \mathbf{I}_2 = \mathbf{E}^2 \chi_e^0 + \mathbf{E}^1 \chi_e^1 + \mathbf{E}^0 \chi_e^2;$$

and so on. More accurately is to apply a piecewise linear approximation of the electric field in each time interval $[n\delta t, (n+1)\delta t]$

$$\mathbf{E}(t) = \mathbf{E}^n + \frac{\mathbf{E}^{n+1} - \mathbf{E}^n}{\delta t}(t - n\delta t). \tag{3.157}$$

Then convolution integral for the electric field at any new time step can be expressed as accumulating sums of the susceptibility function over previous times steps. More about recursive convolution schemes can be read in [2].

Note: Recursive convolution schemes are not the most efficient algorithms due to accumulative sums, and for Drude, Debye, or Lorentz, dispersion models are practically overridden by ADE methods. However, their advantage is that they can be applied to whatever complicated law of dispersion, where the principle part of the ADE formalism – formal conversion of the polarization current from the frequency domain to time domain – does not work.

Note: One more approach for treatment of frequency dispersive materials in FDTD is the Z-transform method [5].

3.6 FDTD METHOD FOR NONLINEAR MATERIALS, MATERIALS WITH GAIN, AND LASING

We give here a short excursus in possible extensions of the FDTD method on nonlinear materials, materials with gain, and lasing systems. Here comes the universality of the FDTD schemes, which serve for the direct solution of the Maxwell's equations in the time domain. Therefore, dynamical quantum and classical processes of different physical nature (heat transfer, mass diffusion, acoustic waves propagation, atoms excitation and emission, tunnelling, etc.) happening in time can be directly linked with the electromagnetic fields evolution scheme.

In case of accounting for the nonlinear optical effects, they can be described by the nonlinear polarization term in Equation (2.6). In general, nonlinear optics comprises a lot of contributions from the second- and third-order susceptibilities. However, we limit ourselves only to the third-order nonlinearities in connection with the Kerr and Raman effects.

3.6.1 NONLINEAR POLARIZATION IN FDTD

The FDTD method by its construction is very convenient for describing linear and nonlinear optical properties of materials. As we mentioned earlier, nonlinear properties are taken into account by adding a nonlinear polarization $\mathbf{P}^{NL}$ in the material equation:

$$\mathbf{D} = \varepsilon_0\varepsilon_\infty\mathbf{E} + \mathbf{P} + \mathbf{P}^{NL}. \tag{3.158}$$

Three models of linear polarization contribution $\mathbf{P}$ are described in Section 3.5.3. Nonlinear polarization $\mathbf{P}^{NL}$ has much more complex structure. There are plenty of optical nonlinear effects, which are described by different terms of the susceptibility function. The most frequent in use are the second- and third-order nonlinearities. We will further show an approach for the third-order susceptibility, when the polarization vector is proportional to the electric field in the cubic power. In the time domain, it leads to a triple time-domain integration:

$$\mathbf{P}^{NL}(\mathbf{r}, t) = \varepsilon_0 \iiint \chi^{(3)}(t - t_1, t - t_2, t - t_3)\mathbf{E}(\mathbf{r}, t_1)\mathbf{E}(\mathbf{r}, t_2)\mathbf{E}(\mathbf{r}, t_3)dt_1dt_2dt_3. \tag{3.159}$$

Here a contraction of $\mathbf{E}$ vectors with the fourth-order susceptibility tensor $\chi^{(3)}$ is assumed.

For a simple model of the electronic response accounting for nonresonant incoherent (i.e., not amplitude, but intensity dependent) nonlinear effects, the third-order nonlinear polarization can be described by the Born–Oppenheimer approximation [23] with susceptibility $\chi_0^{(3)}$ treated as a constant, that is,

$$\mathbf{P}^{NL}(\mathbf{r}, t) = \varepsilon_0\chi_0^{(3)}\mathbf{E}(t) \int g(t - \tau)|\mathbf{E}(\tau)|^2 d\tau. \tag{3.160}$$

Function $g(t)$ is the nonlinear response function normalized in a manner similar to the delta function [24]. The upper limit of integration in (3.160) extends only up to

t because the response function $g(t - \tau)$ must be zero for $\tau > t$ to ensure causality. Such approximation is applied for an instant electronic response like the Kerr effect and phonon–photon interactions (Raman effect). We weigh contributions from these two additive sources by a positive constant $0 \leq \alpha \leq 1$ [24]: the Kerr nonlinearity $g_K(t)$ with weight α and Raman effect $g_R(t)$ with weight $1 - \alpha$:

$$g(t) = \alpha g_K(t) + (1 - \alpha) g_R(t), \tag{3.161}$$

where $g_K(t)$ and $g_{Ra}(t)$ are some functions describing the Kerr and Raman effects, respectively. Having in mind almost instant Kerr electronic response, that is, $g_K(t) = \delta(t)$ – a Dirac delta function – the nonlinear Kerr polarization is readily reduced to [24]

$$\mathbf{P}_{Kerr}(t) = \varepsilon_0 \chi_0^{(3)} \mathbf{E}(t) \int \alpha \delta(t - \tau) |\mathbf{E}(\tau)|^2 d\tau = \alpha \varepsilon_0 \chi_0^{(3)} |\mathbf{E}(t)|^2 \mathbf{E}(t). \tag{3.162}$$

An updating scheme can be taken either straightforwardly,

$$\mathbf{P}_{Kerr}^{n+1} = \alpha \varepsilon_0 \chi_0^{(3)} (\mathbf{E}^{n+1})^3, \tag{3.163}$$

or with some simplification acknowledging appearance of fields in formula (3.162); see, for example, [25]:

$$\mathbf{P}_{Kerr}^{n+1} = \alpha \varepsilon_0 \chi_0^{(3)} |\mathbf{E}^n|^2 \mathbf{E}^{n+1}. \tag{3.164}$$

The first scheme leads to the third power of updated electric field in Ampere's law. It requires an iterative approach to solve the cubic equation for $\mathbf{E}^{n+1}$ [2]. Such situation can be circumvent if we use the displacement field $\mathbf{D}^{n+1}$, which is updated through linear Ampere's law. Then electric field $\mathbf{E}^{n+1}$ can be found analytically by solving a cubic equation derived directly from the constitutive equation $\mathbf{D}^{n+1} = \varepsilon_0 (\varepsilon + \alpha \chi_0^{(3)} |\mathbf{E}^{n+1}|^2 \mathbf{E}^{n+1})$ [26].

The FDTD scheme involving approach (3.164) is still completely explicit. Direct comparison of different schemes for Kerr nonlinearities in the 1D FDTD algorithm is presented in paper [26].

The Raman nonlinear polarization update is more complicated due to the *history* of the field interaction with some material parameters and possible anisotropy of the susceptibility response. We can basically define a sketch of the updating scheme. The Raman polarization is exposed through some function $S(t)$ quadratic in electric field:

$$\mathbf{P}_{Ra}(t) = \varepsilon_0 \mathbf{E}(t) \int (1 - \alpha) \chi_0^{(3)} g_{Ra}(\tau) \mathbf{E}^2(t - \tau) d\tau = \varepsilon_0 \mathbf{E}(t) S(t). \tag{3.165}$$

Assume that it is possible to find an updating scheme for function $S(t)$ [2], then

$$\mathbf{P}_{Ra}^{n+1} = \varepsilon_0 \mathbf{E}^{n+1} S^{n+1}. \tag{3.166}$$

Now, we substitute all updated terms (3.150), (3.153), (3.163), and (3.166) into (3.158)

$$\mathbf{D}^{n+1} = \varepsilon_0 \varepsilon_\infty \mathbf{E}^{n+1} + \mathbf{P}_D^{n+1} + \mathbf{P}_L^{n+1} + \mathbf{P}_{Kerr}^{n+1} + \mathbf{P}_{Ra}^{n+1}. \tag{3.167}$$

Then the FDTD cycle is performed in the sequence mentioned in Section 3.5.3.

Note: However, Equation (3.167) is now the nonlinear equation regarding $\mathbf{E}^{n+1}$. In most cases, it cannot be solved analytically and, hence, must be solved by an iteration procedure in analogy with reported in [2]. Following the simplified approach for the Kerr nonlinearity as discussed in [26], it is possible to avoid the iteration scheme and keep the updating scheme explicit. However, the stability of such approach and its accurateness especially in 2D and 3D realization is still the matter of investigation.

We refer here to Section 5.3, where more detailed analysis of how to tackle with the nonlinear polarization and third-order susceptibility can be read. Discussion there regards numerical methods used for simulation of nonlinear optical processes in fibres.

3.6.2 MEDIUM WITH GAIN: PHENOMENOLOGICAL APPROACH IN FDTD

We can compute the electromagnetic field propagation in a medium with gain employing the macroscopic approach, for example, through the complex electric conductivity σ in Ohm's law with the changed sign, which shows switching from the loss to gain model. A more sophisticated model of a gain medium with the saturation effect can be described by equation [2]:

$$\sigma(\omega) = \frac{1}{1 + I/I_s} \left(\frac{\sigma_0/2}{1 + i(\omega - \omega_0)T} + \frac{\sigma_0/2}{1 + i(\omega + \omega_0)T} \right), \tag{3.168}$$

where some parameters like resonance frequency ω_0, saturation intensity I_s, and resonant period T are acquired from experimental data. The following actions are to extract the real and imaginary parts of $\sigma(\omega)$, transform them into the time domain, and substitute into the FDTD scheme; see [2] for details.

3.6.3 LASING IN FDTD

In this section, we briefly analyze one of the situations originated from the light–matter interactions, namely, the lasing process in a four-level atomic system.

We consider a 4-level atom in the quasiclassic approach (Figure 3.8) following the recipe from [2]. Assuming N_0, N_1, N_2, and N_3 to be electron population density probabilities of the ground, and three excited levels, the polarization densities in the interband photon-absorbing $0 \to 3$ $\mathbf{P}_{30}$ and photon-emitting $2 \to 1$ $\mathbf{P}_{21}$ transitions driven by electric field $\mathbf{E}$ obey the second-order differential equations:

$$\frac{d^2\mathbf{P}_{21}}{dt^2} + \gamma_{21}\frac{d\mathbf{P}_{21}}{dt} + \omega_{21}^2\mathbf{P}_{21} = \zeta_{21}(N_2 - N_1)\mathbf{E},$$

$$\frac{d^2\mathbf{P}_{30}}{dt^2} + \gamma_{30}\frac{d\mathbf{P}_{30}}{dt} + \omega_{30}^2\mathbf{P}_{30} = \zeta_{30}(N_3 - N_0)\mathbf{E}, \tag{3.169}$$

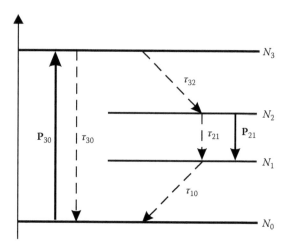

FIGURE 3.8 Four-level atom scheme.

where γ_{21} is the dephasing rate (or nonradiative losses) for transition $(2 \rightarrow 1)$, ω_{21} is the transition resonant frequency, and $\zeta_{21} = 6\pi\varepsilon_0 c^3/\omega_{21}^2\tau_{21}$, τ_{21} is the electron life time on level 2 and corresponding designations for the transition $(0 \rightarrow 3)$.

The electron population density probabilities are governed by the rate equations:

$$\frac{dN_0}{dt} = \frac{N_1(1-N_0)}{\tau_{10}} + \frac{N_3(1-N_0)}{\tau_{30}} - \frac{1}{\hbar\omega_{30}}\mathbf{E}\frac{d\mathbf{P}_{30}}{dt},$$

$$\frac{dN_1}{dt} = -\frac{N_1(1-N_0)}{\tau_{10}} + \frac{N_2(1-N_1)}{\tau_{21}} - \frac{1}{\hbar\omega_{21}}\mathbf{E}\frac{d\mathbf{P}_{21}}{dt},$$

$$\frac{dN_2}{dt} = \frac{N_3(1-N_2)}{\tau_{32}} - \frac{N_2(1-N_1)}{\tau_{21}} + \frac{1}{\hbar\omega_{21}}\mathbf{E}\frac{d\mathbf{P}_{21}}{dt},$$ (3.170)

$$\frac{dN_3}{dt} = -\frac{N_3(1-N_2)}{\tau_{32}} - \frac{N_3(1-N_0)}{\tau_{30}} + \frac{1}{\hbar\omega_{30}}\mathbf{E}\frac{d\mathbf{P}_{30}}{dt},$$

where two main reasons of varying population densities N_i are
By pumping (or stimulated emission) with rate $\mathbf{E}(d\mathbf{P}/dt)$
Via spontaneous emission decay from level i to j with rate N_i/τ_{ij}

Accounting for the Fermi distribution rule reduces the rates of population changes by factor $1 - N$ [2]. The simplified approaches without such accounting are presented in papers [27,28].

Note: Instead of using dipole moments in transition $0 \rightarrow 3$ (optical pumping), we can use just the pumping rate W, which can be attributed for the electrical pumping. Then vector $\mathbf{P}_{30}$ should be neglected in all expressions, and, correspondingly, it will be only one equation (3.169).

To finish the chain of the light–matter interactions, we should consider Ampere's law with an additional polarization current due to the variation of the polarization vectors $n\mathbf{P}_{21}$ and $n\mathbf{P}_{30}$:

$$\frac{\partial \mathbf{E}}{\partial t} = \frac{1}{\varepsilon_0} \nabla \times \mathbf{H} - \frac{n}{\varepsilon_0} \left(\frac{d\mathbf{P}_{21}}{dt} + \frac{d\mathbf{P}_{30}}{dt} \right), \tag{3.171}$$

where n is the population density of the atoms.

Fields Update Algorithm for Lasing Scheme

The updating scheme is performed successively for each time step:

1. By solving Equation (3.169), the polarization densities of two transitions 21 and 30 are updated from $\mathbf{P}^n$ to $\mathbf{P}^{n+1}$ with electric field $\mathbf{E}^n$ and populations probabilities $N_i^n, i = 0, 1, 2, 3$.
2. The electric field is updated to $\mathbf{E}^{n+1}$ through Equation (3.171) with new $\mathbf{P}^{n+1}$ and old $\mathbf{P}^n$ dipole moments involved for the derivatives approximation.
3. The rate equations (3.170) are solved to produce new population density probabilities $N_i^{n+1}, i = 0, 1, 2, 3$ starting from N_3^{n+1} down to N_1^{n+1}. Then N_0^{n+1} is acquired from the conservation rule for electronic probabilities justified in each time moment: $N_0 + N_1 + N_2 + N_3 = 1$.
4. The magnetic vector is updated through Faraday's law (2.2).

3.7 CONCLUSION

We have just considered the fundamentals of the FDTD – a modern computational scheme. The power of the FDTD methods is in its universality – they directly solve Maxwell's equations in the time domain. Of course, we have not been able to cover all FDTD-related material thoroughly, unveiled remain such popular topics as a *total field/scattered field* numerical scheme, *subgridding* and *non-uniform meshing*, and *near field–far field transformation*. As a separate subject stands *parallelization* – by its nature – the FDTD is nicely suited for parallel computing, giving manifold speeding up in the calculations. You can read about these and many other subjects in the cited literature.

EXERCISES

3.1 Derive a fourth-order central difference scheme for $\partial u / \partial x$ using the Taylor series expansion of function u around point x.

3.2 Derive a second-order central difference scheme for $\partial^2 u / \partial x^2$ Equation (3.10). Use $h/2$ discretization and central difference approximations for function $\partial u / \partial x$.

3.3 Consider 1D propagation of an electromagnetic wave in vacuum. Solve the 1D dispersion equation (3.44) regarding the wave number analytically. Find the conditions for the wave number to be real or complex. What kind of waves do such wave numbers describe? Use Equation (3.52) if needed.

3.4 In this exercise, we study superluminal ($v_p > c_0$) and subluminal ($v_p < c_0$) numerical light propagation in vacuum. Show that $\omega \delta t = 2\pi Q/N_\lambda$, where N_λ grid sampling (number of grid points per wavelength in vacuum) is defined by formula (3.60). Derive the expression for the phase velocity of a plane wave on a 1D grid. Plot phase velocity normalized by the speed of light in vacuum in dependence of the grid sampling. Take Courant stability factors: $Q = 0.4; 0.6; 0.8; 0.9; 0.99; 1.0$ and grid sampling from 1.0 (extremely course mesh) to 20 (moderate-to-fine meshing). Use analytical results from Exercise 3.3. Define grid samplings for the superluminal and subluminal propagation.

3.5 In this exercise, we study numerical losses of light in vacuum. If the wave number k is complex, then the plane wave is attenuated, unless an unstable FDTD algorithm is used with $Q > 1$. Wave's amplitude is exponentially decreasing during the propagation. Find an analytical formula for the attenuation constant, which characterizes such decay, in dependence of the grid sampling N_λ. Take Courant stability factors: $Q = 0.4; 0.6; 0.8; 0.9; 0.99; 1.0$ and grid sampling from 1.0 (extremely course mesh) to 20 (moderate-to-fine meshing). Use analytical results from Exercise 3.3.

3.6 Analyze numerical dispersion in the case of 1D wave propagation with the Courant stability factor $Q = 1.0$. Explain why the time step $\delta t = h/c_0$ is called the magic step.

3.7 Show that the finite-difference scheme applied on the 3D Yee grid is divergence-free for the computed magnetic fields.

3.8 Prove that in the limit of infinitesimal spatial grid meshing ($h \to 0$) and time stepping ($\delta t \to 0$), the dispersion equation (3.41) converges to the conventional light dispersion equation in a uniform dielectric (3.32).

3.9 Derive 2D numerical dispersion equation in uniform dielectric:

$$\frac{\varepsilon\mu}{c_0^2 \delta t^2} \sin^2 \frac{\omega \delta t}{2} = \left(\frac{1}{\delta x} \sin \frac{k_x \delta x}{2}\right)^2 + \left(\frac{1}{\delta y} \sin \frac{k_y \delta y}{2}\right)^2 \qquad (3.172)$$

by completing the steps of substituting TM travelling waves into the 2D FDTD scheme

$$\frac{H_x|_{i,j+1/2}^{n+1/2} - H_x|_{i,j+1/2}^{n-1/2}}{\delta t} = -\frac{1}{\mu_0 \mu_{i,j+1/2}} \left(\frac{E_z|_{i,j+1}^n - E_z|_{i,j}^n}{\delta y}\right) \qquad (3.173)$$

$$\frac{H_y|_{i+1/2,j}^{n+1/2} - H_y|_{i+1/2,j}^{n-1/2}}{\delta t} = \frac{1}{\mu_0 \mu_{i+1/2,j}} \left(\frac{E_z|_{i+1,j}^n - E_z|_{i,j}^n}{\delta x}\right) \qquad (3.174)$$

$$\frac{E_z|_{i,j}^{n+1} - E_z|_{i,j}^n}{\delta t} = \frac{1}{\varepsilon_0 \varepsilon_{i,j}} \left(\frac{H_y|_{i+1/2,j}^{n+1/2} - H_y|_{i-1/2,j}^{n+1/2}}{\delta x} - \frac{H_x|_{i,j+1/2}^{n+1/2} - H_x|_{i,j-1/2}^{n+1/2}}{\delta y}\right).$$

$$(3.175)$$

3.10 Solving numerically 2D dispersion equation (3.172) on a square grid, prove that even an empty grid-defined space exhibits anisotropy for light propagation, when phase velocity depends on the direction of propagation. Choose appropriate numerical parameters on your own.

3.11 Assume the 1D numerical domain from $i = 1$ to $i = 11$. All initial electric and magnetic $H_{i+1/2}^0$ fields are nullified except $E_6^{1/2} = 1$. Calculate analytically electric and magnetic fields in the nine consequent time steps $\delta t/2$, and fill up the table with the field amplitudes in space points (coordinates in rows) in different time moments (time steps in columns). Take $Q = 1.0$.

3.12 Write down a script implementing the numerical scheme of a 1D FDTD method. Make your choice of parameter Q and the size of the domain. Implement a monochromatic source (CW) as the right end and a PMC as the left BCs terminating the domain. Test the script for a homogeneous medium, $\varepsilon = \mu = 1$ observing

(a) Fields E_i^n, H_i^n in all coordinate points at several time moments, for example, $n = 100, 200, 500$

(b) Fields E_i^n, H_i^n at all time steps in several points, for example, $i = 20, 50, 200$, at all time steps

(c) Choose $Q \geq 1.0$ and observe unlimited growth of the fields

3.13 Repeat Exercise 3.12 but with a hard CW source positioned in the middle of the domain with $\varepsilon = 9, \mu = 1$ and PEC as the right end BC. Describe the changes in the field pictures you observe? Try to violate the Courant stability condition. Does the limit value depend on ε?

3.14 Implement a pulse current source from the very right end of the numerical domain, generating a pulse like $\sin(\omega t) \sin(\Omega t)$. Take two cases: $\varepsilon = \mu = 1$ and $\varepsilon = 9, \mu = 1$:

(a) Trace the pulse propagation in time and space.

(b) Check the peak value and pulse shape.

(c) Determine the group velocity of the pulse, defined as the speed of the pulse peak propagation.

3.15 Implement a 1D PML from one side of the domain opposite to the position of the CW source. Observe propagation of the wavefront before and after reflection from the PML. Wait for the steady-state conditions. Quantify the PML performance as an ABC by comparing amplitudes of the reflected waves with the perfect conductor BC and PML.

3.16 Somewhere in the middle of the domain, make a Bragg grating with parameters: 10 periods, thicknesses of the layers obey condition, $\sqrt{\varepsilon_1}l_1 = \sqrt{\varepsilon_2}l_2 = \lambda_0/4$ with $\lambda_0 = 1200$ nm, $\sqrt{\varepsilon_1} = 1.45$, and $\sqrt{\varepsilon_2} = 2.6$. Put a CW source at one side of the grating. Observe and describe transmission and reflection from the grating at different wavelengths of the source.

3.17 Implement PMLs from both sides of the domain. Put your source in the middle between two Bragg gratings with the same parameters. Trace CW and pulse propagation in the frequency ranges with good (passband) and poor (stopband) transmission.

3.18 Tune the left Bragg mirror on wavelength $\lambda_0 = 1500$ nm. Generate a CW, short (broadband) and long (narrowband) pulses with the (central) frequencies matching with either the left or right Bragg gratings stopbands. Observe and describe its propagation through the structures.

REFERENCES

1. K. S. Kunz and R. J. Luebbers, *The Finite Difference Time Domain Method for Electromagnetics*, Boca Raton, FL: CRC Press, 1993.
2. A. Taflove and S. C. Hagness, *Computational Electrodynamics: The Finite-Difference Time-Domain Method*, 3rd edn. Boston, MA: Artech House, 2005.
3. A. Elsherbeni and V. Demir, *The Finite-Difference Time-Domain Method for Electromagnetics with MATLAB Simulations*, 3rd edn. Raleigh, NC: SciTech Publishing, 2008.
4. W. Yu, X. Yang, Y. Liu, R. Mittra, and A. Muto, *Advanced FDTD Methods. Parallelization, Acceleration, and Engineering Applications*. Boston, MA: Artech House, 2011.
5. D. M. Sullivan, *Electromagnetic Simulation Using the FDTD Method*, 2nd edn. Hoboken, NJ: IEEE Press, 2013.
6. A. Taflove, Application of the finite-difference time-domain method to sinusoidal steady-state electromagnetic penetration problems, *IEEE Trans. Electromagn. Compat.*, 22, 191–202 (1980).
7. K. S. Yee, Numerical solution of initial boundary value problem involving Maxwell's equations in isotropic media, *IEEE Trans. Ant. Propag.*, 14, 302–307 (1966).
8. A. J. Ward and J. B. Pendry, Calculating photonic Green's functions using a nonorthogonal finite-difference time-domain method, *Phys. Rev. B*, 58, 7252–7259 (1998).
9. F. R. Gantmacher, *The Theory of Matrices*. Providence, RI: AMS Chelsea Publishing, 2000.
10. D. M. Shyroki, Modeling of sloped interfaces on a Yee grid, *IEEE Trans. Ant. Propag.*, 59, 3290–3295 (2011).
11. W. H. Press, S. A.Teukolsky, W. T. Vetterling, and B. P. Flannery, *Numerical Recipes: The Art of Scientific Computing*, 3rd edn. New York: Cambridge University Press, 2007.
12. B. Engquist and A. Majda, Absorbing boundary conditions for the numerical simulation of waves, *Math. Comput.*, 31, 629–651 (1977).
13. G. Mur, Absorbing boundary conditions for the finite-difference approximation of the time-domain electromagnetic field equations, *IEEE Trans. Electromagn. Compat.*, 23, 377–382 (1981).
14. J. P. Berenger, A perfectly matched layer for the absorption of electromagnetic waves, *J. Comput. Phys.*, 114, 185–200 (1994).
15. S. D. Gedney, An anisotropic perfectly matched layer-absorbing medium for the truncation of FDTD lattices, *IEEE Trans. Ant. Propag.*, 44, 1630–1639 (1996).
16. P. G. Petropoulos, L. Zhao, and A. C. Cangellaris, A reflectionless sponge layer absorbing boundary condition for the solution of Maxwell's equations with high-order staggered finite difference schemes, *J. Comput. Phys.*, 139, 184–208 (1998).
17. A. Lavrinenko, P. I. Borel, L. H. Frandsen, M. Thorhauge, A. Harpoth, M. Kristensen, T. Niemi, and H. M. H. Chong, Comprehensive FDTD modelling of photonic crystal waveguide components, *Opt. Express*, 12, 234–248 (2004).
18. D. M. Shyroki and A. V. Lavrinenko, Perfectly matched layers in the finite-difference time-domain and frequency domain methods, *Phys. Stat. Sol. (b)*, 244, 3506–3514 (2007).
19. C. Kittel, *Introduction to Solid State Physics*. New York: John Wiley, 1971.
20. W. H. P. Pernice, F. P. Payne, and D. F. G. Gallagher, A general framework for the finite-difference time-domain simulation of real metals, *IEEE Trans. Ant. Propag.*, 55, 916–923 (2007).

21. T. Kashiwa and I. Fukai, A treatment by FDTD method of dispersive characteristics associated with electronic polarization, *Microw. Opt. Tech. Lett.*, 3, 203–205 (1990).

22. R. M. Joseph, S. C. Hagness, and A. Taflove, Direct time integration of Maxwell's equations in linear dispersive media with absorption for scattering and propagation of femtosecond electromagnetic pulse, *Opt. Lett.*, 16, 1412–1414 (1991).

23. R. W. Hellwarth, Third-order optical susceptibilities of liquid and solids, *J. Prog. Quant. Electron.*, 5, 1–68 (1977).

24. G. P. Agrawal, *Nonlinear Fiber Optics*, 3rd edn. San Diego, CA: Academic Press, 2001.

25. R. W. Ziolkowski, The incorporation of microscopic material models into the FDTD approach for ultrafast optical pulse simulations, *IEEE Trans. Ant. Propag.*, 45, 375–391 (1997).

26. I. S. Maksymov, A. A. Sukhorukov, A. V. Lavrinenko, and Y. S. Kivshar, Comparative study of FDTD-adopted numerical algorithms for Kerr nonlinearities, *IEEE Ant. Wireless Propag. Lett.*, 10, 143–146 (2011).

27. A. S. Nagra and R. A. York, FDTD analysis of wave propagation in nonlinear absorbing and gain media, *IEEE Trans. Ant. Propag.*, 46, 334–340 (1998).

28. X. Jiang and C. M. Soukoulis, Time dependent theory for random lasers, *Phys. Rev. Lett.*, 85, 70–73 (2000).

4 Finite-Difference Modelling of Straight Waveguides

4.1 INTRODUCTION

This chapter describes how light propagation in straight dielectric waveguides can be modelled in the frequency domain (FD) by the finite-difference approach. By a *straight waveguide*, we understand a waveguide structure which is invariant in the direction of light propagation. Important examples of such waveguides are optical fibres, including microstructured fibres, and (the straight sections of) various slab and planar waveguide structures. Understanding and designing the properties of such waveguides are important in areas such as signal transmission (telecommunication), nonlinear signal processing, fibre lasers, and waveguide-based sensors. Therefore, efficient tools for numerical modelling are of great importance for both university and industry researchers in these fields. The finite-difference methods are simple and versatile techniques, which allow to expose the tricks and approximations involved in modelling on a real-space grid without getting lost in details of mathematics and coding. They rely on discretizing the electromagnetic fields of the waveguide modes on a uniform real-space grid, using simple interpolation formulas to evaluate numerical derivatives.

This chapter is structured in the following way: Section 4.2 contains some general considerations on the finite-difference method and the rationale for using the FD. In Section 4.3, the discretization of these equations in one dimension using finite differences is discussed, with emphasis on the treatment of discontinuities in the dielectric function (and therefore in the fields). This section also contains an introduction to the slab waveguide, which is used as a simple test case for the derived finite-difference method in the exercises. Section 4.4 outlines the MATLAB® procedures, in particular eigensolvers, needed to solve the discretized equations. In Section 4.5, the discretization of 2D waveguide problems on the so-called *Yee mesh* is described, in conjunction with the important technique of dielectric function averaging. Also, the exploitation of mirror symmetries is outlined in this section.

4.2 GENERAL CONSIDERATIONS

4.2.1 Time Domain versus Frequency Domain

When attacking a given electromagnetic problem numerically, one must make a choice whether to work in the time domain or FD. In Chapter 3, the *time-domain* approach was outlined: An initial field distribution is assumed, and the time evolution of the fields is tracked by propagating Maxwell's equations forward in time. This method is useful for many kinds of problems. An example is the modelling of signal processing components, in which the light undergoes some complicated transformation over relatively short distances.

However, one of the most important applications of waveguides is as transmission media for light signals over long distances, most notably optical fibres for long-distance telecommunication. Modelling of pulse propagation through a long waveguide (metres or kilometres) is prohibitive by a technique such as finite-difference time domain (FDTD), due to the very large number of time steps required. A better alternative may be to go into the FD, using the Fourier transforms (2.15) and (2.16), to express the time-domain fields. The FD fields can be found by solving wave equations like (2.36) and (2.37), which take the form of (generalized) eigenvalue problems for the squared angular frequency. If all relevant eigenstates are known, along with the expansion of the initial fields on these eigenstates, the time evolution of the fields is trivially determined by Equations (2.15) and (2.16). If the fields appearing in the problem at hand can, at all times, be expressed in terms of only a few frequency eigenstates, and/or the fields are to be propagated over a long time interval, the FD approach will be preferable over the time-domain approach. On the other hand, if the fields are well described only by a superposition of many eigenmodes, working in the time domain is a more appropriate strategy. If many eigenmodes *and* a prolonged time propagation are required, the problem will be difficult regardless of the chosen strategy.

In this chapter, the focus is on calculating FD eigenmode properties of straight waveguides. Examples of how the information about propagation constants and eigenmode fields can be useful for simulating pulse propagation over extended distances can be found in Chapter 5. For the Fourier expansions in Equations (2.15) and (2.16), we will adopt the positive sign in the exponents.

4.2.2   Finite-Difference Methods for Straight Waveguides

The finite-difference method samples electric and/or magnetic fields at a finite set of points in space and evaluates the derivatives appearing in the wave equations from the function values in the sampled points only. The advantages of the method are its simplicity and its generality: No a priori assumptions are made about the structure of the solutions; other than that, they should be representable on the chosen grid of sampling points. Also, the formulation of the wave equations in terms of a finite set of field values is fairly straightforward as we shall see. This is especially the case if we restrict ourselves to grids with a uniform point spacing. It is possible to let the density of gridpoints vary, so that the computational effort can be concentrated in regions of high importance or particularly intricate field structures. This technique does, however, require some care to avoid that the variations in point spacing influence the results, and it will not be discussed further in this chapter.

For straight waveguides, the most serious challenge for the finite-difference method is the fact that it a priori assumes continuous field distributions in order for the finite-difference expressions for first and second derivatives to be valid.

However, most interesting problems in photonics involve discontinuous refractive-index profiles, which lead to discontinuities in the electric field components perpendicular to the index steps or, equivalently, in the parallel components of the displacement field. Examples of such problems include most optical fibres, both the standard step-index fibres and the more recent microstructured fibres, where large index steps are present at the boundaries between silica and airholes. Also integrated waveguide technology is mostly concerned with piecewise constant index distributions, since the fabrication of such waveguides usually involves depositing layer upon layer of various materials, with subsequent position-dependent etch processes forming the structures in the lateral dimension. One might think that the adoption of a magnetic field formulation for the wave equation would resolve the issue, because the magnetic field will be continuous in the absence of magnetic materials, which is usually, though not always, a viable approximation. However, since Ampere's law, Equation (2.3), relates the displacement field to the rotation of the magnetic field, a discontinuous displacement field will imply discontinuities in the first derivative of **H**. Since the finite-difference approach also relies on continuity of the first derivative (see Section 4.3.1), the problems implied by discontinuous dielectric functions in connection with finite-difference methods are real and require extensions of the most straightforward finite-difference formulations of, for example Equation (2.64).

An early attempt of extending finite-difference formulas to account for discontinuities was made by Stern [1]. In this work, a piecewise continuous refractive-index profile was discretized by utilizing continuity and differentiability of the fields in uniform sections, which are local solutions to the Helmholtz equation. Index steps were assumed to lie halfway between gridpoints, and field discontinuities at the index steps were handled using the appropriate electromagnetic boundary conditions (BCs). Other derivations have followed similar ideas, but placing index steps at the gridpoints instead of between them, which implies that a single gridpoint can be associated with up to four refractive-index values in a 2D waveguide problem [2]. An important extension to the formalism of Stern was made by Vassallo, who derived a discretization formula for arbitrary locations of the index steps, while also accounting for discontinuities in the second derivatives of the fields [3]. In these approaches, index steps in a 2D waveguide structure must conform to the chosen (Cartesian) grid, that is a curved or tilted waveguide surface must be approximated by the *staircase* method, in which the curved interface is replaced by a stepwise variation, as will be further discussed in Section 4.5.2. The method has later been generalized to cover arbitrarily curved and tilted interfaces using coordinate-transformation techniques [4,5]. While leading to a method of great generality, this approach is formally somewhat cumbersome. In this chapter, we illustrate this class of techniques by applying a slight extension of Vassallo's formalism to the simple case of a 1D slab waveguide.

An alternative approach to finite-difference modelling is to stay with the conventional finite-difference formulas, while *smoothing* the discontinuous dielectric profiles by an averaging procedure. In a simple approach, one might set the dielectric constant at a boundary point to be the average of the values in the neighbouring regions. However, a much better performance is obtained by using a tensorial averaging scheme, which accounts for the orientation of the dielectric surface. This scheme was first introduced for plane-wave-based methods [6], building on earlier studies of effective-medium theory [7]. However, the technique has also been used for Yee-mesh-based finite-difference modelling in both time domain [8] and FD [9]. The advantage of this method is that a general finite-difference formalism ensures reasonably good convergence results for arbitrary waveguide structures. The finite-difference routines can be formulated once and for all, so all one has to do in order to study a particular waveguide geometry is to perform a suitable dielectric function averaging of the structure. In this chapter, we present a formulation based on the work of Zhu and Brown [9] to study arbitrary 2D waveguide structures.

4.3 MODIFIED FINITE-DIFFERENCE OPERATORS

In this section, the approach of modifying the finite-difference operators to account for field discontinuities are outlined in one dimension. As a prequel, we start out by discretizing the scalar wave equation, Equation (2.66) for the slab waveguide, in order to introduce the general finite-difference terminology. The resulting formalism will then be modified to account for discontinuities.

4.3.1 Discretizing the Scalar Wave Equation

The first step in applying the finite-difference method is to choose an appropriate *discretization grid* for the waveguide structure. As noted previously, we will only consider uniform discretizations, where the grid spacing is the same between all the gridpoints. To formulate the wave equations in terms of the field values on the gridpoints, we need to discretize the second-order differential operator. We shall only concern ourselves with the so-called three-point formulas, that is we want to estimate the derivative at a given gridpoint in terms of the field values at the point itself and its neighbouring points. To be more specific, suppose that we know H_x at the gridpoints x and $x \pm \Delta x$, as illustrated in Figure 4.1. We can write the Taylor expansions as

$$H_x(x + \Delta x) = H_x(x) + H'_x(x)\Delta x + \frac{1}{2}H''_x(x)\Delta x^2 + \frac{1}{6}H'''_x(x)\Delta x^3 + O(\Delta x^4), \quad (4.1)$$

$$H_x(x - \Delta x) = H_x(x) - H'_x(x)\Delta x + \frac{1}{2}H''_x(x)\Delta x^2 - \frac{1}{6}H'''_x(x)\Delta x^3 + O(\Delta x^4), \quad (4.2)$$

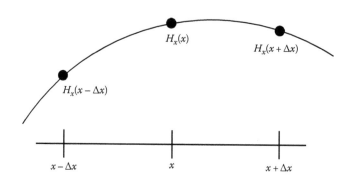

FIGURE 4.1   Three points on a discretization grid equally spaced by the distance Δx.

where H'_x, H''_x, and H'''_x denote the first, second, and derivatives of H_x, etc. We can *invert* these equations to solve for $H'_x(x)$ and $H''_x(x)$:

$$H_x(x + \Delta x) - H_x(x - \Delta x) = 2H'_x(x)\Delta x + \frac{1}{3}H'''_x(x)\Delta x^3 + O(\Delta x^4) \tag{4.3}$$

$$H_x(x + \Delta x) + H_x(x - \Delta x) = 2H_x(x) + H''_x(x)\Delta x + O(\Delta x^4) \tag{4.4}$$

$$\Downarrow$$

$$H'_x(x) = \frac{H_x(x + \Delta x) - H_x(x - \Delta x)}{2\Delta x} + O(\Delta x^2) \tag{4.5}$$

$$H''_x(x) = \frac{H_x(x + \Delta x) + H_x(x - \Delta x) - 2H_x(x)}{\Delta x^2} + O(\Delta x^2). \tag{4.6}$$

The second-order error terms imply that the accuracy of the first- and second-order derivatives is improved four times if Δx is made twice as small.

Note: The aforementioned derivations assume existence (and finiteness) of the field derivatives. If the field or its first or second derivative is discontinuous, the results for the accuracy of the derivative estimates cannot be expected to hold.

In the scalar approximation discussed in Section 2.6, the different vector components of the electromagnetic field are decoupled, so we need only consider one of them, H_x, say

$$\partial_x^2 H_x(x) + \varepsilon(x)k_0^2 H_x(x) = \beta^2 H_x(x). \tag{4.7}$$

In Figure 4.2, an arbitrary $\varepsilon(x)$ distribution and its discretization into five points are shown. At each point in the discretization grid, a permittivity value is stored as well as a value of the field, H. While the permittivity values are known from the outset, we seek to calculate H values such that the scalar eigenvalue equation is fulfilled. We can

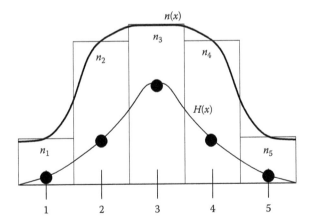

FIGURE 4.2 An arbitrary waveguide geometry discretized in five points. Each point contains both a refractive index, n, and field value, H.

do this by setting up linear equations for the field values at the different gridpoints. Consider first the field at gridpoint 2, H_2. From Equations (4.6) and (4.7), we obtain

$$\frac{1}{(\Delta x)^2}H_1 + \left[\varepsilon_2 k_0^2 - \frac{2}{(\Delta x)^2}\right]H_2 + \frac{1}{(\Delta x)^2}H_3 = \beta^2 H_2. \qquad (4.8)$$

The second-order differential relates H_2 to H_1 and H_3. To get an equation for H_3, we may repeat our procedure at gridpoint 3, to find

$$\frac{1}{(\Delta x)^2}H_2 + \left[\varepsilon_3 k_0^2 - \frac{2}{(\Delta x)^2}\right]H_3 + \frac{1}{(\Delta x)^2}H_4 = \beta^2 H_3. \qquad (4.9)$$

So, now there are two equations and four unknowns. As we add more gridpoints, we will continue to get an extra unknown for each added equation, and the system will remain underdetermined. This situation is, however, saved by the fact that we must eventually restrict our calculations to a finite space.

In the analytical treatment of optical waveguides, it is possible to consider dielectric structures that are infinite in extent. This cannot be done in a numerical finite-difference treatment, since only a finite number of variables can be stored in the computer. Therefore, the computational domain must be truncated, and we need to specify how the second-order differential should be evaluated at the endpoints. As discussed in Section 2.7.3, this is known as the BCs of the problem. In this chapter, we consider only the most simple-minded BC: We assume that the field outside the computational domain is everywhere zero. In setting up an equation for the field at gridpoint 1, H_1, we could then pretend that there exists a further gridpoint to the left, where the field is known to be zero. Use of Equation (4.6) then yields

$$\left[\varepsilon_1 k_0^2 - \frac{2}{(\Delta x)^2}\right]H_1 + \frac{1}{(\Delta x)^2}H_2 = \beta^2 H_1. \qquad (4.10)$$

This *zero-field* BC can be justified for a localized (guided) mode whose field strength must fall to zero at large distances from the waveguide core, provided that the boundary of the computational domain is sufficiently far from the core for the field to be insignificant there. As you will see in the exercises, this requirement on the extent of the computational domain may be dependent on the modal properties, for example the wavelength under consideration. A discussion of other possible BCs may be found in Sections 2.7.3 and 3.4.

Since Equation (4.10) only involves H_1 and H_2, which were already present in Equation (4.8), we have succeeded in adding an equation without increasing the number of unknowns. Applying a similar procedure at the other endpoint, gridpoint 5, we obtain a system of five equations with five unknowns, which can be written in matrix form as

$$\begin{bmatrix} X_1 & a & 0 & 0 & 0 \\ a & X_2 & a & 0 & 0 \\ 0 & a & X_3 & a & 0 \\ 0 & 0 & a & X_4 & a \\ 0 & 0 & 0 & a & X_5 \end{bmatrix} \begin{bmatrix} H_1 \\ H_2 \\ H_3 \\ H_4 \\ H_5 \end{bmatrix} = \beta^2 \begin{bmatrix} H_1 \\ H_2 \\ H_3 \\ H_4 \\ H_5 \end{bmatrix}. \qquad (4.11)$$

Here, the symbols X_i and a are defined as

$$X_i = \frac{-2}{(\Delta x)^2} + \varepsilon_i k_0^2, \tag{4.12}$$

$$a = \frac{1}{(\Delta x)^2}. \tag{4.13}$$

Thus, by applying the finite-difference approximation, we have converted our analytical wave equation into a numerical eigenvalue problem of the following form:

$$\Phi H = \beta^2 H. \tag{4.14}$$

Here, Φ is commonly referred to as the *discretization matrix* of the finite-difference problem.

4.3.2 Inclusion of Discontinuities: General Formalism

After these preliminaries, we are ready to tackle the important problem of a discontinuous dielectric function. We will consider a dielectric distribution, which is piecewise constant, but possibly with refractive-index steps between each point in the finite-difference grid (recall that permittivity and refractive index are related by $\varepsilon = n^2$). A graded-index distribution can thus be approximated by a *staircase* of constant refractive-index levels. The situation is outlined in Figure 4.3: Dielectric boundaries are present at positions $x_i + \xi_i \Delta x$, where ξ_i is a number between 0 and 1. Thus, in addition to information about the value of the dielectric constant at each gridpoint, we also need to specify the exact positions of the dielectric boundaries. Alternatively, one could decide that boundaries are always placed in certain positions, for example at the gridpoints or at the midpoint between two gridpoints. The advantage of the more general scheme is that the waveguide structure can be specified independently of the chosen grid. For generality, the field at gridpoint i is denoted by ψ_i. For a transverse electric (TE) mode, ψ will be taken to be E_y, whereas for a transverse magnetic (TM) mode, ψ will be E_x. Furthermore, we will use the shorthand notation ψ_i', ψ_i'', and ψ_i''' for the first, second, and third derivatives, respectively, of ψ_i with x.

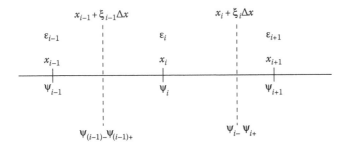

FIGURE 4.3 One-dimensional finite-difference grid with arbitrary dielectric boundaries between the gridpoints. The field just left of the i'th boundary is denoted ψ_{i-}, the field just right of this boundary ψ_{i+}, etc.

Note: One can also formulate magnetic field equations by a similar procedure, but of course with a different final result. Here, we focus on the electric field formulation to demonstrate that the method works well even for a discontinuous field.

In the piecewise constant sections of the structure, that is *almost everywhere*, the Helmholtz equation is locally valid. Our aim is therefore to find an expression for the second derivative of the field at gridpoint i, ψ_i'', expressed in terms of the field values at i and $i \pm 1$, just as we did in the previous section. Once again, the strategy is to do the opposite – specify the $\psi_{i\pm 1}$ in terms of ψ_i and the derivatives at the different gridpoint, then invert the resulting equations to solve for ψ_i''. In between the dielectric boundaries, the fields are continuous and differentiable, which means they can be expressed by the Taylor expansions, which we will initially carry through to third order. At the dielectric boundary to the right of gridpoint i (see Figure 4.3), we have

$$\psi_{i+} = \theta_i \psi_{i-}; \quad \psi_{i+}' = \psi_{i-}'; \quad \theta_i = \frac{\varepsilon_i}{\varepsilon_{i+1}} (\text{TM}); \quad \theta_i = 1 (\text{TE}). \tag{4.15}$$

In the TE case, ψ represents E_y, which is parallel to the dielectric boundary and therefore continuous. The continuity of ψ' in this case follows from the z-component of Faraday's law, Equation (2.2), and the continuity of all components of the magnetic field. In the TM case, ψ represents E_x, whose behaviour at the boundary is determined by the continuity of the x-component of the displacement field, that is $D_x = \varepsilon E_x$. The continuity of ψ' in this case follows from the fact that

$$\nabla \cdot \mathbf{E} = \partial_x E_x - i\beta E_z = 0 \tag{4.16}$$

to the left and right of the dielectric boundary and that E_z is continuous across the boundary.

We can find BCs for ψ'' and ψ''' by considering the wave equation on both sides of the boundary. For the second-order derivative, we have

$$\psi_{i-}'' + k_0^2 \varepsilon_i \psi_{i-} = \beta^2 \psi_{i-} \tag{4.17}$$

$$\psi_{i+}'' + k_0^2 \varepsilon_{i+1} \psi_{i+} = \psi_{i+}'' + \theta_i k_0^2 \varepsilon_{i+1} \psi_{i-} = \beta^2 \psi_{i+} = \theta_i \beta^2 \psi_{i-} \tag{4.18}$$

$$\Downarrow$$

$$\psi_{i+}'' = \theta_i \left[\psi_{i-}'' + k_0^2 (\varepsilon_i - \varepsilon_{i+1}) \psi_{i-} \right]. \tag{4.19}$$

Taking the derivative of the wave equation and using the fact that ψ' is continuous, one finds in a similar way that

$$\psi_{i+}''' = \psi_{i-}''' + k_0^2 (\varepsilon_i - \varepsilon_{i+1}) \psi_{i-}'. \tag{4.20}$$

We can now *track* the field from gridpoint i to gridpoint $i+1$ by using the Taylor expansions in combination with the derived BCs. The first step is to write

$$\psi_{i-} = \psi_i + \psi_i' \xi_i \Delta x + \frac{1}{2} \psi_i'' (\xi_i \Delta x)^2 + \frac{1}{6} \psi_i''' (\xi_i \Delta x)^3, \tag{4.21}$$

$$\psi_{i-}' = \psi_i' + \psi_i'' \xi_i \Delta x + \frac{1}{2} \psi_i''' (\xi_i \Delta x)^2, \tag{4.22}$$

$$\psi_{i-}'' = \psi_i'' + \psi_i''' \xi_i \Delta x, \tag{4.23}$$

$$\psi_{i-}''' = \psi_i''', \tag{4.24}$$

neglecting all field derivatives higher than the third order. Working from the right side of the boundary, we can similarly write

$$\psi_{i+1} = \psi_{i+} + \psi'_{i+}(1 - \xi_i)\Delta x + \frac{1}{2}\psi''_{i+}((1 - \xi_i)\Delta x)^2 + \frac{1}{6}\psi'''_{i+}((1 - \xi_i)\Delta x)^3. \quad (4.25)$$

The final step is to relate the field and its derivatives across the boundary, using Equations (4.15), (4.19), and (4.20). This leads to the following result:

$$\psi_{i+1} = \theta_i\psi_{i-} + \psi'_{i-}(1 - \xi_i)\Delta x + \frac{1}{2}\left(\theta_i\left[\psi''_{i-} + k_0^2(\varepsilon_i - \varepsilon_{i+1})\psi_{i-}\right]\right)((1 - \xi_i)\Delta x)^2$$

$$+ \frac{1}{6}\left(\psi'''_{i-} + k_0^2(\varepsilon_i - \varepsilon_{i+1})\psi'_{i-}\right)((1 - \xi_i)\Delta x)^3$$

$$= \psi_i\theta_i\left[1 + \frac{\Delta x^2}{2}(1 - \xi_i)^2k_0^2(\varepsilon_i - \varepsilon_{i+1})\right]$$

$$+ \psi'_i\Delta x\left[\theta_i\xi_i + (1 - \xi_i) + \frac{\Delta x^2}{6}(1 - \xi_i)^2k_0^2(\varepsilon_i - \varepsilon_{i+1})(3\xi_i\theta_i + (1 - \xi_i))\right]$$

$$+ \psi''_i\frac{\Delta x^2}{2}\left[\theta_i\xi_i^2 + 2(1 - \xi_i)\xi_i + (1 - \xi_i)^2\theta_i\right]$$

$$+ \psi'''_i\frac{\Delta x^3}{6}\left[\theta_i\xi_i^3 + 3(1 - \xi_i)\xi_i^2 + 3(1 - \xi_i)^2\xi_i\theta_i + (1 - \xi_i)^3\right]. \quad (4.26)$$

In this equation, all terms of higher order than Δx^3 have been dropped, and the terms multiplying the different derivatives of ψ_i have been grouped.

To obtain expressions for ψ''_i in terms of ψ_i and $\psi_{i\pm1}$, we need to carry through the same program for expressing ψ_{i-1} in terms of ψ_i and its derivatives. This involves the same series of steps as presented earlier, but this time expanding backwards from ψ_i, using the BCs relevant for the dielectric boundary at $x_i - (1 - \xi_i)\Delta x$. You can easily convince yourself that this leads to a formula similar in structure to Equation (4.26), replacing θ_i by θ_{i-1}^{-1}, ξ_i by $-(1 - \xi_{i-1})$, $(1 - \xi_i)$ by $-\xi_{i-1}$, and $(\varepsilon_i - \varepsilon_{i+1})$ by $(\varepsilon_i - \varepsilon_{i-1})$. Thus, the expression for ψ_{i-1} becomes

$$\psi_{i-1} = \psi_i\theta_{i-1}^{-1}\left[1 + \frac{\Delta x^2}{2}\xi_{i-1}^2k_0^2(\varepsilon_i - \varepsilon_{i-1})\right] - \psi'_i\Delta x$$

$$\times\left[\theta_{i-1}^{-1}(1 - \xi_{i-1}) + \xi_{i-1} + \frac{\Delta x^2}{6}\xi_{i-1}^2k_0^2(\varepsilon_i - \varepsilon_{i-1})(3(1 - \xi_{i-1})\theta_{i-1}^{-1} + \xi_{i-1})\right]$$

$$+ \psi''_i\frac{\Delta x^2}{2}\left[\theta_{i-1}^{-1}(1 - \xi_{i-1})^2 + 2(1 - \xi_{i-1})\xi_{i-1} + \xi_{i-1}^2\theta_{i-1}^{-1}\right]$$

$$- \psi'''_i\frac{\Delta x^3}{6}\left[\theta_{i-1}^{-1}(1 - \xi_{i-1})^3 + 3(1 - \xi_{i-1})^2\xi_{i-1} + 3(1 - xi_{i-1})\xi_{i-1}^2\theta_{i-1}^{-1} + xi_{i-1}^3\right]. \quad (4.27)$$

Equations (4.26) and (4.27) form the basis for deriving finite-difference expressions. In the following, this will be carried out for the TE and TM cases in turn.

4.3.3   Inclusion of Discontinuities: TE Case

In the case of a TE mode, $\theta_i = 1$ for all i, and the aforementioned expressions are considerably simplified:

$$\psi_{i+1} = \psi_i \left[1 + \frac{\Delta x^2}{2} (1 - \xi_i)^2 k_0^2 (\varepsilon_i - \varepsilon_{i+1}) \right]$$

$$+ \psi_i' \Delta x \left[1 + \frac{\Delta x^2}{6} (1 - \xi_i)^2 k_0^2 (\varepsilon_i - \varepsilon_{i+1}) (2\xi_i + 1) \right] + \psi_i'' \frac{\Delta x^2}{2} + \psi_i''' \frac{\Delta x^3}{6},$$

(4.28)

$$\psi_{i-1} = \psi_i \left[1 + \frac{\Delta x^2}{2} \xi_{i-1}^2 k_0^2 (\varepsilon_i - \varepsilon_{i-1}) \right]$$

$$- \psi_i' \Delta x \left[1 + \frac{\Delta x^2}{6} \xi_{i-1}^2 k_0^2 (\varepsilon_i - \varepsilon_{i-1}) (3 - 2\xi_{i-1}) \right] + \psi_i'' \frac{\Delta x^2}{2} - \psi_i''' \frac{\Delta x^3}{6}$$

(4.29)

Adding these two equations, the third-order derivative of ψ is eliminated:

$$\psi_{i+1} + \psi_{i-1} = \psi_i \left(2 + \frac{\Delta x^2}{2} k_0^2 \left[(1 - \xi_i)^2 (\varepsilon_i - \varepsilon_{i+1}) + \xi_{i-1}^2 (\varepsilon_i - \varepsilon_{i-1}) \right] \right)$$

$$+ \psi_i' \frac{\Delta x^3}{6} k_0^2 \left[(1 - \xi_i)^2 (\varepsilon_i - \varepsilon_{i+1})(2\xi_i + 1) - \xi_{i-1}^2 (\varepsilon_i - \varepsilon_{i-1})(3 - 2\xi_{i-1}) \right]$$

$$+ \psi_i'' \Delta x^2.$$

(4.30)

Since ψ_i' is continuous across the interfaces, whereas, from Equation (4.19), ψ_i'' is not, we can write

$$\psi_i' = \frac{\psi_{i+1} - \psi_{i-1}}{2\Delta x} + O(\Delta x)$$

(4.31)

(note that the accuracy is $O(\Delta x)$, not $O(\Delta x^2)$ because ψ_i'' is not continuous). Since the ψ_i' term in Equation (4.30) is already of order Δx^3, we can introduce this approximation with no additional error (remember that we have cast away Δx^4 terms from the outset). This yields the desired expression for ψ_i'':

$$\psi_i'' = \frac{1}{\Delta x^2} \left(\psi_{i+1} + \psi_{i-1} - \psi_i \left(2 + \frac{\Delta x^2}{2} k_0^2 \left[(1 - \xi_i)^2 (\varepsilon_i - \varepsilon_{i+1}) + \xi_{i-1}^2 (\varepsilon_i - \varepsilon_{i-1}) \right] \right) \right)$$

$$- (\psi_{i+1} - \psi_{i-1}) \frac{1}{12} k_0^2 \left[(1 - \xi_i)^2 (\varepsilon_i - \varepsilon_{i+1})(2\xi_i + 1) - \xi_{i-1}^2 (\varepsilon_i - \varepsilon_{i-1})(3 - 2\xi_{i-1}) \right]$$

$$+ O(\Delta x^2) \equiv \alpha_{i-} \psi_{i-1} + \alpha_i \psi_i + \alpha_{i+} \psi_{i+1} + O(\Delta x^2),$$

(4.32)

$$\alpha_i = -\frac{2}{\Delta x^2} - \frac{1}{2} k_0^2 \left[(1 - \xi_i)^2 (\varepsilon_i - \varepsilon_{i+1}) + \xi_{i-1}^2 (\varepsilon_i - \varepsilon_{i-1}) \right]; \quad \alpha_{i\pm} = \frac{1}{\Delta x^2} \pm \delta_i,$$

(4.33)

$$\delta_i = -\frac{1}{12} k_0^2 \left[(1 - \xi_i)^2 (\varepsilon_i - \varepsilon_{i+1})(2\xi_i + 1) - \xi_{i-1}^2 (\varepsilon_i - \varepsilon_{i-1})(3 - 2\xi_{i-1}) \right].$$

(4.34)

Note: These expressions reduce to Equation (4.6) in the case where there is no dielectric boundaries, so that $\varepsilon_{i-1} = \varepsilon_i = \varepsilon_{i+1}$. On the other hand, even though both the field and its first derivative are continuous in the TE case, correction terms appear when accounting for refractive-index steps. This is because the second-order derivative of the field is not continuous.

Looking at Equation (4.34), one realizes that the complicated finite-difference expression for discontinuous dielectric structures can in fact be discretized in the same way as was outlined in Section 4.3.1 for the scalar wave equation. In the discontinuous TE case, the discretization matrix for a five-point structure will appear like

$$
\begin{bmatrix}
\beta_1 & \alpha_{1+} & 0 & 0 & 0 \\
\alpha_{2-} & \beta_2 & \alpha_{2+} & 0 & 0 \\
0 & \alpha_{3-} & \beta_3 & \alpha_{3+} & 0 \\
0 & 0 & \alpha_{4-} & \beta_4 & \alpha_{4+} \\
0 & 0 & 0 & \alpha_{5-} & \beta_5
\end{bmatrix}
\begin{bmatrix}
E_1 \\ E_2 \\ E_3 \\ E_4 \\ E_5
\end{bmatrix}
= \beta^2
\begin{bmatrix}
E_1 \\ E_2 \\ E_3 \\ E_4 \\ E_5
\end{bmatrix},
\tag{4.35}
$$

$$
\beta_i = \alpha_i + k_0^2 \varepsilon_i.
\tag{4.36}
$$

This should be compared to Equation (4.11). The off-diagonal coefficients are now position dependent, but otherwise the matrix structure is unchanged.

4.3.4 Inclusion of Discontinuities: TM Case

In the case of TM modes, we have $\theta_i = \varepsilon_i/\varepsilon_{i+1}$, and the general equations (4.26), (4.27) cannot be readily simplified. To proceed, we must drop the third-order terms in the expansions from the outset, accepting a lower-order accuracy of the resulting expressions. We also need to be more careful about the specification of ψ'. In the TE case, the simple $O(\Delta x)$ estimate in Equation (4.31) was acceptable, because all ψ' terms were multiplied by Δx^3. In the TM case, we will need a more accurate estimate. Starting from Equation (4.27), and dropping all terms of higher order than Δx^2, we obtain

$$
\psi_{i-1} = \psi_i \theta_{i-1}^{-1} \left[1 + \frac{\Delta x^2}{2} \xi_{i-1}^2 k_0^2 (\varepsilon_i - \varepsilon_{i-1}) \right] - \psi_i' \Delta x \left[\theta_{i-1}^{-1}(1 - \xi_{i-1}) + \xi_{i-1} \right]
$$
$$
+ \psi_i'' \frac{\Delta x^2}{2} \left[\theta_{i-1}^{-1}(1 - \xi_{i-1})^2 + 2(1 - \xi_{i-1})\xi_{i-1} + \xi_{i-1}^2 \theta_{i-1}^{-1} \right]
\tag{4.37}
$$

$$
\Downarrow
$$

$$
\psi_i' \Delta x = \frac{1}{\theta_{i-1}^{-1}(1 - \xi_{i-1}) + \xi_{i-1}} \left\{ \psi_i \theta_{i-1}^{-1} \left[1 + \frac{\Delta x^2}{2} \xi_{i-1}^2 k_0^2 (\varepsilon_i - \varepsilon_{i-1}) \right] \right.
$$
$$
\left. + \psi_i'' \frac{\Delta x^2}{2} \left[\theta_{i-1}^{-1}(1 - \xi_{i-1})^2 + 2(1 - \xi_{i-1})\xi_{i-1} + \xi_{i-1}^2 \theta_{i-1}^{-1} \right] - \psi_{i-1} \right\}
\tag{4.38}
$$

From this point, the derivation is straightforward, albeit complicated: We insert the result for ψ_i', Equation (4.38), into Equation (4.26) without third-order terms and solve for ψ_i''. The result of this calculation is

$$\psi_i'' = \gamma_{i+}\psi_{i+1} + \gamma_{i-}\psi_{i-1} - \gamma_i\psi_i + O(\Delta x), \tag{4.39}$$

$$\gamma_{i-} = \gamma_{i+}\frac{\theta_i\xi_i + 1 - \xi_i}{\theta_{i-1}^{-1}(1 - \xi_{i-1}) + \xi_{i-1}}; \quad \gamma_{i+} = \frac{2}{\Delta x^2 D}, \tag{4.40}$$

$$D = \theta_i\xi_i^2 + 2(1 - \xi_i)\xi_i + (1 - \xi_i)^2\theta_i + \left(\theta_{i-1}^{-1}\left(\xi_{i-1}^2 + (1 - \xi_{i-1})^2\right)\right)$$

$$+ 2\xi_{i-1}(1 - \xi_{i-1}))\frac{\theta_i\xi_i + 1 - \xi_i}{\theta_{i-1}^{-1}(1 - \xi_{i-1}) + \xi_{i-1}}, \tag{4.41}$$

$$\gamma_i = \gamma_{i+}\left[\theta_i\left(1 + \frac{\Delta x^2}{2}(1 - \xi_i)^2 k_0^2(\varepsilon_i - \varepsilon_{i+1})\right)\right.$$

$$\left. + \theta_{i-1}^{-1}\left(1 + \frac{\Delta x^2}{2}\xi_{i-1}^2 k_0^2(\varepsilon_i - \varepsilon_{i-1})\right)\frac{\theta_i\xi_i + 1 - \xi_i}{\theta_{i-1}^{-1}(1 - \xi_{i-1}) + \xi_{i-1}}\right]. \tag{4.42}$$

Because we had to leave out third-order terms in the field expansions, the result for ψ_i'' in this case has error terms of order Δx. The structure of the discretization matrix is similar to what we found for the TE case, with α replaced by γ. Thus, both the TE and TM problems can be discretized by matrices having elements only in the diagonal and the nearest sub-diagonals.

4.4   NUMERICAL LINEAR ALGEBRA IN MATLAB

In the previous sections, we showed how we could use the finite-difference method to convert the wave equation into a numerical matrix problem. In this course, we use MATLAB as a numerical tool for solving such problems. This section briefly introduces some of the important concepts from numerical linear algebra and discusses how different matrix operations are carried out in MATLAB.

4.4.1   Sparse Matrices

The finite-difference discretization matrices derived in the previous section had the property that only elements in the diagonal and the nearest two sub-diagonals were non-zero. For a matrix like the one given in Equation (4.35), this means that 13 out of 25 elements are non-zero. However, for a larger matrix, the fraction of non-zero elements becomes much lower, for example a 1000×1000 matrix would have 2998 out of 10^6 elements non-zero. Matrices with a low fraction of non-zero elements are commonly denoted *sparse matrices*. Since we only need to store and manipulate the non-zero elements when working with sparse matrices, huge gains in computational efficiency are possible by taking advantage of this property.

MATLAB features built-in functions for storing and manipulating sparse matrices. To set up a sparse matrix in MATLAB, we must first declare the matrix, by telling MATLAB about its full dimension, and the maximal number of non-zero elements that storage must be reserved for. This is done using the *SPALLOC* command. To set up an $N \times N$ matrix with a maximum of $3N$ non-zero elements, we would issue the command

```
spmat=spalloc(N,N,3N);
```

Hereafter, elements can be added just like for an ordinary matrix:

```
spmat(2,3) = 10;
```

etc. However, an error will occur if one tries to add more than the $3N$ non-zero elements specified by the *SPALLOC* command.

4.4.2 Direct and Iterative Eigensolvers

By an eigensolver, we understand a numerical algorithm that is capable of solving a numerical eigenvalue problem of the following form:

$$\overline{\overline{A}}\mathbf{x} = \lambda\mathbf{x}, \tag{4.43}$$

where

  $\overline{\overline{A}}$ is a quadratic matrix
  $\mathbf{x}$ is the eigenvector
  λ is the eigenvalue

Usually, an eigenvalue and its corresponding eigenvector are referred to as an eigensolution of the matrix. The total number of eigensolutions corresponds to the number of rows (or columns) of the considered matrix. In order to find all the eigenvectors and eigenvalues, we can use the *EIG* command in MATLAB:

```
[eigvec eigval]=eig(mat);
```

The routine takes a full matrix as input argument and returns one matrix, *eigvec*, containing the eigenvectors and another, *eigval*, containing the eigenvalues in the diagonal. The *EIG* command is an example of a so-called *direct* eigensolver, which solve eigenvalue problems by general methods suitable for arbitrary matrices. Such eigensolvers are unsuitable for the very large matrices arising in finite-difference problems for two reasons: Firstly, they calculate all eigensolutions of a given matrix, whereas typically we are only interested in a few of them. Storage of all eigenvectors requires as much memory as storage of the input matrix in its full, non-sparse, format, which is typically infeasible. Secondly, the direct eigensolvers work in the same way for all matrices and thus cannot take advantage of a property like sparsity. In particular, the time complexity for a direct eigensolver scales with the cube of the matrix dimension, which rapidly makes the calculation of eigensolutions intractable for typical finite-difference matrices, having dimensions on the order of 10^4–10^6.

A class of eigensolvers which are better suited for sparse-matrix problems are the so-called *iterative* eigensolvers. As the name implies, these solvers work by improving an initial trial vector in an iterative process, until it converges towards an eigenvector. Thus, eigensolutions are found one by one, a clear advantage when one only needs a few solutions out of a large set. The iterative procedure typically involves one or more multiplications of the matrix with the trial vector. While the multiplication of an arbitrary $N \times N$ matrix with an N-element vector is an N^2 process, multiplication with a sparse matrix scales linearly with N, a very significant improvement for large N. It is these two properties that make iterative eigensolvers advantageous for sparse-matrix problems.

As a simple example of an iterative eigensolver, consider the following algorithm: Given a matrix, $\overline{\overline{A}}$, and a (clever or random as preferred) starting guess for an eigenvector, $\mathbf{V}_0$, let the sequence of vectors $\mathbf{V}_i$ be given by

$$\mathbf{V}_{i+1} = \mathbf{V}_i - \delta \overline{\overline{A}} \mathbf{V}_i, \tag{4.44}$$

where δ is some small parameter. What this algorithm really does is to numerically propagate the following differential equation:

$$\partial_t \mathbf{V} = -\overline{\overline{A}} \mathbf{V}_i, \tag{4.45}$$

along a fictitious time axis. If the eigenvectors of $\overline{\overline{A}}$ form a complete set, we can write $\mathbf{V}_0$ as a linear combination of such eigenvectors, and in that case, we can easily write up a formal solution to Equation (4.45):

$$\mathbf{V}_0 = \sum_n a_n \mathbf{V}^{(n)}, \quad \overline{\overline{A}} \mathbf{V}^{(n)} = \lambda_n \mathbf{V}^{(n)}, \tag{4.46}$$

$$\mathbf{V}(t) = \sum_n a_n \exp(-\lambda_n t) \mathbf{V}^{(n)}, \tag{4.47}$$

which shows that the eigenvector with the lowest eigenvalue will have increasing relative weight in the sum as t becomes larger. So, after a large number of steps in the iterative scheme, $\mathbf{V}_i$ will be (approximately) equal to the eigenvector with the lowest eigenvalue.

While the aforementioned example of an iterative eigensolver is very unsophisticated and not recommended for practical use, it does illustrate some general features of iterative eigensolvers. For one, we see from Equation (4.47) that the iterative solution takes more steps if many closely spaced eigenvalues are present. Another important point is that the iterative eigensolver described is only able to find the lowest eigenvalue of $\mathbf{A}$ and its corresponding eigenvector. It could easily, though, be modified to find the *highest* eigenvalue (how?), but finding an *interior* eigenvalue, as is often desired, is more difficult. For instance, we might want to find the eigenvector whose eigenvalue is closest to some target eigenvalue, λ_T. The typical strategy employed for such problems is to search for the eigenvalues of a different matrix, where the eigenvalue closest to λ_T becomes an extremal eigenvalue. For instance, we could consider finding the lowest eigenvalue of the matrix

$$\overline{\overline{A}} = \left(\overline{\overline{A}} - \lambda_T\right)^2, \tag{4.48}$$

which will obviously be given by

$$(\lambda^{(k)} - \lambda_T)^2, \tag{4.49}$$

where $\lambda^{(k)}$ is the eigenvalue of $\overline{\overline{A}}$ closest to λ_T. However, the solution of such *squared* eigenproblems turns out to have a much slower convergence than the original problem. A better approach has been found to be using a matrix of the following kind:

$$\overline{\overline{A}} = \left(\overline{\overline{A}} - \lambda_T\right)^{-1}. \tag{4.50}$$

The iterative eigensolver in MATLAB solves targeted eigenvalue problems using this approach. It is a distinct advantage of the finite-difference approach that the finite-difference matrix can be explicitly written down and inverted. In other numerical approaches, such as the plane-wave method, the matrix of the problem cannot in practice be written down and inverted explicitly, which means that iterative inversion methods must be used, which typically increase the computational effort considerably.

Iterative eigensolvers can find multiple solutions by first finding the lowest (or highest) eigenvalue, as described earlier, then repeating the procedure in a reduced vector space, which is orthogonal to the solution just found. In this way, one obtains the second lowest (highest) eigenvalue, and the process can be repeated. As the number of desired eigenvalues grows, the orthogonalization process becomes increasingly time consuming, and iterative eigensolvers are typically found to be inefficient if a substantial fraction of the total number of eigensolutions is needed.

In order to solve an eigenvalue problem iteratively in MATLAB, we can use the *EIGS* command:

```
[eigvec eigval]=eigs(spmat);
```

The command takes a sparse matrix as input argument, and again it returns one matrix, *eigvec*, containing the eigenvectors and another, *eigval*, containing the eigenvalues in the diagonal. In the aforementioned example, MATLAB calculates the six numerically largest eigenvalues of the matrix. If we are interested in finding the *neig* solution with eigenvalue closest to some target eigenvalue, *eigt*, we add the following parameters to the *EIGS* command:

```
[eigvec eigval]=eigs(spmat,neig,eigt);
```

4.5   2D WAVEGUIDES AND THE YEE MESH

In this section, we present an approach for the modelling of arbitrary 2D waveguides using tensorial ε-averaging on a Yee mesh [10]. The development of the finite-difference discretization follows that of Zhu and Brown [9]. The circular step-index fibre geometry is used as a test case for the numerical accuracy, since semi-analytical solutions are available for this waveguide structure.

4.5.1   Yee Mesh

To appreciate the idea behind the Yee mesh, we start by taking a deeper look at the FD dynamic Maxwell's equations (2.27) and (2.28) (note that similar arguments were presented in Chapter 3). Written out in individual vector components, these equations look like this:

$$\partial_y E_z - \partial_z E_y = -i\omega\mu_0 H_x; \quad \partial_y H_z - \partial_z H_y = j\omega\varepsilon_0\varepsilon E_x, \tag{4.51}$$

$$\partial_z E_x - \partial_x E_z = -i\omega\mu_0 H_y; \quad \partial_z H_x - \partial_x H_z = j\omega\varepsilon_0\varepsilon E_y, \tag{4.52}$$

$$\partial_x E_y - \partial_y E_x = -i\omega\mu_0 H_z; \quad \partial_x H_y - \partial_y H_x = j\omega\varepsilon_0\varepsilon E_z. \tag{4.53}$$

Let us first consider Equation (4.53), which requires us to evaluate the derivatives $\partial_x E_y$ and $\partial_y E_x$ to find H_z. If we would do the partial derivatives by a centred difference formula like Equation (4.31), we would need to know E_y at positions to the left and right of H_z and E_x at positions above and below H_z (this terminology assumes that the x-axis runs left–right and the y-axis up–down). But we do not need to know E_x and E_y right at the point where we want to evaluate H_z. So, it makes sense to set up a grid where E_x, E_y, and H_z are specified at different spatial points. If we now consider Equation (4.51) as another example, we see that finite-difference discretization of this equation would require us to know E_z at points above and below H_x. However, the z-derivative of E_y can be replaced by $-i\beta$ in a waveguide, so E_y should be specified at the same point as H_x. By similar arguments, one can show that all the six Maxwell's equations (4.51 through 4.53) may be conveniently discretized on the grid illustrated in Figure 4.4. Each elementary *cell* of the grid has side lengths Δx and Δy and consists of four points where different field components are computed.

If we imagine that dielectric discontinuities are present only on lines running through the gridpoints where electric fields are represented, one notes that all electric field components are parallel to these interfaces and therefore have well-defined values at the interface. This was originally the main rationale for introducing the Yee mesh (in fact, Yee's original work was concerned with electromagnetic fields in metallic structures). The restriction of discontinuities to such grid lines is somewhat inconvenient for, for example modelling of dielectric object with circular boundaries. Such structures must then be approximated by right-angled steps, the so-called *stair-case* approximation. It is also an unpleasant feature that the discretized dielectric structure is connected to the grid size and shape, which means that arbitrarily small changes of the structure (smaller than the grid spacing) cannot be studied. It has been found that these issues can be resolved by using a so-called *dielectric function averaging* scheme, which will be described in Section 4.5.2 (see also Equation 3.63).

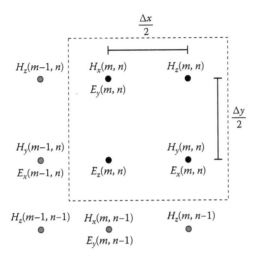

FIGURE 4.4 The 2D Yee mesh. The dashed box illustrates a fundamental cell in the grid, of side lengths Δx and Δy containing four interior points with different field components.

For the moment, all we need to know is that the scheme will require us to use different ε-values for the different electric field components, and the formulation of finite-difference equations on the Yee mesh will proceed accordingly.

It is possible to discretize Equation (2.64) directly, using three-point derivatives (or *stencils*, as introduced in Section 3.1.1) for the first and second derivatives of the magnetic fields and ε functions. However, these equations are formulated for a scalar ε function (meaning that the same ε value is used for all components of the electric field at a given point), and so are not directly useful for the dielectric function averaging to be implemented later. Furthermore, one finds that equations discretized in this way have significant zero-order error terms, that is even if a calculation is converged with respect to the discretization stepsize, it still does not produce the right answer! A better procedure turns out to be discretizing Maxwell's equations already at the first-order level and then combining the discretized first-order equations into second-order equations for the transverse components of **H**.

The discretized versions of Equations (4.51) through (4.53) look like this:

$$\frac{E_z(m, n+1) - E_z(m, n)}{\Delta y} + j\beta E_y(m, n) = -j\omega \mu_0 H_x(m, n), \qquad (4.54)$$

$$\frac{H_z(m, n) - H_z(m, n-1)}{\Delta y} + j\beta H_y(m, n) = j\omega \varepsilon_0 \varepsilon_x(m, n) E_x(m, n), \qquad (4.55)$$

$$-j\beta E_x(m, n) - \frac{E_z(m+1, n) - E_z(m, n)}{\Delta x} = -j\omega \mu_0 H_y(m, n), \qquad (4.56)$$

$$-j\beta H_x(m, n) - \frac{H_z(m, n) - H_z(m-1, n)}{\Delta x} = j\omega \varepsilon_0 \varepsilon_y(m, n) E_y(m, n), \qquad (4.57)$$

$$\frac{E_y(m+1,n) - E_y(m,n)}{\Delta x} - \frac{E_x(m,n+1) - E_x(m,n)}{\Delta y} = -j\omega\mu_0 H_z(m,n), \quad (4.58)$$

$$\frac{H_y(m,n) - H_y(m-1,n)}{\Delta x} - \frac{H_x(m,n) - H_x(m,n-1)}{\Delta y} = j\omega\varepsilon_0\varepsilon_z(m,n)E_z(m,n).$$

$$(4.59)$$

To understand the appearance of the discretization terms, it is useful to refer to Figure 4.4. For example $H_x(m,n)$ is situated between $E_z(m,n+1)$ and $E_z(m,n)$, so it is these E_z values that should be used in evaluating the finite-difference derivative $\partial_y E_z$. On the other hand, $E_x(m,n)$ is situated between $H_z(m,n)$ and $H_z(m,n-1)$, so $\partial_y H_z$ should be evaluated from these H_z-values. Note also that different ε-values have been used for the different components of the electric fields. These ε-values should be evaluated at the gridpoints where the corresponding electric field components are defined, as detailed in the next section.

From this point onwards, the derivation of a second-order finite-difference eigenvalue equation for β^2 is more or less similar to the derivation of Equation (2.64), even though it takes place with discretized field equations. First, the expressions for the electric field components, Equations (4.55), (4.57), and (4.59), are substituted into Equations (4.54), (4.56), and (4.58). Then H_z is eliminated by the use of Equation (2.65) in its discretized form. The algebraic manipulations are somewhat complicated and will not be shown here. The final discretized equations for H_x and H_y become

$$\begin{aligned}
\beta^2 H_x(m,n) &= \left[k_0^2 \varepsilon_y(m,n) - \frac{1}{\Delta y^2}\left(\frac{\varepsilon_y(m,n)}{\varepsilon_z(m,n)} + \frac{\varepsilon_y(m,n)}{\varepsilon_z(m,n+1)} \right) - \frac{2}{\Delta x^2} \right] H_x(m,n) \\
&\quad - \frac{1}{\Delta x \Delta y}\left[1 - \frac{\varepsilon_y(m,n)}{\varepsilon_z(m,n)} \right] H_y(m,n) + \frac{1}{\Delta x^2}\left[H_x(m+1,n) + H_x(m-1,n) \right] \\
&\quad + \frac{\varepsilon_y(m,n)}{\Delta y^2}\left[\frac{1}{\varepsilon_z(m,n)} H_x(m,n-1) + \frac{1}{\varepsilon_z(m,n+1)} H_x(m,n+1) \right] \\
&\quad + \frac{1}{\Delta x \Delta y}\left[1 - \frac{\varepsilon_y(m,n)}{\varepsilon_z(m,n)} \right] H_y(m-1,n) \\
&\quad + \frac{1}{\Delta x \Delta y}\left[1 - \frac{\varepsilon_y(m,n)}{\varepsilon_z(m,n+1)} \right] \left(H_y(m,n+1) - H_y(m-1,n+1) \right),
\end{aligned}$$

$$(4.60)$$

$$\begin{aligned}
\beta^2 H_y(m,n) &= \left[k_0^2 \varepsilon_x(m,n) - \frac{1}{\Delta x^2}\left(\frac{\varepsilon_x(m,n)}{\varepsilon_z(m,n)} + \frac{\varepsilon_x(m,n)}{\varepsilon_z(m+1,n)} \right) - \frac{2}{\Delta y^2} \right] H_y(m,n) \\
&\quad - \frac{1}{\Delta x \Delta y}\left[1 - \frac{\varepsilon_x(m,n)}{\varepsilon_z(m,n)} \right] H_x(m,n) \\
&\quad + \frac{1}{\Delta y^2}\left[H_y(m,n+1) + H_y(m,n-1) \right] \\
&\quad + \frac{\varepsilon_x(m,n)}{\Delta x^2}\left[\frac{1}{\varepsilon_z(m,n)} H_y(m-1,n) + \frac{1}{\varepsilon_z(m+1,n)} H_y(m+1,n) \right]
\end{aligned}$$

$$+ \frac{1}{\Delta x \Delta y} \left[1 - \frac{\varepsilon_x(m,n)}{\varepsilon_z(m,n)} \right] H_x(m, n-1)$$

$$+ \frac{1}{\Delta x \Delta y} \left[1 - \frac{\varepsilon_x(m,n)}{\varepsilon_z(m+1,n)} \right] (H_x(m+1, n) - H_x(m+1, n-1)).$$

$$(4.61)$$

Complex as these equations may seem, they can be treated numerically as a sparse-matrix eigenvalue problem in the same way as was outlined for the case of a 1D waveguide in the previous section. The bookkeeping involved in setting up the sparse matrix is somewhat more intricate in the 2D case, due to the appearance of two spatial dimensions and two field components, but the underlying principles are completely similar.

4.5.2 Dielectric Function Averaging

As discussed in the previous section, the electric field components on the Yee mesh have well-defined values if dielectric discontinuities are assumed to follow lines connecting points where electric fields are represented. This implies the problem that the shape of an arbitrary (e.g. a curved) dielectric discontinuity will be misrepresented on a rectangular finite-difference grid, as illustrated in Figure 4.5. Furthermore, even though the electric and magnetic fields are continuous at the mesh points, it does not necessarily imply that they are differentiable. For instance, Equation (4.53) and the continuity of E_z imply that either the x-derivative of H_y or the y-derivative of H_x or both must be discontinuous at a dielectric interface. Thus, we cannot a priori be sure that the finite-difference equations formulated previously will be accurate.

An immediate, physically motivated idea for solving this problem would be to replace the index discontinuities by continuous functions with a rapid variation in the boundary region. Surely, this is the way things must really be in the physical world: Dopant profiles are smeared by diffusion, material surfaces by roughness, and so on. However, this physical argument does not, in fact, help us a lot. The characteristic

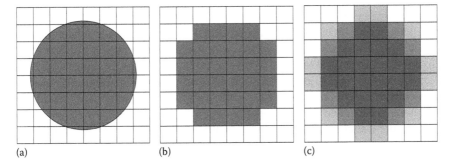

(a)   (b)   (c)

FIGURE 4.5 (a) Representing a circle on a square grid. In (b), a simple *staircase* approximation is used, where the index of each square is put equal to the index of the circle if the square centre falls within the circle radius. In (c), an index-averaging approach is sketched.

scale of variations for index profiles smeared by these mechanisms can easily be as small as a few nanometres or even angstroms. To resolve these functions, we would need finite-difference grids with mesh spacings of a few angstroms, making the description of micron-sized waveguide structures hopelessly inefficient. If we apply smearing on a larger scale, we must ensure that the smearing does not crucially affect the results we are interested in and this can in itself be a quite cumbersome task.

Instead of modifying the ε-profile, we could think of modifying the finite-difference grid to conform to the structure, so all discontinuities lie on boundaries between elementary grid cells. However, while a rectangular refractive-index profile would be quite easy to treat on a rectangular finite-difference grid, other shapes, such as circles, would be difficult because they do not conform to the shape of the grid. If we had complete freedom to choose the positions of the gridpoints, we could increase the resolution just at the index steps and shape the grid so as to closely follow the structure of the discontinuities. These techniques are in fact heavily used in finite-element schemes; however, the evaluation of derivatives on non-rectangular grids is far from trivial, and automatic routines for *meshing* an arbitrary structure in an intelligent way are also quite complicated to set up. Therefore, this approach is by no means a straightforward extension of the finite-difference scheme outlined earlier.

As mentioned previously, a simpler way of improving the finite-difference description of index steps is to use the dielectric function averaging techniques. Here, the idea is to introduce an averaging procedure to define the index distribution on the grid. A simple 1D approach for this was given in Equation (3.63). In two dimensions, the approach is illustrated in Figure 4.5c: Imagine that we place the gridpoints in the middle of the square boxes and consider the ε-value at a particular point to represent the average of ε over the surrounding box rather than the value just at the gridpoint. Several simple arguments favour this kind of approach. Firstly, there is no reason why the ε-value exactly at a particular gridpoint should have special significance compared to the values at all other points in the surrounding box. Secondly, the averaging procedure produces a certain rounding of the ε-distribution, but without introducing a smearing function with free parameters that would need to be determined. Thirdly, this procedure means that the integral, $\int d\mathbf{r}\varepsilon(\mathbf{r})$, is preserved and that the calculation in principle is sensitive to arbitrarily small changes in the circle radius in Figure 4.5c, which is not necessarily the case for the *staircase* approximation in Figure 4.5b.

Although the averaging procedure outlined does not introduce free parameters, it is not unique. For instance, if we calculate the average of the *inverse* ε over a single cell, it will not in general be equal to the inverse of the averaged ε. Since both ε and $1/\varepsilon$ enter the wave equations, one may ask whether it is better to average one quantity or the other. In fact, it turns out that the optimal averaging procedure is a combination of the two, as explained in the following.

Consider a single point in some finite-difference grid (Yee or not) and the box defined by points in the plane, which are closer to this gridpoint than any other gridpoint. Assume further that a refractive-index step boundary traverses this box, as illustrated in Figure 4.6. Let us denote the field of an *exact* solution to the wave equation as $\mathbf{E}(\mathbf{r})$. In the finite-difference scheme, we do not find the full $\mathbf{E}(\mathbf{r})$ function but only get a value at the gridpoint, which we could call $\mathbf{E}_i$. Since the gridpoint itself

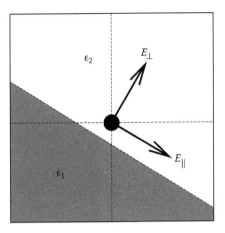

FIGURE 4.6 A single gridpoint and the box of points closest to this gridpoint. A dielectric boundary traverses the box and defines directions parallel and perpendicular to the dielectric interface.

does not have special significance, the best we can hope for is that $\mathbf{E}_i$ will represent the average value of the electric field vector in the surrounding box. While reasonable, this requirement in itself is not very illuminating with respect to ε-averaging. However, we could equally well argue that the electric field energy density, $\mathbf{E} \cdot \mathbf{D}$, also a physical quantity, should obey a similar rule. So, ideally, we might want a finite-difference scheme which works such that $\mathbf{E}_i$ fulfils the following relations:

$$\mathbf{E}_i = \langle \mathbf{E}(\mathbf{r}) \rangle, \tag{4.62}$$

$$\mathbf{E}_i \cdot \mathbf{D}_i \equiv \mathbf{E}_i \cdot \varepsilon_i \mathbf{E}_i = \langle \mathbf{E}(\mathbf{r}) \cdot \mathbf{D}(\mathbf{r}) \rangle, \tag{4.63}$$

where
   ε_i is our chosen *average* value for ε at the gridpoint
   $\langle ... \rangle$ denotes an average over the box

It seems appropriate to further assume that Equation (4.63) should be fulfilled independently by the components of $\mathbf{E}$ and $\mathbf{D}$ parallel and perpendicular to the index step, rather than through some fortuitous cancellation, that is we should have

$$\mathbf{E}_{i\parallel} \cdot \varepsilon_i \mathbf{E}_{i\parallel} = \langle \mathbf{E}_{\parallel}(\mathbf{r}) \cdot \mathbf{D}_{\parallel}(\mathbf{r}) \rangle, \tag{4.64}$$

$$E_{i\perp} \varepsilon_i E_{i\perp} = \langle E_{\perp}(\mathbf{r}) \cdot D_{\perp}(\mathbf{r}) \rangle. \tag{4.65}$$

Since the component of $\mathbf{E}(\mathbf{r})$, which is parallel to the index step, $\mathbf{E}_{\parallel}(\mathbf{r})$, is continuous across the interface, if the box is sufficiently small (i.e. the grid spacing is sufficiently fine), $\mathbf{E}_{\parallel}(\mathbf{r})$ should be approximately constant over the box. Similarly, the component of $\mathbf{D}(\mathbf{r})$ perpendicular to the interface, $\mathbf{D}_{\perp}(\mathbf{r})$, should be continuous over the interface and therefore approximately a constant. So we have

$$\mathbf{E}_{\parallel}(\mathbf{r}) \approx \mathbf{E}_{i\parallel}, \tag{4.66}$$

$$\mathbf{D}_{\perp}(\mathbf{r}) \approx \mathbf{D}_{i\perp}. \tag{4.67}$$

With this, we can rewrite Equation (4.64) as follows:

$$|\mathbf{E}_{i\parallel}|^2 \, \varepsilon_i = \langle \mathbf{E}_{\parallel}(\mathbf{r}) \cdot \mathbf{D}_{\parallel}(\mathbf{r}) \rangle \approx \mathbf{E}_{i\parallel} \cdot \langle \varepsilon(\mathbf{r}) \mathbf{E}_{\parallel}(\mathbf{r}) \rangle \approx |\mathbf{E}_{i\parallel}|^2 \, \langle \varepsilon(\mathbf{r}) \rangle \tag{4.68}$$

From this, we have directly that $\varepsilon_i = \langle \varepsilon(\mathbf{r}) \rangle$, that is the averaging scheme described previously. However, from Equation (4.65), we get

$$E_{i\perp} \varepsilon_i E_{i\perp} = \frac{1}{\varepsilon_i} D_{i\perp}^2 = \langle E_{\perp}(\mathbf{r}) D_{\perp}(\mathbf{r}) \rangle \approx D_{i\perp} \left\langle \frac{1}{\varepsilon(\mathbf{r})} D_{\perp}(\mathbf{r}) \right\rangle \approx D_{i\perp}^2 \left\langle \frac{1}{\varepsilon(\mathbf{r})} \right\rangle, \tag{4.69}$$

which suggests that we should rather use the averaging $\varepsilon_i^{-1} = \left\langle \dfrac{1}{\varepsilon(\mathbf{r})} \right\rangle$.

With these conflicting results, it seems that we cannot devise a finite-difference scheme that would fulfil Equations (4.62), (4.64), and (4.65) simultaneously. However, there is a way out of the dilemma: Since one requirement on the averaged ε came from considering $\mathbf{E}_{\parallel}, \mathbf{D}_{\parallel}$, and the other from $\mathbf{E}_{\perp}, \mathbf{D}_{\perp}$, we can satisfy them both if we replace the *scalar*-averaged ε function by a *tensor* average of the following form:

$$\overline{\overline{\varepsilon}}_i = \hat{P}_{\parallel} \langle \varepsilon \rangle \hat{P}_{\parallel} + \hat{P}_{\perp} \frac{1}{\langle \varepsilon^{-1} \rangle} \hat{P}_{\perp}, \tag{4.70}$$

where $\hat{P}_{\parallel}$ and $\hat{P}_{\perp}$ are, respectively, projection operators onto the directions parallel and perpendicular to the dielectric boundary cutting through the box surrounding our gridpoint. So, if the angle between the normal vector to the dielectric interface and the x-axis is θ, the projection operators act as follows:

$$\hat{P}_{\parallel}\hat{\mathbf{x}} = \sin\theta \left(\hat{\mathbf{x}} \sin\theta - \hat{\mathbf{y}} \cos\theta \right); \quad \hat{P}_{\parallel}\hat{\mathbf{y}} = -\cos\theta \left(\hat{\mathbf{x}} \sin\theta - \hat{\mathbf{y}} \cos\theta \right), \tag{4.71}$$

$$\hat{P}_{\perp}\hat{\mathbf{x}} = \cos\theta \left(\hat{\mathbf{x}} \cos\theta + \hat{\mathbf{y}} \sin\theta \right); \quad \hat{P}_{\perp}\hat{\mathbf{y}} = \sin\theta \left(\hat{\mathbf{x}} \cos\theta + \hat{\mathbf{y}} \sin\theta \right). \tag{4.72}$$

Note that these projection operators are *local* entities, which depend on the orientation of the dielectric interface at the gridpoint in question. Written out in the global coordinate system, with axes parallel to the edges of the box, our averaged dielectric tensor reads

$$\overline{\overline{\varepsilon}}_i = \begin{bmatrix} \langle \varepsilon \rangle \sin^2\theta_i + \dfrac{\cos^2\theta_i}{\langle \varepsilon^{-1} \rangle} & \left(\dfrac{1}{\langle \varepsilon^{-1} \rangle} - \langle \varepsilon \rangle \right) \cos\theta_i \sin\theta_i & 0 \\ \left(\dfrac{1}{\langle \varepsilon^{-1} \rangle} - \langle \varepsilon \rangle \right) \cos\theta_i \sin\theta_i & \langle \varepsilon \rangle \cos^2\theta_i + \dfrac{\sin^2\theta_i}{\langle \varepsilon^{-1} \rangle} & 0 \\ 0 & 0 & \langle \varepsilon \rangle \end{bmatrix}, \tag{4.73}$$

where θ_i is the angle between the normal vector to the dielectric interface at gridpoint i and the x-axis.

The aforementioned arguments do not actually prove that the use of tensor averaging will give results complying with Equations (4.62) and (4.63) but only that it is a necessary condition. However, as we shall see from the numerical exercises, this form of averaging in fact leads to significant convergence improvements.

The introduction of off-diagonal dielectric tensor components, in this case ε_{xy}, is problematic on the Yee mesh since such components connect electric field components x and y, which are represented on different gridpoints. Our rather simple-minded solution will be to ignore the off-diagonal components and only work with the diagonals. These will be averaged over a box centred on the gridpoint where the relevant field component is represented, that is ε_x is calculated from $\langle \varepsilon \rangle$ and $1/\langle \varepsilon^{-1} \rangle$ averaged around the E_x, H_y gridpoint.

How large is the difference between $\langle \varepsilon \rangle$ and $1/\langle \varepsilon^{-1} \rangle$? If we denote by f_1, f_2 the fractions of the box area occupied by material with $\varepsilon = \varepsilon_1, \varepsilon_2$, respectively, we can work out the averages to be

$$\langle \varepsilon \rangle = \varepsilon_2 + f_1(\varepsilon_1 - \varepsilon_2), \tag{4.74}$$

$$\langle \varepsilon^{-1} \rangle = \frac{1}{\varepsilon_2} + f_1 \left(\frac{1}{\varepsilon_1} - \frac{1}{\varepsilon_2} \right) = \frac{1 + f_1 \dfrac{\varepsilon_2 - \varepsilon_1}{\varepsilon_1}}{\varepsilon_2}, \tag{4.75}$$

where we used that $f_1 + f_2 = 1$. The difference between $\langle \varepsilon \rangle$ and $1/\langle \varepsilon^{-1} \rangle$ is then given by

$$\langle \varepsilon \rangle - \frac{1}{\langle \varepsilon^{-1} \rangle} = \varepsilon_2 + f_1(\varepsilon_1 - \varepsilon_2) - \frac{\varepsilon_2}{1 + f_1 \dfrac{\varepsilon_2 - \varepsilon_1}{\varepsilon_1}} = \frac{f_1 f_2(\varepsilon_1 - \varepsilon_2)^2}{\varepsilon_1 - f_1(\varepsilon_1 - \varepsilon_2)} \tag{4.76}$$

after a few manipulations. This result shows that the difference between the two averaging schemes is of second order in $(\varepsilon_1 - \varepsilon_2)$. For the weak index contrasts in doped optical fibres (on the order of 10^{-2}), this difference is negligible, and the averaged ε-tensor reduces to a scalar. For a silica/air structure, the difference between the two averages is around 15% for $f_1 = f_2 = 0.5$. For a silicon/air structure, the difference is very large and cannot be neglected without serious errors.

4.5.3 Use of Mirror Symmetries

Consider the three waveguide structures in Figure 4.7. The rectangular waveguide in Figure 4.7a is symmetric under a mirror reflection in the y-axis. If $n_2 = n_3$, the structure is also mirror symmetric in the x-axis. The circular and microstructured fibres in Figure 4.7b are mirror symmetric in both x- and y-axes. In fact, most practical waveguide structures exhibit (at least approximately) a mirror symmetry in at least one of these axes. As will be shown, this fact allows a reduction of the computational domain, thus increasing the efficiency of any numerical method. The proof will be done from Equation (2.64) but of course holds equally well for the first-order-discretized equations on the Yee mesh.

Imagine we have some structure which is mirror symmetric in the x-axis, meaning that the relation

$$\hat{M}_x \varepsilon(x, y) = \varepsilon(x, -y) = \varepsilon(x, y) \tag{4.77}$$

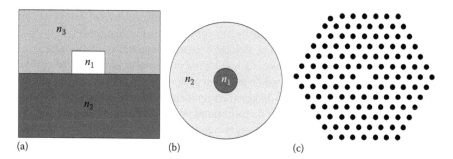

FIGURE 4.7 Refractive-index distributions of three common dielectric waveguides: (a) a planar waveguide structure, (b) a step-index optical fibre, and (c) a microstructured optical fibre.

holds. Here, the operator $\hat{M}_x$ was defined to effect a reflection in the x-axis for the scalar function ε. For the vector of TM field components, (H_x, H_y), we define the action of $\hat{M}_x$ to be

$$\hat{M}_x H_x(x, y) = H_x(x, -y); \qquad \hat{M}_x H_y(x, y) = -H_y(x, -y). \tag{4.78}$$

This seems natural, since a reflection of the $\mathbf{H}$ vector in the x-axis would imply a change of sign of the y-component. Note also that

$$f'(x, y) = \partial_y f(x, y) \Rightarrow f'(x, -y) = -\partial_y f(x, -y), \tag{4.79}$$

where f is any scalar function.

Consider now applying the $\hat{M}_x$ operator to both sides of Equation (2.64). For the x-component, we have

$$\hat{M}_x \left[\left(\nabla_\perp^2 + k_0^2 \varepsilon(x, y) \right) H_x(x, y) + \frac{1}{\varepsilon(x, y)} \partial_y \varepsilon(x, y) \left(\partial_x H_y(x, y) - \partial_y H_x(x, y) \right) \right]$$

$$= \left(\nabla_\perp^2 + k_0^2 \varepsilon(x, -y) \right) H_x(x, -y)$$

$$- \frac{1}{\varepsilon(x, -y)} \partial_y \varepsilon(x, -y) \left(\partial_x H_y(x, -y) + \partial_y H_x(x, -y) \right)$$

$$= \left(\nabla_\perp^2 + k_0^2 \varepsilon(x, y) \right) \hat{M}_x H_x(x, y) + \frac{1}{\varepsilon(x, y)} \partial_y \varepsilon(x, y) \left(\partial_x \hat{M}_x H_y(x, y) - \partial_y \hat{M}_x H_x(x, y) \right)$$

$$= \beta^2 \hat{M}_x H_x(x, y). \tag{4.80}$$

A similar result is readily derived for the y-component of Equation (2.64). Writing the wave equations in the operator form $\hat{\theta} \mathbf{H}_\perp = \beta^2 \mathbf{H}_\perp$, we have shown that

$$\hat{M}_x \hat{\theta} \mathbf{H}_\perp = \hat{\theta} \hat{M}_x \mathbf{H}_\perp \tag{4.81}$$

that is $\hat{\theta}$ and $\hat{M}_x$ are commuting operators. So, a reflection operation leaving $\varepsilon(\mathbf{r})$ invariant will also leave the wave equation invariant, which intuitively was to be expected.

Consider now an eigenstate of the wave equation. We have

$$\beta^2 \hat{M}_x \mathbf{H}_\perp = \hat{M}_x \hat{\Theta} \mathbf{H}_\perp = \hat{\Theta} \hat{M}_x \mathbf{H}_\perp, \qquad (4.82)$$

so that $\hat{M}_x \mathbf{H}_\perp$ is also an eigenstate of the wave equation. If this eigenstate is non-degenerate (no other eigenstates have the same eigenvalue), it follows immediately that application of $\hat{M}_x$ can only amount to multiplication by a constant. In other words, the eigenstate of the wave equation is also an eigenstate of $\hat{M}_x$. The frequently encountered case of degenerate eigenstates is more complicated, but one may still show that eigenstates of $\hat{M}_x$ may be obtained by linear combinations of the degenerate wave equation eigenstates, which will still be eigenstates of the wave equation. Thus, the wave equation eigenstates may always be chosen to also be eigenstates of $\hat{M}_x$. Since $\hat{M}_x^2 = 1$, the only possible eigenvalues of $\hat{M}_x$ are ± 1. This leads to the important conclusion that the H_x field of a wave equation eigenstate must be either even or odd upon reflection in the x-axis provided that $\varepsilon(\mathbf{r})$ is symmetric under this reflection. If H_x is even, H_y is odd by Equation (4.78) and vice versa. Of course, if $\varepsilon(\mathbf{r})$ is symmetric under reflection in the y-axis, a completely similar result applies.

Let us return to the optical fibre in the middle panel of Figure 4.7 or any other waveguide with mirror symmetries in both x- and y-axes of some suitable coordinate system. If the field distribution in the first quadrant of this coordinate system is known, along with the eigenvalues of the mirror operators, the field in the second quadrant can be determined by reflection in the y-axis, and the field in the remaining quadrants can then be determined by reflection in the x-axis. Only the first quadrant contains unique information. This allows us to restrict our calculations to the first quadrant by applying appropriate BCs at the mirror symmetry axes. The situation is illustrated in Figure 4.8. Here, the E_z gridpoint has been placed at the origin,

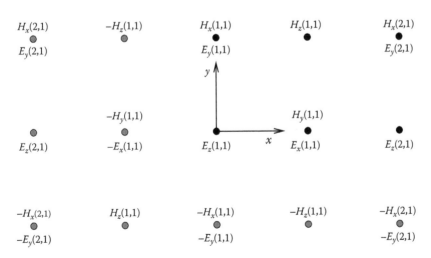

FIGURE 4.8 Illustration of gridpoints close to mirror symmetry lines. The field values at the grey points are fully determined by the values at the black points if the eigenvalues of the mirror operators are known.

and it has been assumed that the H_y field is even under mirror reflections in the x-axis and odd under mirror reflections in the y-axis. It then follows that H_x has the opposite parities under these reflections. The parities of the other field components can be deduced from the Maxwell's equations (4.51 through 4.53). For example since H_y is odd under $x \rightarrow -x$ (reflection in the y-axis), $\partial_x H_y$ is even under this reflection. Therefore, E_z must also be even under this operation and could thus be nonvanishing at the origin. This is also consistent with H_x and therefore $\partial_y H_x$, being even under reflection in the y-axis. Similar arguments show that E_z is also even under reflections in the x-axis and the symmetries of the other field components can be derived correspondingly. Note that if we had chosen another symmetry, which required E_z to be odd under one of the reflection operations, it would have been forced to zero at the origin.

Now, if we consider Equation (4.60) for the H_x field at $m, n = (1,1)$, that is at the origin, this field value is connected to, among others, the field values $H_x(0, 1)$, $H_x(1, 0)$, $H_y(0, 1)$, and $H_y(0, 2)$, neither of which falls within our symmetry-reduced finite-difference grid. In the equation, we must therefore substitute those fields by fields inside the reduced domain by using the symmetry relations. Referring to the symmetries illustrated in Figure 4.8, we would have $H_x(0, 1) = H_x(2, 1)$, $H_x(1, 0) = -H_x(1, 1)$, $H_y(0, 1) = -H_y(1, 1)$, and $H_y(0, 2) = -H_y(1, 2)$. This modifies the structure of the finite-difference matrix and therefore affects the solutions found. To find all solutions of all possible mirror symmetries, we should consider four sets of BCs, corresponding to H_x being either even or odd in the x- and y-axes, respectively. One can easily show that solutions having different mirror symmetries will also be orthogonal with respect to the usual inner products where either $[\mathbf{E} \times \mathbf{H}]_z$ or $\mathbf{E} \cdot \mathbf{D}$ is integrated over the waveguide cross section.

Thus, the price of reducing the computational domain to quarter size is that we only get a fourth of the eigenvalues, so it seems we must do four times as many calculations. However, since the numerical effort scales with the cube of the number of gridpoints, this still implies a reduction of the numerical complexity. Furthermore, one may consider it an advantage that attention can be focussed on eigenstates of a predetermined symmetry. In a single-mode waveguide, one is perhaps only interested in the fundamental mode, which will usually be symmetric in both axes. Therefore, only one of the symmetry combinations needs to be investigated in this case.

EXERCISES

4.1 This exercise is concerned with setting up a mode solver for TE and TM modes in a general 1D waveguide, discretized on a grid of N points with resolution Δx.

A template MATLAB script called *discmat_te.m* is provided, which sets up the discretization matrix for TE modes in a 1D waveguide, as specified in Equations (4.32) through (4.34). Read and understand this routine and create a sister routine to set up the discretization matrix for TM modes, according to Equations (4.40) through (4.42).

Set up ε and ξ for a symmetric slab waveguide of width 1.2345, a core refractive index of 1.45, and a cladding refractive index of 1 (a glass membrane in air). Take $\Delta x = 0.01$, $W = 5$, where W is the total width of the computational domain. Set up the TE discretization matrix for a wavelength $\lambda = 1.55$. Find the highest β^2-eigenvalue using the following command:

```
[eigvec eigval]=eigs(mat,1,(2*pi*1.45/1.55)^2);
```

The MATLAB function *eigs* is described in Section 4.4.2. *mat* is the discretization matrix in sparse format that you just calculated. The second entry is the desired number of eigenstates. The third entry is a *target* eigenvalue, that is the eigensolver finds the eigenvalue closest to this target. On output, the *eigvec* array will contain the eigenvector, with 500 elements. Similarly, *eigval* contains the β^2-eigenvalues.

We define the *effective index* of a waveguide mode as

$$n_{eff} = \frac{\beta}{k_0}. \tag{4.83}$$

In an ordinary dielectric waveguide, such as this one, the guided modes will have the highest effective indices (and therefore the highest β^2-eigenvalues), and these effective indices will be bounded from above by the refractive index of the core. Therefore, we can search the guided modes by putting the target eigenvalue to $(k_0 n_{core})^2$, as indicated earlier.

Calculate the effective index of the TE mode from the β value you just found and repeat the whole procedure for the TM mode also. If your routine is correct, you should find the results 1.37823 and 1.35294 (omitting the following digits). Plot the fields stored in *eigvec*, can you see where the refractive-index discontinuities are?

Finally, bind all of these steps together in a single MATLAB function with the following syntax:

```
[neff,field]=slab_solver(eps,xi,dx,lambda,ntar,neig,mode);
```

So, the inputs to this function should be the ε distribution *eps*, the ξ function *xi*, the resolution Δx, the wavelength *lambda*, the target effective index *ntar*, the desired number of eigenstates *neig*, and a flag specifying whether the TE or TM mode should be calculated (*mode*). The function should set up the appropriate discretization matrix, determine the target eigenvalue from the target effective index, find the eigenstates using *eigs*, and calculate the effective indices from the β^2-eigenvalues. The returned variables are then the effective indices of the eigenstates and the eigenvector, that is the E_y field for TE modes and the E_x field for TM modes.

4.2 As a starting point for this exercise, the script *neff_slab.m* is provided. The syntax of this script is

```
[neff_te,neff_tm]=neff_slab(lambda,n1,n2,nc,d);
```

Here, *lambda* is the wavelength, *n1* is the cladding refractive index on one side of the slab, *n2* is the cladding index on the other side, *nc* is the core index, and *d* is the width of the core. The script returns exact values for the effective indices of the fundamental TE and TM modes of the asymmetric slab. In the exercises, we will only consider symmetric cases, $n1 = n2 \equiv ncl$.

We start from the symmetric structure introduced in Exercise 4.1. From now on, we quantify the wavelength in terms of the *normalized frequency* or *V* parameter, which, for the slab waveguide, is defined as

$$V = \frac{\pi d}{\lambda} \sqrt{n_c^2 - n_{cl}^2}. \tag{4.84}$$

Take $V = \pi/2$ (which is the single-mode cutoff of the slab) and set the width of the computational domain to 15. Investigate the convergence of the finite-difference results for TE and TM modes. How does the deviation from the exact result scale with Δx? Use the MATLAB function *polyfit* to extrapolate your results for n_{eff} to $\Delta x = 0$ (this technique is known as the *Richardson extrapolation*). Repeat the investigations for the following cases:
- $nc = 1.45$, $nl = nr = 1.44$ (a waveguide in doped silica)
- $nc = 3.45$, $nl = nr = 1$ (a silicon membrane in air)

4.3 Now that you have a feel for the convergence behaviour with Δx, it is time to investigate the importance of the domain width. Starting from the structure in the previous exercise with $nc = 1.45$, $ncl = 1$, and $V = \pi/2$, reduce the domain width, and plot the finite-difference error versus this parameter (use the Richardson extrapolation to get results accurately converged in Δx). How does the error depend on the domain width? Can you explain why? Repeat the investigation for $V = 1$ and $V = 2$.

4.4 In this and the following exercises, you will apply the 2D finite-difference scheme on the Yee grid to the problem of a circular step-index fibre. Three basic scripts to start from are provided: The first is called *epsfiber.m* and has the syntax

```
[eps,width]=epsfiber(w,dx,a,ncore,nclad);
```

This function sets up a dielectric function for a step-index fibre core of radius *a*, using the simple *staircase* approach sketched in Figure 4.5b. The input variable *w* is a target width, and the output *width* is a two-component vector, whose values may deviate slightly from *w*. The variable *dx* is the discretization step. The output is a quarter of the step-index fibre structure, in the first quadrant of a Cartesian coordinate system with the origin in the core centre.

The second script is *discmat_yee.m*, which is a full-fledged MATLAB routine for setting up the 2D discretization matrix corresponding to Equations (4.60) and (4.61). Its syntax is

```
mat=discmat(eps,width,lambda,symx,symy);
```

Here, *lambda* is the wavelength, whereas *eps* and *width* are the dielectric function and domain width vector output from *epsfiber.m*, that is the two scripts are compatible. The variables *symx* and *symy* determine the BCs on the *inner* boundaries (left and bottom edges of the domain). *symx* = 0 gives a zero-field BC on the x-axis. This option is useful for structures where the guided mode is in the centre of the computational domain and the field at the domain boundaries is assumed to be vanishingly small. However, if mirror symmetries of the waveguide are to be utilized, other BCs are needed, as explained in Section 4.5.3. *symx* = 1 gives an H_x field even under reflections in the x-axis, while *symx* = −1 gives an H_x field, which is odd under this reflection. The variable *symy* determines the BCs for H_x on the y-axis in a similar way.

The *discmat* routine accepts an ε function which can be either of the format *eps(nx, ny)* or *eps(nx, ny, 3)*. In the former case, the ε function in a given cell of the Yee mesh will have the same value at E_x, E_y, and E_z points. In the latter case, *eps(nx, ny, 1)* contains ε_x, *eps(nx, ny, 2)* contains ε_y, and *eps(nx, ny, 3)* contains ε_z.

The final script provided transforms the raw eigenvector, obtained from solving the eigenvalue problem with the matrix output by *discmat_yee*, to 2D H_x, H_y fields. The syntax of the script is

```
[hx,hy]=fields(eigvec,nx,ny);
```

where *eigvec* is the eigenvector output by the *eigs* routine and nx and ny are the Yee grid dimensions. The output variable $hx(m, n)$ gives the H_x field in the Yee cell of index m, n. The variable $hy(m, n)$ gives the H_y field in the same cell. Note that the fields are given at positions which are shifted relative to each other, as shown in Figure 4.4.

Set up the dielectric structure of a glass rod in air (*ncore* = 1.45, *nclad* = 1) with a = 0.5 μm. Take λ = 1.0 μm, dx = 0.04 μm. Calculate the first few eigenmodes for each combination of *symx* = ±1 and *symy* = ±1. How many guided modes are there? Plot the H_x and H_y fields and describe their polarization properties. Use Equation (2.65) to calculate the z-component of the magnetic field. Plot the results and show that there exists a TM mode, with vanishing H_z field in the step-index fibre.

4.5 In this exercise, you will study the convergence of the effective index of the fundamental mode in the step-index fibre, using the staircase approximation ε function given by the *epsfiber* script. Using the same dielectric structure as in Exercise 4.4, take λ = 1.55 μm, dx = 0.05 μm initially and investigate the convergence with the domain width, as you did earlier for the 1D slab waveguide.

Next, vary dx and investigate the error relative to the exact solution, which you can get from the script *neff_sif.m*. The script has the syntax

```
neff=neff_sif(lambda,ncore,nclad,a);
```

This script calculates the effective index of the fundamental fibre mode in the step-index fibre, that is the mode with the highest effective index, using an essentially exact mode-matching technique.

How does the finite-difference error scale with Δx? As discussed earlier, use the Richardson extrapolation to estimate results for $\Delta x \to 0$. Do you obtain the correct limit?

Redo these investigations also for a cladding index of 1.44 and for a core index of 3.45 with cladding index of 1 to see how the accuracy of the FD scheme evolves with refractive-index contrast. As presented earlier, rescale the wavelength to keep the fibre V parameter constant – it is defined in a similar way as the slab V parameter, with $d = 2a$ being the diameter of the fibre core.

4.6 Based on the *epsfiber* function, create two other functions that set up the step-index distribution using either simple averaging of ε over a grid cell centred on an E_z point or tensorial averaging as in Equation (4.73). In the latter case, only the diagonal components of the dielectric tensor should be specified. The averages for ε_{xx} should be calculated in a box centred on the Yee gridpoint where E_x is represented (refer to Figure 4.4 and remember that the E_z point is at the origin), and similarly ε_{yy} should be calculated at the E_y point, and ε_{zz} at the E_z point.

Repeat the investigations of Δx convergence from Exercise 4.5 and discuss the improvements in accuracy from the different averaging techniques.

If time permits, check also the dependence of the FD accuracy on wavelength, for example in the range from 0.5 to 2 µm. Remember that you may have to change your domain width to ensure convergence for the longer wavelengths.

REFERENCES

1. M. S. Stern, Rayleigh quotient solution of semivectorial field problems for optical waveguides with arbitrary index profiles, *IEE Proc. J. Optoelectron.*, 138, 123–127 (1991).
2. P. Lusse, P. Stuwe, J. Schule, and H.-G. Unger, Analysis of vectorial mode fields in optical waveguides by a new finite difference method, *J. Lightwave Technol.*, 12, 487–494 (1994).
3. C. Vassallo, Improvement of finite difference methods for step-index optical waveguides, *IEE Proc. J. Optoelectron.*, 139, 137–142 (1992).
4. Y.-C. Chiang, Y.-P. Chiou, and H.-C. Chang, Improved full-vectorial finite-difference mode solver for optical waveguides with step-index profiles, *J. Lightwave Technol.*, 20, 1609–1618 (2002).
5. Y.-C. Chiang, Higher order finite-difference frequency domain analysis of 2-D photonic crystals with curved dielectric interfaces, *Opt. Express* 17, 3305–3315 (2009).
6. R. D. Meade, A. M. Rappe, K. D. Brommer, J. D. Joannopoulos, and O. L. Alerhand, Accurate theoretical analysis of photonic band-gap materials, *Phys. Rev. B*, 48, 8434–8437 (1993).
7. D. E. Aspnes, Local-field effects and effective-medium theory—A microscopic perspective, *Am. J. Phys.*, 50, 704–709 (1982).
8. A. Farjadpour, D. Roundy, A. Rodriguez, M. Ibanescu, P. Bermel, J. D. Joannopoulos, S. G. Johnson, and G. W. Burr, Improving accuracy by subpixel smoothing in the finite-difference time domain, *Opt. Lett.*, 31, 2972–2974 (2006).
9. Z. Zhu and T. G. Brown, Full-vectorial finite-difference analysis of microstructrured optical fibres, *Opt. Express*, 10, 853–864 (2002).
10. K. S. Yee, Numerical solution of initial boundary value problems involving Maxwell's equations in isotropic media, *IEEE Trans. Antennas Propag.*, AP14, 302–307 (1966).

5 Modelling of Nonlinear Propagation in Waveguides

5.1 INTRODUCTION

In this chapter, we outline how nonlinear propagation in the guided modes of optical waveguides can be described efficiently in a so-called 1+1D propagation formalism. This terminology implies that the transverse degrees of freedom in the waveguide cross section are integrated out and a first-order differential equation describing the evolution of the temporal, or spectral, profile of an optical pulse along the length of the waveguide is obtained. The two *dimensions* involved in the calculation are therefore the spatial coordinate along the length axis of the waveguide, and the time, or equivalently the frequency, coordinate.

Nonlinear optics in waveguides is an important scientific and industrial subject, because a waveguide allows light to propagate in one or more guided modes which do not diffract. Thereby, a high optical intensity can be sustained over extended distances, which can lead to substantial effects of optical nonlinearities, even if the optical nonlinearity of the waveguide base material is quite low. An extreme example of this is silica-based optical fibres, where light may propagate over many metres or even kilometres without significant attenuation. This can lead to very pronounced nonlinear effects even though the optical nonlinearity of silica is at least two orders of magnitude lower than those of most semiconductors. In telecommunication links, as well as in high-power fibre-based lasers, fibre-optic gyroscopes, and other components, it is important to mitigate such effects. In other applications, such as all-optical signal processing and frequency conversion, nonlinear effects are desired and utilized. Efficient modelling tools for nonlinear propagation are therefore increasingly important for waveguide designers.

For materials with inversion symmetry, such as glasses or high-symmetry crystals, the leading nonlinear contribution to the dielectric polarization is of third order in the electric field. This contribution can be associated with an intensity-dependent change in the material refractive index. One typically finds that the index changes that can be induced at optical intensities below the breakdown threshold are very small, on the order of 10^{-4} or smaller. Therefore, the nonlinearity hardly changes the profile of the guided modes, because the cross-sectional index contrasts shaping these modes are typically one-to-three orders of magnitude larger. On the other hand, since the waveguide is invariant, or at least slowly varying, in time and along the axial coordinate, even a small index change in these dimensions can be significant. These observations form the basis of the modelling scheme to be developed in this chapter. We assume that the light propagates with a transverse profile dictated by the guided modes of the waveguide; however, the temporal/spectral profile may change, as may the distribution of light between different waveguide modes.

Many texts on nonlinear propagation modelling start out by deriving the simplest reasonable propagation equation, which is then gradually extended to take various physical effects into account. The approach of the present chapter is the opposite: we start out in Section 5.2 by deriving a quite general formalism, which we specialize in Section 5.3 to the case of an isotropic nonabsorbing dielectric. In addition, we introduce a couple of assumptions (most notably one of single-mode guidance), which facilitate the formulation of a propagation equation that is numerically tractable. In Section 5.4, we introduce further approximations which allow to derive an equation, the so-called *nonlinear Schrödinger equation* (NLS), which is also amenable to analytical treatment. Section 5.5 discusses issues relating to the numerical solution of the propagation equations and their implementation in MATLAB®. In the exercises, a numerical simulation tool is developed, tested against the analytical results for the case of silica fibres, and used to study how these solutions are modified as the approximations leading to the NLS are relaxed.

5.2   FORMALISM

5.2.1   GENERAL PROPAGATION EQUATION

The starting point is Maxwell's equations with the displacement term separated into a linear term describing the ideal waveguide and a small perturbation to the polarization, $\delta \mathbf{P}$:

$$\nabla \times \mathbf{E} = -\mu_0 \partial_t \mathbf{H}, \tag{5.1}$$

$$\nabla \times \mathbf{H} = \varepsilon_0 \varepsilon(\mathbf{r}_\perp) \partial_t \mathbf{E} + \partial_t \delta \mathbf{P}. \tag{5.2}$$

The **E** and **H** fields are expanded into modal fields:

$$\mathbf{H}(\mathbf{r}, t) = \frac{1}{\sqrt{2\pi}} \sum_m \int d\omega A_m(z, \omega) \mathbf{h}_m(\mathbf{r}, \omega; t), \tag{5.3}$$

$$\mathbf{E}(\mathbf{r}, t) = \frac{1}{\sqrt{2\pi}} \sum_m \int d\omega A_m(z, \omega) \mathbf{e}_m(\mathbf{r}, \omega; t), \tag{5.4}$$

$$\mathbf{h}_m(\mathbf{r}, \omega; t) = \mathbf{h}_m(\mathbf{r}_\perp, \omega) e^{i(\omega t - \beta_m(\omega) z)}, \tag{5.5}$$

$$\mathbf{e}_m(\mathbf{r}, \omega; t) = \mathbf{e}_m(\mathbf{r}_\perp, \omega) e^{i(\omega t - \beta_m(\omega) z)}, \tag{5.6}$$

$$A_m(z, -\omega) = A_m^*(z, \omega); \quad \mathbf{e}_m(\mathbf{r}_\perp, -\omega) = \mathbf{e}_m^*(\mathbf{r}_\perp, \omega). \tag{5.7}$$

where the ω integration extends over both positive and negative values, and the m-index denotes different mode orders. Note that we adopt the positive sign convention from Equations (2.15) and (2.16) in these expressions. The modal fields

$\mathbf{h}_m(\mathbf{r}_\perp, \omega)$ and $\mathbf{e}_m(\mathbf{r}_\perp, \omega)$ are solutions of the linear Maxwell's equations for the waveguide, with propagation constants $\beta_m(\omega)$. These modal fields are orthogonal and normalized according to

$$\int d\mathbf{r}_\perp \left[\mathbf{e}_m(\mathbf{r}_\perp, \omega) \times \mathbf{h}_n^*(\mathbf{r}_\perp, \omega) - \mathbf{h}_m(\mathbf{r}_\perp, \omega) \times \mathbf{e}_n^*(\mathbf{r}_\perp, \omega) \right]_z = N_m(\omega)\delta_{mn}. \quad (5.8)$$

The choice of normalization $N_m(\omega)$ is discussed in Section 5.2.2.

The fundamental assumption of the following derivations is that $\delta\mathbf{P}$ is sufficiently small that the modal amplitudes $A_m(z, \omega)$ are slowly varying in z compared to $\exp[-i\beta_m(\omega)z]$. For silica glass, as well as most other dielectric materials, this is a well-justified assumption, since the optical intensities needed to violate it are typically high above the damage thresholds. The derivation of a 1+1D propagation equation then proceeds by projecting Maxwell's equations (5.1) and (5.2) onto the modal fields as follows:

$$\mathbf{e}_m^*(\mathbf{r}, \omega; t) \cdot [\partial_t\delta\mathbf{P} + \partial_t\mathbf{D}] = \mathbf{e}_m^*(\mathbf{r}, \omega; t) \cdot \nabla \times \mathbf{H}, \quad (5.9)$$

$$-\mu_0\mathbf{h}_m^*(\mathbf{r}, \omega; t) \cdot \partial_t\mathbf{H} = \mathbf{h}_m^*(\mathbf{r}, \omega; t) \cdot \nabla \times \mathbf{E}, \quad (5.10)$$

$$\Downarrow$$

$$\mathbf{e}_m^*(\mathbf{r}, \omega; t) \cdot [\partial_t\delta\mathbf{P} + \partial_t\mathbf{D}] = \nabla \cdot \left[\mathbf{H} \times \mathbf{e}_m^*(\mathbf{r}, \omega; t)\right] + \mathbf{H} \cdot \nabla \times \mathbf{e}_m^*(\mathbf{r}, \omega; t), \quad (5.11)$$

$$-\mu_0\mathbf{h}_m^*(\mathbf{r}, \omega; t) \cdot \partial_t\mathbf{H} = \nabla \cdot \left[\mathbf{E} \times \mathbf{h}_m^*(\mathbf{r}, \omega; t)\right] + \mathbf{E} \cdot \nabla \times \mathbf{h}_m^*(\mathbf{r}, \omega; t), \quad (5.12)$$

where we used the mathematical result $\mathbf{V}_1 \cdot \nabla \times \mathbf{V}_2 = \nabla \cdot [\mathbf{V}_2 \times \mathbf{V}_1] + \mathbf{V}_2 \cdot \nabla \times \mathbf{V}_1$. Using Equations (5.1) and (5.2) again, these equations become

$$\mathbf{e}_m^*(\mathbf{r}, \omega; t) \cdot [\partial_t\delta\mathbf{P} + \partial_t\mathbf{D}] = \nabla \cdot \left[\mathbf{H} \times \mathbf{e}_m^*(\mathbf{r}, \omega; t)\right] + i\omega\mu_0\mathbf{H} \cdot \mathbf{h}_m^*(\mathbf{r}, \omega; t), \quad (5.13)$$

$$-\mu_0\mathbf{h}_m^*(\mathbf{r}, \omega; t) \cdot \partial_t\mathbf{H} = \nabla \cdot \left[\mathbf{E} \times \mathbf{h}_m^*(\mathbf{r}, \omega; t)\right] - i\omega\varepsilon_0\varepsilon(\mathbf{r}_\perp, \omega)\mathbf{E} \cdot \mathbf{e}_m^*(\mathbf{r}, \omega; t). \quad (5.14)$$

Integrating over time and transverse spatial coordinates and doing a partial integration assuming that all fields vanish at infinity, one obtains

$$i\omega \int d\mathbf{r}_\perp dt \mathbf{e}_m^*(\mathbf{r}, \omega; t) [\delta\mathbf{P} + \mathbf{D}] = \partial_z \int d\mathbf{r}_\perp dt \left[\mathbf{H} \times \mathbf{e}_m^*(\mathbf{r}, \omega; t)\right]_z$$

$$+ i\omega\mu_0 \int d\mathbf{r}_\perp dt \mathbf{H} \cdot \mathbf{h}_m^*(\mathbf{r}, \omega; t), \quad (5.15)$$

$$-i\omega\mu_0 \int d\mathbf{r}_\perp dt \mathbf{h}_m^*(\mathbf{r}, \omega; t) \cdot \mathbf{H} = \partial_z \int d\mathbf{r}_\perp dt \left[\mathbf{E} \times \mathbf{h}_m^*(\mathbf{r}, \omega; t)\right]_z$$

$$- i\omega\varepsilon_0 \int d\mathbf{r}_\perp dt \varepsilon(\mathbf{r}_\perp, \omega)\mathbf{E} \cdot \mathbf{e}_m^*(\mathbf{r}, \omega; t). \quad (5.16)$$

The next step is to introduce the modal field expansions (Equations 5.3 and 5.4) for the $\mathbf{E}$ and $\mathbf{H}$ fields. The expression of $\delta\mathbf{P}$ in terms of the modal fields is deferred for the moment. This leads to the expressions

$$i\omega \int d\mathbf{r}_\perp dt e_m^*(\mathbf{r}, \omega; t)\delta\mathbf{P}$$

$$+ i\omega\varepsilon_0\sqrt{2\pi}\sum_n \int d\mathbf{r}_\perp A_n(z, \omega)\varepsilon(\mathbf{r}_\perp, \omega)\mathbf{e}_n(\mathbf{r}, \omega; t) \cdot \mathbf{e}_m^*(\mathbf{r}, \omega; t)$$

$$= \sqrt{2\pi}\partial_z \sum_n \int d\mathbf{r}_\perp \left[A_n(z, \omega)\mathbf{h}_n(\mathbf{r}, \omega; t) \times \mathbf{e}_m^*(\mathbf{r}, \omega; t) \right]_z$$

$$+ i\omega\mu_0\sqrt{2\pi}\sum_n \int d\mathbf{r}_\perp A_n(z, \omega)\mathbf{h}_n(\mathbf{r}, \omega; t) \cdot \mathbf{h}_m^*(\mathbf{r}, \omega; t) \qquad (5.17)$$

$$i\omega\mu_0\sqrt{2\pi}\sum_n \int d\mathbf{r}_\perp A_n(z, \omega)\mathbf{h}_n(\mathbf{r}, \omega; t) \cdot \mathbf{h}_m^*(\mathbf{r}, \omega; t)$$

$$= -\sqrt{2\pi}\partial_z \sum_n \int d\mathbf{r}_\perp A_n(z, \omega) \left[\mathbf{e}_n(\mathbf{r}, \omega; t) \times \mathbf{h}_m^*(\mathbf{r}, \omega; t) \right]_z$$

$$+ i\omega\varepsilon_0\sqrt{2\pi}\sum_n \int d\mathbf{r}_\perp A_n(z, \omega)\varepsilon(\mathbf{r}, \omega)\mathbf{e}_n(\mathbf{r}, \omega; t) \cdot \mathbf{e}_m^*(\mathbf{r}, \omega; t). \qquad (5.18)$$

In deriving these equations, the integral representation of the Dirac delta function

$$\delta(\omega - \omega') = \frac{1}{2\pi} \int dt e^{i(\omega - \omega')t} \qquad (5.19)$$

was used. Adding Equations (5.17) and (5.18) yields

$$N_m(\omega)\partial_z A_m(z, \omega) = -\frac{i\omega}{\sqrt{2\pi}} \int d\mathbf{r}_\perp dt e_m^*(\mathbf{r}, \omega; t)\delta\mathbf{P}. \qquad (5.20)$$

The choice of normalization factor is still arbitrary, so the propagation equation may be simplified by choosing, for example $N_m(\omega) = 1$. Note, however, that other derivations of the nonlinear propagation equations may lead to formulations corresponding to other choices of $N_m(\omega)$. In the next section, we examine how physical quantities such as power and energy are expressed for arbitrary choices of $N_m(\omega)$.

5.2.2 PULSE POWER AND PULSE ENERGY

The instantaneous power of the propagating fields can be written as

$$P(t) = \frac{1}{2} \int d\mathbf{r}_\perp \left[\mathbf{E}(\mathbf{r}, t) \times \mathbf{H}^*(\mathbf{r}, t) - \mathbf{H}(\mathbf{r}, t) \times \mathbf{E}^*(\mathbf{r}, t) \right]_z$$

$$= \frac{1}{4\pi} \sum_{mn} \int d\omega_1 d\omega_2 A_m(z, \omega_1) A_n^*(z, \omega_2) e^{i((\omega_1 - \omega_2)t - \beta_m(\omega_1) + \beta_n(\omega_2))}$$

$$\times \int d\mathbf{r}_\perp \left[\mathbf{e}_m(\mathbf{r}_\perp, \omega_1) \times \mathbf{h}_n^*(\mathbf{r}_\perp, \omega_2) - \mathbf{h}_m(\mathbf{r}_\perp, \omega_1) \times \mathbf{e}_n^*(\mathbf{r}_\perp, \omega_2) \right]_z. \qquad (5.21)$$

Neglecting the frequency dependence of the modal fields and averaging over several optical cycles, this expression becomes

$$P(t) = \sum_m N_m \, |\tilde{A}_m(z,t)|^2;$$

$$\tilde{A}_m(z,t) = \frac{1}{\sqrt{2\pi}} \int_0^\infty d\omega e^{i\omega t} \tilde{A}_m(z,\omega); \quad \tilde{A}_m(z,\omega) = A_m(z,\omega)e^{-i\beta_m(\omega,z)}. \tag{5.22}$$

Equation (5.7) has been used to restrict the frequency integral in the definition of $\tilde{A}_m(z,t)$ to positive frequencies. Under the stated assumptions, the squared norm of the amplitudes $\tilde{A}_m(z,t)$ can then be directly interpreted as a temporal power profile if one chooses $N_m = 1$. If the mode profiles have a significant frequency variation, this interpretation becomes less straightforward. For the diagonal terms in the sum over modal fields, one can write

$$\int d\mathbf{r}_\perp \left[\mathbf{e}_m(\mathbf{r}_\perp, \omega_1) \times \mathbf{h}_m^*(\mathbf{r}_\perp, \omega_2) - \mathbf{h}_m(\mathbf{r}_\perp, \omega_1) \times \mathbf{e}_m^*(\mathbf{r}_\perp, \omega_2) \right]_z$$

$$= \int d\mathbf{r}_\perp \left[\mathbf{e}_m(\mathbf{r}_\perp, \omega_1) \times \mathbf{h}_m(\mathbf{r}_\perp, \omega_2) - \mathbf{h}_m(\mathbf{r}_\perp, \omega_1) \times \mathbf{e}_m(\mathbf{r}_\perp, \omega_2) \right]_z$$

$$= N_m(\omega_1) + \int d\mathbf{r}_\perp \left[\mathbf{e}_m(\mathbf{r}_\perp, \omega_1) \times \partial_{\omega_1} \mathbf{h}_m(\mathbf{r}_\perp, \omega_1) - \mathbf{h}_m(\mathbf{r}_\perp, \omega_1) \times \partial_{\omega_1} \mathbf{e}_m(\mathbf{r}_\perp, \omega_1) \right]_z$$

$$\times (\omega_2 - \omega_1) + O\left((\omega_2 - \omega_1)^2\right) = N_m(\omega_1) + \frac{1}{2} \partial_{\omega_1} N_m(\omega_1)(\omega_2 - \omega_1)$$

$$+ O\left((\omega_2 - \omega_1)^2\right). \tag{5.23}$$

The first equality stems from the fact that the transverse components of the modal fields $\mathbf{e}_m(\mathbf{r}_\perp, \omega)$ and $\mathbf{h}_m(\mathbf{r}_\perp, \omega)$ may be taken real. It is seen that if a frequency-independent $N_m(\omega)$ is chosen, such as $N_m(\omega) = 1$, the interpretation of $|\tilde{A}_m(z,t)|^2$ as temporal modal power profiles holds to first order in the frequency dispersion of the mode profiles, provided that off-diagonal terms can be neglected. This will often be the case when modes belong to different symmetry classes or between different polarization states of modes of the same symmetry.

Regardless of mode profile dispersion, the total pulse energy is given by

$$E_p = \int dt P(t) = \frac{1}{2} \sum_m \int d\omega \, |\tilde{A}_m(z,\omega)|^2 N_m(\omega) \tag{5.24}$$

meaning that for $N_m(\omega) = 1$, the positive-frequency norm-squared amplitudes $A_m(z,\omega)$ integrate directly to the total pulse energy. In the following, we will stick to this normalization convention.

5.3 NONLINEAR POLARIZATION

The formalism derived so far is of a very general and abstract nature, since we have not made any assumptions about the nature and properties of the polarization $\delta\mathbf{P}$.

It could take any reasonable form, even including linear perturbations such as absorption, perturbations to the waveguide structure, and external time-dependent modulations of the waveguide properties. Thus, Equation (5.20) can be used as a starting point for deriving propagation equations for a wide variety of physical problems. In this section, we begin by giving a brief overview of the most commonly encountered nonlinear processes, after which we formulate propagation equations for a particular case, namely that of an isotropic and nonabsorbing dielectric. We will do this with an eye to the case of silica-based optical fibres since the exercises will be based on this particular example. For a more comprehensive introduction to nonlinear optics in general, we refer the reader to specialized books on this subject, such as the book of Boyd [1]. For an in-depth introduction to nonlinear fibre optics, one may consult the book of Agrawal [2].

5.3.1 NONLINEAR PROCESSES

Since in most materials nonlinear optical processes are of a weak nature, it is customary to write the nonlinear polarization as a power series expansion:

$$\delta P_i(\mathbf{r}, t) = \sum_{j=2}^{\infty} \sum_{k_1 \cdots k_j} \int dt_1 \cdots dt_j \, \chi^{(j)}_{ik_1 \cdots k_j}(\mathbf{r}, \mathbf{r}_1, \ldots, \mathbf{r}_j; \, t, t_1, \ldots, t_j) E_{k_1}(\mathbf{r}_1, t_1) \cdots E_{k_j}(\mathbf{r}_j, t_j).$$

(5.25)

Here, j counts the order of the nonlinear process, $\chi^{(j)}$ is the nonlinear susceptibility tensor of order j, and the i and k_j indices label Cartesian coordinates. In addition to this expansion, there are nonlinear effects associated with field derivatives and material interfaces [3,4]. While such effects can be important in special cases, they are usually weak and will not be considered further in this book. Apart from these effects, Equation (5.25) is a quite general expression, allowing for the nonlinear response to be nonlocal in both time and space, as well as having an overall time and space variations. Common simplifications are assumptions of homogeneity in time and space, so that the t_k and $\mathbf{r}_k$ dependencies of the $\chi^{(n)}$ can be expressed in terms of $t - t_k$ and $\mathbf{r} - \mathbf{r}_k$, respectively.

Note: Especially, the assumption of spatial homogeneity may break down in a number of relevant cases involving waveguides composed of materials having widely different nonlinear properties. For spatially local nonlinear effects, the $\mathbf{r}_k$ dependence is given by $\delta(\mathbf{r} - \mathbf{r}_k)$, and in this case, it is not very difficult to correct for inhomogeneity effects.

Since both $\delta \mathbf{P}$ and $\mathbf{E}$ are vectors, one expects that a change of sign in $\mathbf{E}$ should lead to a change of sign in $\delta \mathbf{P}$ as well if the nonlinear material has *inversion symmetry*, that is if it is invariant under a change of sign of the atomic position vectors. This implies that all $\chi^{(j)}$ coefficients in Equation (5.25) with even j must vanish, meaning that for such materials, $\chi^{(3)}$ is the lowest-order nonlinear coefficient [5]. Examples of (macroscopically) inversion-symmetric materials are amorphous structures, such as glasses, and many high-symmetry crystal structures. It is important to note, however, that there are also many single-crystal materials which do not have inversion symmetry, and where one can therefore utilize second-order nonlinearities.

Various physical effects may contribute to the nonlinear susceptibility tensors. Typically, nonlinear terms in the response of the electron clouds to an external field, the so-called *Kerr nonlinearities*, are the most important ones [5]. Due to the low electron mass and the strong electrostatic fields governing electronic motions in solids, this response is usually very rapid, on the order of a femtosecond (fs) or so. There are also important responses from motion of the atomic nuclei making up the nonlinear material [5]. These can be divided into the *Raman scattering* [6], associated with the excitation of local atomic vibrations (or *optical phonons* in a quantum-mechanical picture), and *electrostriction* effects, associated with the excitation of sound waves (or *acoustic phonons*) [7]. The latter effect, which is also known as the *stimulated Brillouin scattering*, is characterized by a response time on the order of a nanosecond, and therefore does not affect pulses significantly shorter than that. On the other hand, the Raman response usually has sub-picosecond and, in many cases few fs response time, and therefore affects all but the very shortest pulses. All of these responses can usually be assumed to be spatially local. The implementation of the Kerr and Raman nonlinearities into the FDTD method was discussed in Section 3.6.

Absorption effects may give rise to several kinds of nonlinearity. *Multiphoton absorption* takes place when two or more photons are absorbed simultaneously to match the energy of some quantum-mechanical transition in the material [8]. The effect of multiphoton absorption will be intensity dependent and is instantaneous in nature, unless some resonant intermediate state is involved. For sufficiently large intensity and/or fluence, one may see effects of *saturable absorption* [9], where the absorption coefficient goes down with intensity/fluence because absorption centres already excited cannot absorb further. The magnitude and temporal response of this effect depend on both the number density of the absorption centres and the lifetime of the excited level, both of which can vary over many orders of magnitude. Such absorption effects are described by the imaginary parts of the nonlinear $\chi^{(j)}$ tensors. The excitation of absorption centres also affects the real dielectric constant of the material (i.e. the real parts of the $\chi^{(j)}$ tensors) by way of the Kramers–Kronig relations between the real and imaginary parts of the dielectric constant [10]. In semiconductors, the excitation of carriers to the conduction band can lead to electron-gas screening of the optical fields [11]. Finally, absorption effects will typically lead to heating of the material, which can again lead to a change in dielectric constant through thermo-optic and thermo-elastic effects [12,13]. Thermally induced nonlinearities are often very strong, dominating all other nonlinear effects in magnitude [14], but at the same time have very slow response times, which means that they mainly contribute to changes in the transverse beam profile, possibly with low-frequency modulations [15]. Whereas most other absorption effects are spatially local, the thermo-optic effect may be strongly nonlocal due to heat diffusion.

In the following, we will develop a specific propagation model for $\chi^{(3)}$ non-linearities in an isotropic material, including electronic and the Raman effects, as these are the nonlinearities of most general relevance. As a specific example, we will parametrize the case of silica glass, which is a widely used base material in both fibre and integrated optics.

5.3.2 $\chi^{(3)}$ Nonlinear Processes

The third-order nonlinear polarization can in general be written as

$$\delta P_i(\mathbf{r},t) = \sum_{jkl} \int dt_1 dt_2 dt_3 E_j(\mathbf{r},t_1)\chi_{ijkl}^{(3)}(\mathbf{r},t-t_1,t-t_2,t-t_3)E_k(\mathbf{r},t_2)E_l(\mathbf{r},t_3). \quad (5.26)$$

This is basically the same equation as (3.159), except that Equation (5.26) allows for a spatial variation of the nonlinear susceptibility. For an isotropic material such as silica glass, this form can be reduced to [16]

$$\delta P_i(\mathbf{r}) = \sum_{jkl} E_j(\mathbf{r},t)\chi^{(3)}(\mathbf{r}) \int dt' r_{ijkl}(t-t')E_k(\mathbf{r},t')E_l(\mathbf{r},t'). \quad (5.27)$$

The response tensor r has an electronic contribution which is proportional to $\delta(t-t')$ δ_{ijkl} and a delayed-response (Raman) term, which can be expressed as

$$r_{ijkl}^R(t) = a(t)\delta_{ij}\delta_{kl} + \frac{1}{2}b(t)\left(\delta_{ik}\delta_{jl} + \delta_{il}\delta_{jk}\right). \quad (5.28)$$

Experimental data indicate that for silica glass $b(t) \ll a(t)$, over most of the frequency spectrum of these function [16]. Neglecting $b(t)$, Equation (5.27) can be recast as

$$\delta \mathbf{P}(\mathbf{r},t) = \varepsilon_0 \chi^{(3)}(\mathbf{r})\mathbf{E}(\mathbf{r},t) \int dt' \tilde{R}(\mathbf{r},t-t') \,|\,\mathbf{E}(\mathbf{r},t')\,|^2. \quad (5.29)$$

We will develop our formalism from Equation (5.29), for example neglecting $b(t)$.

Before we go on, it is instructive to discuss the overall implications of having an $\chi^{(3)}$ nonlinearity. Consider the case of a purely electronic nonlinearity, with instantaneous response, and an input field given by the expansion

$$\mathbf{E}(t) = \hat{\mathbf{x}} \sum_k E_k \left(e^{i(\omega_0+k\delta\omega)t} + e^{-i(\omega_0+k\delta\omega)t}\right), \quad (5.30)$$

where we neglect the spatial field structure for the time being. This expansion could describe a train of linearly polarized pulses, with a repetition rate given by $2\pi/\delta\omega$. Under the stated assumptions, the nonlinear polarization is given by

$$\delta P_x \propto E(t)^3 = \sum_{klm} E_k E_l E_m \left(e^{i(3\omega_0+(k+l+m)\delta\omega)t} + 3e^{i(\omega_0+(k-l-m)\delta\omega)t} + c.c.\right) \quad (5.31)$$

where $c.c.$ denotes the complex conjugate of the preceding expression. Assuming that the input spectrum is narrowly centred on ω_0, that is that $|\,k\delta\omega\,| \ll \omega_0$ for all k entering the sum in Equation (5.30), one sees that the spectrum of the nonlinear polarization is like the one sketched in Figure 5.1: components centred on $\pm 3\omega_0$ appear in addition to the spectral peaks around $\pm\omega_0$, which may be modified and widened compared to the input spectrum. The generation of tripled frequencies is denoted *third-harmonic generation*. It is usually a small effect due to the lack of phase matching. Therefore, the modification of the spectrum around ω_0 is the most important effect of $\chi^{(3)}$ nonlinearity and the one that will be the objective of our modelling efforts.

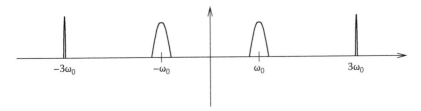

FIGURE 5.1 Illustration of a typical frequency spectrum. The pulse amplitudes at positive and negative frequencies must be complex conjugates of each other in order for the total field to remain a real quantity.

5.3.3 SINGLE-MODE PROPAGATION MODEL

Inserting the modal expansion for **E** into Equation (5.29), one obtains

$$
\frac{i\omega}{\sqrt{2\pi}} \int d\mathbf{r}_\perp dt\, \mathbf{e}_m^*(\mathbf{r}, \omega; t) \cdot \delta\mathbf{P}(\mathbf{r}, t)
$$

$$
= \frac{i\omega\varepsilon_0}{\sqrt{2\pi}} \sum_{npq} \int d\omega_1 d\omega_2 \tilde{A}_n(z, \omega_1) \tilde{A}_p^*(z, \omega_2) \tilde{A}_q(z, \omega - \omega_1 + \omega_2) e^{i\beta_m(\omega)z}
$$

$$
\times \int d\mathbf{r}_\perp \chi^{(3)}(\mathbf{r}) \tilde{R}(\mathbf{r}, \omega - \omega_1) \mathbf{e}_m^*(\mathbf{r}_\perp, \omega) \cdot \mathbf{e}_n(\mathbf{r}_\perp, \omega_1) \mathbf{e}_p^*(\mathbf{r}_\perp, \omega_2)
$$

$$
\cdot \mathbf{e}_q(\mathbf{r}_\perp, \omega - \omega_1 + \omega_2) \tag{5.32}
$$

after a few manipulations. In the following, the sum over modal fields will be restricted to a single state in order to simplify the final derivations a bit. Accordingly, the modal indices will be suppressed from the notation. Such an approach may be justified for the case of a single-mode polarization-maintaining waveguide. It is also quite useful for modelling the propagation of unpolarized light in a single mode, although in that case, the magnitude of $\chi^{(3)}$ should be reduced by a factor of 8/9 [2]. As a further approximation, we will neglect the frequency dependence of the transverse mode profile. This is usually a reasonable approximation, although it may be insufficient when spectrally broad pulses or unusual waveguide structures (e.g. waveguides based on photonic-bandgap effects) are considered [17]. Equation (5.32) can now be recast as

$$
\frac{i\omega}{\sqrt{2\pi}} \int d\mathbf{r}_\perp dt\, \mathbf{e}^*(\mathbf{r}, \omega; t) \cdot \delta\mathbf{P}(\mathbf{r}, t)
$$

$$
= \frac{i\omega\varepsilon_0}{\sqrt{2\pi}} \int d\omega_1 d\omega_2 \tilde{A}(z, \omega_1) \tilde{A}^*(z, \omega_2) \tilde{A}(z, \omega - \omega_1 + \omega_2) e^{i\beta(\omega)z}
$$

$$
\times \int d\mathbf{r}_\perp \chi^{(3)}(\mathbf{r}) \tilde{R}(\mathbf{r}, \omega - \omega_1) |\mathbf{e}(\mathbf{r}_\perp)|^4 . \tag{5.33}
$$

Importantly, the integral over transverse coordinates on the RHS is now frequency independent. Let us assume that the magnitude of $\chi^{(3)}(\mathbf{r})$ and $\tilde{R}$ is independent of position. It is then customary to rewrite the modal overlap integral in terms of an *effective area*, defined as

$$A_{eff} = \frac{\mu_0 \left[\text{Re} \left(\int d\mathbf{r}_\perp \mathbf{e}(\mathbf{r}_\perp) \times \mathbf{h}(\mathbf{r}_\perp) \right) \right]^2}{n_0^2 \varepsilon_0 \int_m d\mathbf{r}_\perp \mid \mathbf{e}(\mathbf{r}_\perp) \mid^4} = \frac{\mu_0 N^2}{4 n_0^2 \varepsilon_0 \int_m d\mathbf{r}_\perp \mid \mathbf{e}(\mathbf{r}_\perp) \mid^4}, \qquad (5.34)$$

where n_0 is a reference refractive index, chosen to be representative of the (frequency-dependent) material refractive index, and N is the normalization factor given by Equation (5.8). This expression can be seen to have dimension of an area when adopting the normalization convention $N = 1$. With this choice of normalization, Equation (5.20) can be written as

$$\partial_z A(z, \omega) = -\frac{i\omega \mu_0 \chi^{(3)} e^{i\beta(\omega)z}}{4 n_0^2 A_{eff} \sqrt{2\pi}}$$

$$\times \int d\omega_1 d\omega_2 \tilde{A}(z, \omega_1) \tilde{A}^*(z, \omega_2) \tilde{A}(z, \omega - \omega_1 + \omega_2) \tilde{R}(\omega - \omega_1)$$

$$\equiv -\frac{i\omega n_2 e^{i\beta(\omega)z}}{3 c A_{eff} \sqrt{2\pi}} \int d\omega_1 d\omega_2 \tilde{A}(z, \omega_1) \tilde{A}^*(z, \omega_2)$$

$$\times \tilde{A}(z, \omega - \omega_1 + \omega_2) \tilde{R}(\omega - \omega_1) \qquad (5.35)$$

$$n_2 \equiv \frac{3 c \mu_0 \chi^{(3)}}{4 n_0^2} = \frac{3 \chi^{(3)}}{4 n_0^2 c \varepsilon_0} \qquad (5.36)$$

where we introduced the nonlinear parameter n_2. The reason for the factor of 3 appearing in the definition of this quantity will become apparent shortly.

In Equation (5.36), the frequency integrals are still assumed to run from minus to plus infinity. However, Equation (5.7) stipulates a relation between amplitudes and fields at positive and negative frequencies, which comes from the fundamental requirement that the physical electric field should be a real quantity. Therefore, we can restrict ourselves to consider, for example the positive-frequency amplitudes without losing information. Rewriting Equation (5.36) with integrals running over the positive half-axis only, we obtain

$$\partial_z A(z, \omega) = -\frac{i\omega n_2 e^{i\beta(\omega)z}}{3 c A_{eff} \sqrt{2\pi}}$$

$$\times \int_+ d\omega_1 d\omega_2 \Big[\tilde{A}(z, \omega_1) \tilde{A}^*(z, \omega_2) \tilde{A}(z, \omega - \omega_1 + \omega_2) \tilde{R}(\omega - \omega_1)$$

$$+ \tilde{A}(z, \omega_1) \tilde{A}^*(z, -\omega_2) \tilde{A}(z, \omega - \omega_1 - \omega_2) \tilde{R}(\omega - \omega_1)$$

$$+ \tilde{A}(z, -\omega_1) \tilde{A}^*(z, \omega_2) \tilde{A}(z, \omega + \omega_1 + \omega_2) \tilde{R}(\omega + \omega_1)$$

$$+ \tilde{A}(z, -\omega_1) \tilde{A}^*(z, -\omega_2) \tilde{A}(z, \omega + \omega_1 - \omega_2) \tilde{R}(\omega + \omega_1) \Big]$$

$$= -\frac{i\omega n_2 e^{i\beta(\omega)z}}{3cA_{eff}\sqrt{2\pi}} \int_+ d\omega_1 d\omega_2$$

$$\times \left[\tilde{A}(z,\omega_1)\tilde{A}^*(z,\omega_2)\tilde{A}(z,\omega-\omega_1+\omega_2)\tilde{R}(\omega-\omega_1) \right.$$

$$+ \tilde{A}(z,\omega_1)\tilde{A}(z,\omega_2)\tilde{A}^*(z,\omega_1+\omega_2-\omega)\tilde{R}(\omega-\omega_1)$$

$$+ \tilde{A}^*(z,\omega_1)\tilde{A}^*(z,\omega_2)\tilde{A}(z,\omega+\omega_1+\omega_2)\tilde{R}(\omega+\omega_1)$$

$$\left. + \tilde{A}^*(z,\omega_1)\tilde{A}(z,\omega_2)\tilde{A}(z,\omega+\omega_1-\omega_2)\tilde{R}(\omega+\omega_1) \right]. \quad (5.37)$$

Here, the symbol $\int_+$ is shorthand for a positive half-axis integral, $\int_0^\infty$. In the last equation, all frequency arguments occurring in the field amplitudes will be positive, provided that the spectrum is not too wide. This may be appreciated by considering again the schematic spectrum in Figure 5.1. Let us write $\omega_i = \omega_0 + \Omega_i$ and assume that third-harmonic generation is negligible. If a frequency argument like $\omega + \omega_1 - \omega_2$ should be negative, we would need to have $\omega_0 < \Omega_2 - \Omega_1 - \Omega$. If we have $|\Omega_i| < \frac{\omega_0}{3}$, this relation cannot be fulfilled, and all frequency arguments of the field amplitudes in Equation (5.37) will be positive. Furthermore, when third-harmonic generation is neglected, the third term in Equation (5.37) falls away, because $\tilde{A}(z, \omega + \omega_1 + \omega_2) \approx 0$. Finally, the second term can be rewritten by the change of variables $\omega_2 \to \omega_1 + \omega_2 - \omega$ and the fourth term by interchanging ω_1 and ω_2, to yield

$$\partial_z A(z,\omega) = -\frac{i\omega n_2 e^{i\beta(\omega)z}}{3cA_{eff}\sqrt{2\pi}} \times \int_+ d\omega_1 d\omega_2 \tilde{A}(z,\omega_1)\tilde{A}^*(z,\omega_2)\tilde{A}(z,\omega-\omega_1+\omega_2)$$

$$\times \left[2\tilde{R}(\omega-\omega_1) + \tilde{R}(\omega+\omega_1) \right]. \quad (5.38)$$

Since the electronic (Kerr) part of the response function is approximated as a delta function in time, it will be a frequency-independent constant in frequency. On the other hand, the delayed response due to the Raman effects will typically be significantly slower than the optical frequency, because it is connected with effects of lattice vibrations. One usually assumes that the Fourier-transformed delayed-response function has negligible weight at the sum frequency $\omega + \omega_1$ in Equation (5.39). We can introduce the Raman response function R in a simple-minded fashion as

$$\tilde{R}(t) = (1-f_R)\delta(t) + f_R R(t) \Leftrightarrow \tilde{R}(\omega) = (1-f_R) + f_R R(\omega); \quad \int\limits_0^\infty dt R(t) = 1.$$

$$(5.39)$$

For optical pulses much longer than the Raman response time, the nonlinear interaction can to a good approximation be described as an instantaneous nonlinearity with $\tilde{R}(\omega) = 1$. On the other hand, for pulses much shorter than the Raman response time, an instantaneous response with $\tilde{R}(\omega) = 1 - f_R$ will be seen. Thus, f_R can be regarded as the fractional Raman contribution to the total response. However, the resulting propagation equation becomes

$$\partial_z A(z,\omega) = -\frac{i\omega n_2 e^{i\beta(\omega)z}}{cA_{eff}\sqrt{2\pi}} \times \int_+ d\omega_1 d\omega_2 \tilde{A}(z,\omega_1)\tilde{A}^*(z,\omega_2)\tilde{A}(z,\omega-\omega_1+\omega_2)$$

$$\times \left[(1-f_R) + f_R \frac{2}{3} R(\omega-\omega_1) \right]. \tag{5.40}$$

To avoid the factor of $2/3$, it has become a common practice, at least in the fibre-optics community, to partition the response function as

$$\tilde{R}(\omega) = (1-f_R) + \frac{3}{2} f_R R(\omega), \tag{5.41}$$

leading to the propagation equation

$$\partial_z A(z,\omega) = -\frac{i\omega n_2 e^{i\beta(\omega)z}}{cA_{eff}\sqrt{2\pi}} \times \int_+ d\omega_1 d\omega_2 \tilde{A}(z,\omega_1)\tilde{A}^*(z,\omega_2)\tilde{A}(z,\omega-\omega_1+\omega_2)$$

$$\times \left[(1-f_R) + f_R R(\omega-\omega_1) \right]. \tag{5.42}$$

Experimentally, one can do experiments to measure effects of instantaneous and delayed response and extract values for $\chi^{(3)}$ (or, equivalently, n_2) and f_R to fit the experimental results. The extracted parameters will depend on whether Equation (5.41) or (5.42) is used for the analysis. As an example, for silica glass, Equation (5.41) leads to $f_R \approx 0.3$ [18], whereas Equation (5.42) leads to $f_R \sim 0.2$ ($f_R = 0.18$ is a commonly used value) [2]. Since the magnitude of the Kerr term $n_2(1-f_R)$ should correspond to experiment in both formulations, this difference in f_R will also lead to different estimates of n_2; however, the difference is on the order of 10%, which is within the spread of published n_2 values for silica. Milam obtained a value of $n_2 = (2.74 \pm 0.17) \cdot 10^{-8}$ $\mu m^2/W$ for linearly polarized light at a wavelength around 1 μm, by averaging a selection of literature results [19]. Measurements at a wavelength of 1.55 μm have yielded values around $2.2 \cdot 10^{-8}$ $\mu m^2/W$ for randomly polarized light in standard telecommunication fibres [20,21]. To compare with measurements for linearly polarized light, this figure should be multiplied by 9/8 [20], leading to a value of $\sim 2.5 \cdot 10^{-8}$ μm^2. The discrepancy may be ascribed to frequency dispersion of the nonlinear coefficient or to measurement uncertainties. Two further considerations are worth mentioning. Firstly, the n_2 of pure silica may be considerably enhanced by Ge doping, and decreased by F doping [20]. These dopants are commonly used to, respectively, raise or lower the refractive index in standard optical fibres. Secondly, the values cited are valid for pulses shorter than about 1 ns. For longer pulses, the effect of *electrostriction* increases the value of n_2 by about 20% [21].

Equation (5.42) is our final result, the propagation equation which will be implemented numerically in the exercises. However, before going further in the analysis of this equation, it is necessary to discuss the issue of temporal reference frame. Although Equation (5.42) is formulated in the frequency domain, it makes implicit use of a time representation, because the convolution integrals appearing on the RHS are most conveniently done by using the numerical fast Fourier transform (FFT) to go into the time domain where the frequency convolutions become products, as displayed in Equation (2.19). In practically relevant nonlinear optics problems, the ratio between the pulse duration and the duration of propagation is often very large. For instance, problems in integrated-optics signal processing, or ultrafast fibre optics, often involve fs pulses propagating over centimetres or even metres of waveguide. In telecommunications, one is typically looking at sub-ns pulses propagating over many kilometres of fibre. Light in a silica-based waveguide travels roughly 0.2 mm in a picosecond, so for such a waveguide, a time grid with sufficient resolution to resolve the pulse, and sufficient width to describe its entire propagation, would have to contain a very large number of points, making the calculations numerically demanding. In fact, it is not necessary to work with such a grid, because at any point z along the waveguide, the pulse only occupies a small part of the total time span. Therefore, it is preferable to transform to a time frame T that moves with the pulse, so that at a given point z along the waveguide, we have

$$T = t - \frac{z}{v_g} = \beta_1 z, \tag{5.43}$$

where $\beta_1 = d\beta/d\omega$ is the inverse group velocity. The concept is illustrated in Figure 5.2. The pulse is always centred around $T = 0$, and one can choose a T-grid that is

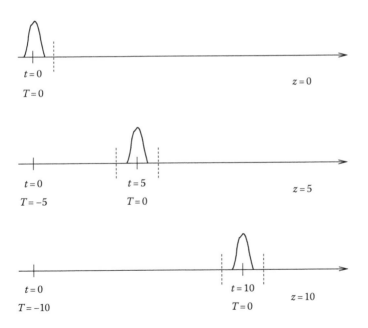

FIGURE 5.2 Illustration of the moving time frame T, relative to the *laboratory* frame t, in which the waveguide is stationary. In the illustration, the pulse group velocity has been set to 1. The vertical dashed lines indicate the boundaries of the moving time grid.

just wide enough to accommodate the pulse width. From Equation (5.4), the electric field can be expressed in terms of T by writing

$$\mathbf{E}(\mathbf{r}, t) = \frac{1}{\sqrt{2\pi}} \int d\omega A(z, \omega) e^{i[\omega_0 t - \beta(\omega_0)z]} \mathbf{e}(\mathbf{r}_\perp) e^{i[(\omega-\omega_0)(T+\beta_1 z)-(\beta(\omega)-\beta(\omega_0))z]}$$

$$= \frac{1}{\sqrt{2\pi}} \sum_m \int d\omega \hat{A}(z, \omega) \mathbf{e}(\mathbf{r}_\perp) e^{i[(\omega-\omega_0)T - \tilde{\beta}(\omega)z]}, \qquad (5.44)$$

$$\hat{A}(z, \omega) = A(z, \omega) e^{i[\omega_0 t - \beta(\omega_0)z]}; \quad \tilde{\beta}(\omega) = \beta(\omega) - \beta(\omega_0) - \beta_1(\omega_0)(\omega - \omega_0). \qquad (5.45)$$

It may be verified that the derivations of a nonlinear propagation equation can be carried through with β substituted by $\tilde{\beta}$ and A substituted by $\hat{A}$. In the following, this substitution will be implicit, that is we will continue to write A and β although they are really replaced by $\hat{A}$ and $\tilde{\beta}$.

5.4 NONLINEAR SCHRÖDINGER EQUATION

In the previous section, a fairly general nonlinear propagation equation was derived. In this section, an approximation to this equation is derived and used to illustrate the most important features of nonlinear propagation in single-mode waveguides. Historically, this equation has been known as the nonlinear Schrödinger, or NLS, equation. The more general forms of nonlinear propagation equations, such as Equation (5.42), are correspondingly known as *generalized nonlinear Schrödinger* (GNLS) equations. In the following, it will be implicitly understood that frequency integrals run over the positive half-axis only.

5.4.1 DERIVATION OF THE NLS EQUATION

Starting out from Equation (5.42), we may arrive at the NLS equation by a series of approximations, which in many cases are reasonable, but in some cases will fail. These are as follows:

1. Second-order dispersion only: The propagation constant $\beta(\omega)$ in Equation (5.45) can be written as a Taylor expansion around the central frequency ω_0:

$$\beta(\omega) = \beta(\omega_0) + \beta_1(\omega_0)(\omega - \omega_0)$$

$$+ \frac{1}{2}\beta_2(\omega_0)(\omega - \omega_0)^2 + \cdots; \quad \beta_n \equiv \frac{d^n\beta}{d\omega^n}. \qquad (5.46)$$

   Going into the moving reference frame, as described earlier, the first two terms fall away. In the NLS equation, all terms higher than the second order are neglected, so only the β_2 term will enter the equation.
2. Neglect of delayed response: We approximate the delayed (Raman) response as an instantaneous response, leaving out the response function and putting the f_R parameter to zero. As discussed in the previous section, this may be a

valid approximation for pulses much longer than the Raman response time. Typically, the approximation is considered justified for pulses longer than a picosecond or so. Note, however, that the Raman effects may also show up in longer pulses or even CW waves. This is because infinitesimal noise fluctuations can be amplified by the Raman effect and the pulse eventually acquires a chaotic substructure on the few-fs level. This amplification process happens after a certain length of waveguide, which is dependent on the pulse power, so the NLS equation is only justified for lengths shorter than that.

3. Neglect of *self-steepening*: Finally, we replace the prefactor ω on the RHS of Equation (5.42) with ω_0. This leads to the neglect of a self-steepening effect which will be discussed later and in the exercises.

The propagation equation now looks like this:

$$\partial_z A(z, \omega) = -\frac{i\omega_0 n_2 e^{i\beta(\omega)z}}{cA_{eff}\sqrt{2\pi}} \int d\omega_1 d\omega_2 \tilde{A}(z, \omega_1)\tilde{A}^*(z, \omega_2)\tilde{A}(z, \omega - \omega_1 + \omega_2).$$

(5.47)

We may transform this to an equation for $\tilde{A}(z, \omega)$ which reads

$$\partial_z \tilde{A}(z, \omega) = -\frac{i}{2}\beta_2(\omega - \omega_0)^2$$

$$- \frac{i\omega_0 n_2}{A_{eff}\sqrt{2\pi}} \int d\omega_1 d\omega_2 \tilde{A}(z, \omega_1)\tilde{A}^*(z, \omega_2)\tilde{A}(z, \omega - \omega_1 + \omega_2), \quad (5.48)$$

where it is from now on implicit that β_2 is evaluated at ω_0. Finally, we can Fourier transform this equation into the time domain, using the shifted frequency $\omega - \omega_0$ as an integration variable. This replaces the term $(\omega - \omega_0)^2$ with a double time derivative (note that this is in the comoving reference frame, as discussed in the previous section). Furthermore, the convolution integrals in frequency become direct products in the time domain, as discussed in Section 2.4. The NLS equation in the time domain then becomes

$$\partial_z \tilde{A}(z, t) = \frac{i}{2}\beta_2 \partial_t^2 \tilde{A}(z, t) - i\gamma \tilde{A}(z, t) \mid \tilde{A}(z, t) \mid^2, \quad (5.49)$$

$$\gamma \equiv \frac{\omega_0 n_2}{A_{eff}}. \quad (5.50)$$

The quantity γ is commonly called the *nonlinear coefficient* and is measured in $(Wm)^{-1}$. Its magnitude is seen to control the magnitude of the only nonlinear term in the equation.

Note: From a numerical point of view, there is not much reason to make these approximations. Solving the full propagation equation (5.42) on the computer is usually only slightly more involved than solving Equation (5.49). However, the NLS equation lends itself much better to formal analysis and allows several analytic solutions which illustrate basic phenomena of nonlinear optics. This will be the subject of the following sections.

5.4.2   DISPERSION AND SELF-PHASE MODULATION

Consider first the limit of a weak pulse, where the nonlinear term in Equation (5.49) can be neglected. We are left with the equation

$$\partial_z \tilde{A}(z, t) = \frac{i}{2} \beta_2 \partial_t^2 \tilde{A}(z, t). \qquad (5.51)$$

Let us consider a Gaussian pulse given by the expression

$$\tilde{A}(z, t) = A_0(z) e^{-a(z) t^2}. \qquad (5.52)$$

It is then straightforward to show that Equation (5.51) is solved by

$$A_0(z) = \frac{A_0(0) t_0}{\sqrt{t_0^2 + i \beta_2 z}}; \quad a(z) = \frac{1}{2(t_0^2 + i \beta_2 z)}. \qquad (5.53)$$

where t_0 is a parameter controlling the initial temporal width of the pulse. The temporal power profile of the pulse is then given by

$$| \tilde{A}(z, t) |^2 = \frac{| A_0(0) |^2 t_0^2}{\sqrt{t_0^4 + (\beta_2 z)^2}} e^{-\frac{t^2}{t_0^2(z)}}; \quad t_0(z) = \sqrt{t_0^2 + \left(\frac{\beta_2 z}{t_0} \right)^2}. \qquad (5.54)$$

Clearly, the temporal width of the pulse increases with z, and for $\beta_2 z \gg t_0^2$, the increase is approximately linearly proportional to $| \beta_2 | z$. The parameter β_2 is therefore called the dispersive parameter, and the first term on the RHS of Equation (5.49) is denoted the dispersive term. In the spectral domain, it is possible to write a more general solution of the propagation equation without the nonlinear part:

$$\tilde{A}(z, \omega) = e^{-i\beta(\omega) z} \tilde{A}(0, \omega). \qquad (5.55)$$

which shows that the amplitudes of the spectral components are unchanged, that is the pulse spectrum remains constant, as we expect for linear propagation.

The β_2 parameter can be interpreted as expressing the frequency variation of the inverse group velocity, v_g, since

$$\frac{1}{v_g(\omega)} = \beta_1(\omega) = \beta_1(\omega_0) + \beta_2(\omega_0)(\omega - \omega_0) + \cdots \qquad (5.56)$$

If β_2 is positive, v_g decreases with frequency, so dispersion will move long wavelengths to the front of the pulse and short wavelengths to the back. The opposite will happen for a negative β_2, but in both cases, pulse broadening will be seen if the starting point is a *transform-limited* pulse, that is one for which all frequency components have the same phase. If the propagation distance z is large, we end up with a highly *chirped* pulse, where the different wavelengths in the spectrum are spread out in time. This is illustrated schematically in Figure 5.3a through c.

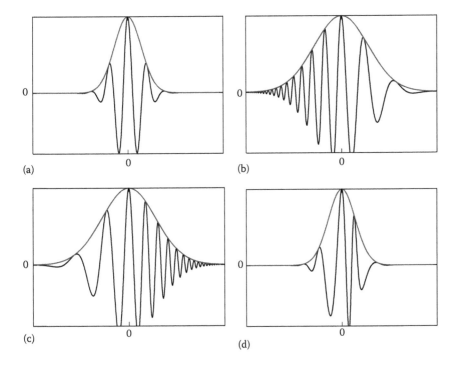

(a)

(b)

(c)

(d)

FIGURE 5.3 **(See colour insert.)** (a) Schematic of an unchirped few-cycle pulse with a Gaussian envelope. (b) The pulse is broadened after passage through a waveguide with negative β_2 – short wavelengths travel faster. (c) The pulse is broadened in a waveguide with positive β_2 – long wavelengths are faster. (d) The pulse is spectrally broadened by self-phase modulation (SPM), but the temporal profile is unchanged.

In the opposite case of a very strong pulse, we may neglect the dispersive term and consider the nonlinear equation

$$\partial_z \tilde{A}(z,t) = -i\gamma \tilde{A}(z,t) \mid \tilde{A}(z,t) \mid^2 . \tag{5.57}$$

Equation (5.57) implies that $\mid \tilde{A}(z,t) \mid^2$ will be invariant as is readily shown:

$$\partial_z \mid \tilde{A}(z,t) \mid^2 = \tilde{A}^*(z,t)\partial_z \tilde{A}(z,t) + \tilde{A}(z,t)\partial_z \tilde{A}^*(z,t)$$

$$= -\gamma \left[\tilde{A}^*(z,t)i\tilde{A}(z,t) - \tilde{A}(z,t)i\tilde{A}^*(z,t) \right] \mid \tilde{A}(z,t) \mid^2 = 0. \tag{5.58}$$

This leads to the straightforward time-domain solution

$$\tilde{A}(z,t) = \tilde{A}(0,t)e^{-i\gamma \tilde{A}(0,t)|^2 z}. \tag{5.59}$$

The invariance of $\mid \tilde{A}(z,t) \mid^2$ implies that the temporal power profile is constant. Only the phase of the temporal profile is changed. For this reason, the effect of a pure Kerr nonlinearity on a pulse in the absence of dispersion and *self-steepening* (see later) is denoted SPM. After a Fourier transform of $\tilde{A}$ given by Equation (5.59), one finds that the spectral amplitudes are *not* constant, that is SPM modifies the spectrum of the pulse. This shows an interesting *duality* between the dispersive and nonlinear terms in the NLS equation: the dispersive term only changes the temporal profile of the pulse, whereas the nonlinear term only changes its spectrum. However, when both terms are present, complicated dynamical phenomena are found to emerge.

As with the dispersive term, the action of SPM also leads to a chirped pulse. We can quantify the chirp by introducing an *instantaneous frequency* $\omega(t)$, given by

$$\omega(t) = \partial_t \Phi(z,t) + \omega_0; \quad \tilde{A}(z,t) \equiv |\tilde{A}(z,t)| \, e^{i\Phi(z,t)}. \tag{5.60}$$

The rationale is that the frequency of a light wave expresses the rate of variation of its phase. The frequency ω_0 must be added because we separated it out in the derivation of the NLS equation. With this definition, it is easy to evaluate the instantaneous frequency of an SPM-chirped pulse:

$$\omega(t) = \omega_0 - \gamma \partial_t \, |\tilde{A}(z,t)|^2 \, z. \tag{5.61}$$

On the front edge of the pulse, the power (and hence $|\tilde{A}(z,t)|^2$) is rising, and SPM gives a negative contribution to $\omega(t)$, whereas on the back edge, the power is decreasing and SPM leads to an increase in frequency. The result is a chirp as sketched in Figure 5.3d. On the front edge of the pulse, the wavelength is longer where the pulse profile is steep, and on the back edge, it is correspondingly shorter. At the centre of the pulse, and in its tails where the power is low, there is little change in the wavelength. Note finally that if we put $\gamma = 0$ so that no SPM is present, we have $\omega(t) = \omega_0$ everywhere, even though a short pulse will have a wide spectrum. Clearly, the concept of instantaneous frequency is only meaningful for highly chirped pulses.

In the tails of the pulse, the power and its derivative eventually decay, and the frequency shift in these parts therefore becomes negligible. So, if we start from the centre of the pulse and go towards its leading edge, we will first see the instantaneous frequency decrease, then increase again to get back to the base frequency. It follows that a given frequency in the spectrum of the SPM-broadened pulse will be generated at least in two different places. When Fourier transforming the temporal pulse to get the spectral function, this will give rise to interference effects, leading to an oscillating structure of the spectra. You will see examples of this in the numerical exercises.

5.4.3 OPTICAL SOLITONS

In the previous section, we showed that SPM effects alone could broaden the spectrum of an optical pulse. This implies that the pulse duration can be shortened if all frequency components are brought into phase. This can in fact be achieved by means of dispersion. Looking again at Figure 5.3, one realizes that the chirp imposed by SPM can be undone by sending the pulse through a waveguide with negative β_2, so that the short wavelengths in the pulse catch up with the long ones. Such compression schemes are indeed being used in ultrafast fibre lasers.

What happens if negative β_2 and SPM are present at the same time? In that case, SPM will continuously decrease the wavelength on the leading edge of the pulse and increase it on the trailing edge, while at the same time, dispersive effects will shift the long wavelengths towards the rear of the pulse and the short wavelengths towards the front. If the pulse shape and power are just right, it is possible to balance these two effects and obtain a pulse which propagates with a constant temporal and spectral shape, even in the presence of dispersion and nonlinearity [22]. Such a pulse is called an *optical soliton*, and its mathematical form is given by

$$\tilde{A}(z,t) = \sqrt{P_s}\,\text{sech}\left(\frac{t}{t_s}\right)e^{iz/(2z_s)}; \quad P_s = \frac{|\beta_2|}{\gamma t_s^2}; \quad z_s = \frac{1}{\gamma P_s}. \tag{5.62}$$

Here, t_s could be taken as a free parameter, which will then determine the peak power P_s, or vice versa. The soliton energy, E_s, is given by $2P_s t_s$, which allows us to write

$$t_s = \frac{2|\beta_2|}{\gamma E_s} \tag{5.63}$$

showing that the soliton duration scales inversely with pulse energy. During propagation, there is an overall phase change with z, but the temporal and spectral intensity distributions remain constant during propagation. In fact, an infinite family of soliton solutions exist, most of which show a periodic evolution of the intensity distribution upon propagation. For this reason, the pulse described by Equation (5.62) is often denoted the *fundamental soliton*. Since the hyperbolic secant is its own Fourier transform, the pulse can be written in the frequency domain as

$$\tilde{A}(z,\omega) = \frac{2\sqrt{P_s}}{\omega_s}\,\text{sech}\left(\frac{\omega}{\omega_s}\right)e^{iz/(2z_s)}; \quad \omega_s = \frac{2}{\pi t_s}. \tag{5.64}$$

The full width at half maximum (FWHM) of the temporal power profile, $|\tilde{A}(z,t)|^2$, is given by $t_F \approx 1.763 t_s$, and similarly, the spectral FWHM of $|\tilde{A}(z,t)|^2$ is given by $\omega_F \approx 1.763\omega_s = 3.526/\pi t_s$. This yields a *time-bandwidth product* for the soliton of

$$t_F \nu_F = t_F \frac{\omega_F}{2\pi} \approx \frac{2 \cdot 1.763^2}{2\pi^2} \approx 0.315. \tag{5.65}$$

For an unchirped pulse, the time-bandwidth product is only dependent on the pulse shape and gives a useful relation between its temporal and spectral widths.

5.4.4 SOLITONS AND RAMAN EFFECTS

It turns out that the fundamental soliton is quite stable against perturbations, such as the terms neglected in deriving the NLS equation from Equation (5.42). Therefore, fundamental solitons are easily observed in experiments and are found to propagate stably over long distances, in spite of the approximate nature of the NLS equation.

On the other hand, higher-order solitons will often break up into one or more fundamental solitons under the influence of, for example Raman effects.

Although Raman scattering does not cause fundamental solitons to break up, it does influence them in a rather interesting way. The inelastic scatterings with the lattice of the waveguide material lead to a continuous downshift in frequency of the photons making up the soliton. As a result, it is found to maintain its shape while being redshifted to longer wavelengths. If the β_2 parameter changes with wavelength (due to higher-order terms in the dispersion), the soliton will be able to adjust its shape provided that the shift is not too fast. This effect is commonly denoted the *soliton self-frequency shift*. A perturbative treatment reveals that the redshift rate is given by [23]

$$\frac{d\omega_s}{dz} = -\gamma P_s f_R \int\limits_{-\infty}^{\infty} dx \operatorname{sech}^2(x)\tanh(x) \int\limits_{0}^{\infty} dy R(yt_s)\operatorname{sech}^2(x-y) \tag{5.66}$$

In the limit where the soliton duration t_s is large compared with the Raman response time, this can be approximated as

$$\frac{d\omega_s}{dz} \approx -\frac{8}{15}\gamma f_R \frac{P_s}{t_s^2} t_R = -\frac{8}{15} \frac{|\beta_2| f_R t_R}{t_s^4}; \quad t_R = \int\limits_{0}^{\infty} dt\, t\, R(t) \tag{5.67}$$

This formula shows that the rate of the frequency shift is strongly dependent on the temporal width of the soliton.

The form of the Raman response function is highly material dependent. While crystalline materials can have very narrow Raman resonances, amorphous materials typically have a broader range of phonon frequencies contributing to the delayed response. In fibre optics, amorphous silica glass is the most important base material. A simple and widely used parametrization for the response function of silica is [2]

$$R(t) = \frac{\tau_1^2 + \tau_2^2}{\tau_1 \tau_2^2} \exp\left[\frac{t}{\tau_2}\right] \sin\left(\frac{t}{\tau_1}\right)\theta(t); \quad \tau_1 = 12.2\text{fs} \quad \tau_2 = 32\text{fs} \tag{5.68}$$

where $\theta(t)$ is the Heaviside step function, which is 0 for $t < 0$ and 1 otherwise. Here, τ_1 represents an inverse average phonon frequency, whereas τ_2 is a characteristic damping time of the phonon. With this response function, the time parameter t_R controlling the soliton self-frequency shift becomes

$$t_R = \int\limits_{0}^{\infty} dt\, t\, R(t) = 2\frac{\tau_1^2 \tau_2}{\tau_1^2 + \tau_2^2} \approx 8.122\text{ fs} \tag{5.69}$$

5.4.5 SELF-STEEPENING

Let us return to Equation (5.48) and reintroduce ω instead of ω_0 in the prefactor of the nonlinear term. The resulting equation can be written as

$$\partial_z \tilde{A}(z, \omega) = -\frac{i}{2}\beta_2(\omega - \omega_0)^2 - \frac{i}{\sqrt{2\pi}}\left[\gamma + \frac{(\omega - \omega_0)n_2}{cA_{eff}}\right]$$

$$\times \int d\omega_1 d\omega_2 \tilde{A}(z, \omega_1)\tilde{A}^*(z, \omega_2)\tilde{A}(z, \omega - \omega_1 + \omega_2). \tag{5.70}$$

Relaxing the approximation $\omega \approx \omega_0$ is seen to add an extra term to the NLS equation, proportional to $\omega - \omega_0$. Now, if we had not made the transformation to a comoving frame by introducing $\tilde{\beta}$ defined in Equation (5.45), we would have had a term of $\beta_1(\omega - \omega_0)$ appearing alongside the β_2 term, and it would determine the propagation velocity of the light pulse in the laboratory frame, as the group velocity $v_g = 1/\beta_1$. Since the instantaneous (Kerr) nonlinearity of the waveguide can be understood as an intensity-dependent modification of its refractive index, we can expect there will be an intensity-dependent correction to the group velocity as well. The last term in Equation (5.71) is actually an expression of this fact.

Consider a pulse having a Gaussian, or similar, shape, with a central intensity peak decaying smoothly as we move to earlier or later times. The Kerr nonlinearity slightly raises the refractive index in the centre of the pulse, and the group velocity correspondingly becomes lower. The light in the pulse centre is delayed relative to the light in the pulse wings and is therefore pushed towards the trailing edge of the pulse. As a result, this edge becomes steeper, whereas the leading edge becomes smoother [24,25]. So, the temporal shape of the pulse gets modified, even if we put β_2 (and higher-order dispersion terms) to zero. In that case, it may be shown that the steepness will become infinite after a finite propagation distance, an effect which is called an *optical shock front*. Since the SPM-induced spectral broadening scales with the steepness of the pulse edges, a strong spectral broadening to the short-wavelength side is seen as the shock front is approached. In a real waveguide, dispersive effects will distort the pulse some time before the limit of infinite steepness is reached, a process sometimes described as *wave breaking*.

5.4.6 CONSERVATION LAWS

The existence of conserved quantities in differential equations is important for analysing their behaviour and useful for validating numerical propagation schemes. If certain conservation laws are expected on physical grounds, it is also interesting to investigate whether these laws are reflected in approximate mathematical equations describing the physics, such as the GNLS equations derived here.

Physically, we can define the so-called *photon number* of a pulse by

$$N_{ph} = \int d\omega \frac{|A(z, \omega)|^2}{\hbar\omega}. \tag{5.71}$$

Since $|A(z, \omega)|^2$ integrates to the total pulse energy, one can think of this expression as summing the energy over different frequencies, while dividing by the energy of a single photon, $\hbar\omega$. Therefore, N_{ph} should be equal to the physical number of photons. Of course, one can leave out the constant $\hbar$ and retain a conserved quantity.

The GNLS in Equation (5.42) can be shown to conserve the photon number in the following way: we start out by considering the z-derivate of N_{ph}, which becomes

$$\partial_z N_{ph} = \int d\omega \frac{A^*(z,\omega)\partial_z A(z,\omega) + c.c.}{\omega} = -\int d\omega \frac{in_2 A^*(z,\omega)e^{i\beta(\omega)z}}{cA_{eff}\sqrt{2\pi}}$$

$$\times \int d\omega_1 d\omega_2 \tilde{A}(z,\omega_1)\tilde{A}^*(z,\omega_2)\tilde{A}(z,\omega-\omega_1+\omega_2)\tilde{R}(\omega-\omega_1) + c.c.$$

$$= -\frac{in_2}{cA_{eff}\sqrt{2\pi}} \int d\omega d\omega_1 d\omega_2 \tilde{A}^*(z,\omega)\tilde{A}(z,\omega_1)\tilde{A}^*(z,\omega_2)$$

$$\times \tilde{A}(z,\omega-\omega_1+\omega_2)\tilde{R}(\omega-\omega_1)$$

$$+ \frac{in_2}{cA_{eff}\sqrt{2\pi}} \int d\omega d\omega_1 d\omega_2 \tilde{A}(z,\omega)\tilde{A}^*(z,\omega_1)\tilde{A}(z,\omega_2)$$

$$\times \tilde{A}^*(z,\omega-\omega_1+\omega_2)\tilde{R}(\omega_1-\omega), \tag{5.72}$$

where $\tilde{R}(\omega-\omega_1) = (1-f_R) + f_R R(\omega-\omega_1)$ was used, along with the fact that $\tilde{R}(-\omega) = \tilde{R}^*(\omega)$ since $\tilde{R}(t)$ is a real function.

The two terms in the last equality cancel each other, which may be seen by changing the integration variables in the second term to $\tilde{\omega} = \omega_1$, $\tilde{\omega}_1 = \omega$, and $\tilde{\omega}_2 = \omega_2 - \tilde{\omega} + \tilde{\omega}_1$. Thus, the z-derivative of N_{ph} vanishes, meaning that this quantity is conserved during propagation.

The nonlinear processes described by Equation (5.42) include elastic photon–photon scattering and inelastic photon–phonon (Raman) scattering. These processes should conserve the photon number, but not the pulse energy, because energy is lost to the material during the Raman scattering. Therefore, the conservation of N_{ph} as defined in Equation (5.71) reflects that the derived propagation equations give a good representation of the underlying physics. Note, however, that the result depends on the restriction of frequency integration to the positive half-axis, which was used to derive Equations (5.37) and (5.39). Without this restriction, the formalism includes the effect of third-harmonic generation, which conserves energy, but not photon number (three photons are converted into one photon at three times the frequency). Mathematically, the definition of photon number in Equation (5.71) would have to be modified if the frequency integrations cover both positive and negative frequencies, since in this case, Equation (5.71) would just give zero. Thus, the most general propagation equation following from the derivations in Section 5.2 conserves neither energy nor photon number. Usually, the neglect of third-harmonic generation is an acceptable approximation though.

If the ω-prefactor in Equation (5.42) is approximated by ω_0, that is if self-steepening is omitted, it is easily seen that this argument can be carried through for the pulse energy $E_p = \int d\omega |A(z,\omega)|^2$ instead of the photon number. So without self-steepening, the propagation equation conserves pulse energy rather than photon number, even in the presence of the Raman effects. This is an unphysical result and shows that one should be wary of using an equation including the Raman but not self-steepening effects.

5.5 NUMERICAL IMPLEMENTATION

5.5.1 FOURIER METHOD

The propagation equation (5.42) is formulated for the amplitudes $A(z, \omega)$ in the frequency domain and involves a double integral of the form

$$\int d\omega_1 \tilde{A}(z, \omega_1) R(\omega - \omega_1) \int d\omega_2 \tilde{A}^*(z, \omega_2) \tilde{A}(z, \omega - \omega_1 + \omega_2). \tag{5.73}$$

If this integral can be evaluated for a given function $\tilde{A}(z, \omega)$, the first-order derivative of this function is given by Equation (5.42), and the equation can be solved by standard numerical methods for ordinary differential equations (ODEs) such as the Runge–Kutta method. If the double integral is done by direct summation on a frequency grid with N points, its evaluation will be an N^3 operation, because the 2D integration must be done at all values of ω.

In fact, the integral can be evaluated much faster by taking advantage of the numerical FFT and the convolution theorem which states that

$$\int d\omega_1 F(\omega - \omega_1) G(\omega_1) = \frac{1}{\sqrt{2\pi}} \int dt e^{-i\omega t} F(t) G(t); \quad F(t) = \frac{1}{\sqrt{2\pi}} \int d\omega e^{i\omega t} F(\omega) \tag{5.74}$$

or, in other words, the convolution of two functions in the frequency domain is given by the Fourier transform of their direct product in the time domain. Inspection of Equation (5.73) reveals that it can be regarded as two successive convolutions: the ω_2 integral convolves $\tilde{A}^*(z, \omega_2)$ with $\tilde{A}(z, \omega - \omega_1 + \omega_2)$ to produce a function of $\omega - \omega_1$. This is then multiplied by $R(\omega - \omega_1)$, and the ω_1 integral represents a convolution of the resulting function with $\tilde{A}(z, \omega_1)$. This implies that Equation (5.73) can be evaluated by means of FFTs with a time complexity of $N\log(N)$ – a vast improvement over the N^3 complexity of the direct-summation approach. This huge gain in computational efficiency means that practically all numerical schemes for solving nonlinear propagation equations in waveguides are based on the Fourier transform methods.

The numerical procedure for evaluating the derivative of $A(z, \omega)$ can then be summarized as follows:

Given an input $A(z, \omega)$, calculate $\tilde{A}(z, t)$ using

$$\tilde{A}(z, \omega) = e^{-i\beta(\omega)z} A(z, \omega); \quad \tilde{A}(z, t) = \frac{1}{\sqrt{2\pi}} \int d\omega e^{i\omega t} \tilde{A}(z, \omega). \tag{5.75}$$

Evaluate the function $F(\omega)$ given by

$$F(\omega) = \frac{1}{\sqrt{2\pi}} \int dt e^{-i\omega t} |\tilde{A}(z, t)|^2 \tag{5.76}$$

(this is the result of the ω_2 integral in Equation 5.73).

Multiply by the Raman response function and go back into the time domain:

$$G(\omega) = [(1 - f_R) + f_R R(\omega)] F(\omega); \quad G(t) = \frac{1}{\sqrt{2\pi}} \int d\omega e^{i\omega t} G(\omega). \quad (5.77)$$

Multiply by $\tilde{A}(z, t)$ and transform back into the frequency domain:

$$I(t) = G(t)\tilde{A}(z, t); \quad I(\omega) = \frac{1}{\sqrt{2\pi}} \int dt e^{-i\omega t} I(t). \quad (5.78)$$

Evaluate the $A(z, \omega)$ derivative as

$$\partial_z A(z, \omega) = -\frac{i\omega n_2 e^{i\beta(\omega)z}}{cA_{eff}\sqrt{2\pi}} I(\omega). \quad (5.79)$$

This procedure involves a total of four FFT operations. Since $R(t)$ and $|\tilde{A}(z, t)|^2$ are real functions, $G(t)$ will also be real, and we will have $F(\omega) = F^*(-\omega)$ and similarly for $G(\omega)$. This can be used to perform the FFTs in Equations (5.76) and (5.77) more efficiently – we say that these are *real FFTs*, whereas the FFT operations in Equations (5.75) and (5.78) are *complex FFTs* since they involve functions which are complex in both time and frequency domains. In MATLAB, a real FFT is evaluated twice as fast as a complex FFT of the same size.

If the nonlinear response is instantaneous ($f_R = 0$), the procedure can be simplified. In that case, one can replace Equation (5.76) with

$$\tilde{F}(\omega) = \frac{1}{\sqrt{2\pi}} \int dt e^{-i\omega t} \tilde{A}(z, t) |\tilde{A}(z, t)|^2, \quad (5.80)$$

and evaluate the derivative of $A(z, \omega)$ as

$$\partial_z A(z, \omega) = -\frac{i\omega n_2 e^{i\beta(\omega)z}}{cA_{eff}\sqrt{2\pi}} \tilde{F}(\omega), \quad (5.81)$$

thus saving the two real FFTs (note that $\tilde{F}$ is complex in both time and frequency domains). So, the nonlinear propagation has less numerical complexity in the absence of delayed response, but the difference is less than a factor of two because real FFTs are faster than complex FFTs.

5.5.2 STEPPING TECHNIQUES

Equation (5.42) is a first-order ODE and can as such be propagated using conventional solvers for such problems. A common variant is the fourth-order Runge–Kutta method (RK4). The advantage of this method is that it is simple to programme, has a fairly high-order accuracy, and allows for an easy implementation of adaptive stepsize, which is usually a requirement for efficient ODE solution. More advanced ODE methods exist, but RK4 is usually found to perform quite well for solving the kind of problems that tend to occur in nonlinear waveguide modelling. In MATLAB, the ODE method is implemented as a built-in function, and we will therefore not go into a discussion of the mathematics and coding of the method.

Another solution method which deserves mentioning is the so-called *split-step* method, which has traditionally been widely used in the nonlinear-optics community. This method is particularly simple and efficient for the NLS equation. Its basis is the existence of analytical solutions to Equation (5.49) in the two cases where either the dispersive term or the nonlinear term is omitted. These analytical solutions were discussed in Section 5.4.2. The idea of the split-step method is to apply the dispersive and nonlinear operators separately, making use of these analytical solutions.

The general solution to the dispersive part of the NLS equation can be written explicitly in the frequency domain, as was done in Equation (5.55). Here, dispersive propagation is simply described by multiplication of a phase factor. On the other hand, the solution for the nonlinear (SPM) part, given by Equation (5.59), is conveniently expressed in the time domain, again as multiplication by a phase factor. In order to apply these two solutions in turn, one must therefore transform between the time and frequency domains by means of the FFT.

A simple version of the split-step scheme is the procedure

$$\tilde{A}(z + \Delta z, \omega) = \hat{F} e^{\Delta z \hat{N}} \hat{F}^{-1} e^{\Delta z \hat{D}} \tilde{A}(z, \omega), \tag{5.82}$$

where $\hat{F}$ denotes a Fourier transform, whereas $\hat{D}$ and $\hat{N}$ are the dispersion and nonlinear operators, respectively:

$$\hat{D}\tilde{A}(z, \omega) = -i\beta(\omega)\tilde{A}(z, \omega); \quad \hat{N}\tilde{A}(z, t) = -i\gamma\tilde{A}(z, t)\,|\tilde{A}(z, t)|^2\,. \tag{5.83}$$

The analytical solutions in Equations (5.55) and (5.59) then tell us that we can write the split-step expression as

$$\tilde{A}(z + \Delta z, \omega) = \hat{F}\tilde{C}(z, t)e^{-i\gamma|\tilde{C}(z,t)|^2\Delta z}; \quad \tilde{C}(z, t) = \hat{F}^{-1}e^{i\beta(\omega)\Delta z}\tilde{A}(z, \omega). \tag{5.84}$$

It may be shown that the leading error term in this stepping scheme is of order Δz^2, which compares poorly to the Δz^5 error of the RK4 approach. On the other hand, the split-step procedure only requires two FFT operations, as opposed to eight for a successful RK4 step. The accuracy of the split-step method may be enhanced to a Δz^3 error by modifying the procedure to

$$\tilde{A}(z + \Delta z, \omega) = e^{\frac{\Delta z}{2}\hat{D}}\hat{F}e^{\int_z^{z+\Delta z}\hat{N}dz}\hat{F}^{-1}e^{\frac{\Delta z}{2}\hat{D}}\tilde{A}(z, \omega). \tag{5.85}$$

For the NLS equation, this is a straightforward modification, because the temporal power profile remains unchanged when $\hat{N}$ is applied. However, the estimates of numerical accuracy also hold for the more general nonlinear operator which includes the Raman and self-steepening effects. For such a case, one in practice needs an ODE scheme to evaluate the exponential of $\hat{N}$. The RK4 method is a common choice, and indeed, it is possible to increase the accuracy of the split-step scheme itself to the same order as RK4 by using a modified stepping procedure. However, compared to the direct use of the RK4 method on Equation (5.42), this approach is more complicated in its implementation, particularly when adaptive stepsize is desired. We will therefore not consider the split-step method further in this book.

5.5.3 DISCRETE FOURIER GRIDS

For a numerical implementation of the nonlinear propagation equations, we must replace the integrals of continuous functions by summations over a finite number of points on a time or frequency grid. This is necessary because we can only store and manipulate a finite number of variables in the computer. In this section, we recapitulate the basic mathematics of discrete Fourier grids and transforms.

As a preliminary, we consider the frequency range of the functions $A(z, \omega)$ and $R(\omega - \omega_1)$ in Equation (5.73). The Raman response function has components at both positive and negative frequencies, including a strong magnitude around zero frequency. Regarding the modal amplitudes, one will typically wish to represent them on a frequency grid containing positive frequencies over some range which is physically reasonable. For a silica-based optical fibre, excessive optical losses are present at wavelengths above 2.3 μm and below ~0.3 μm (these values are somewhat dependent on what one exactly understands by an *excessive* loss). In most nonlinear experiments (with supercontinuum generation as an important exception [26]), the signal bandwidth is in fact much lower than these limits, and if one has a reasonable idea of the bandwidths that can be expected to arise, the frequency grid can be restricted accordingly, in order to save computational time. Thus, from a physical point of view, the modal amplitudes should be defined in a range of positive frequencies, typically on the order of hundreds of THz. Mathematically, however, we may shift the frequency coordinates of the integrals in Equation (5.73) by any constant to write

$$\int d\omega_1 \tilde{A}(z, \omega_1) R(\omega - \omega_1) \int d\omega_2 \tilde{A}^*(z, \omega_2) \tilde{A}(z, \omega - \omega_1 + \omega_2)$$

$$= \int d\Omega_1 \tilde{A}(z, \Omega_1 + \omega_c) R(\Omega - \Omega_1) \int d\Omega_2 \tilde{A}^*(z, \Omega_2 + \omega_c) \tilde{A}(z, \Omega - \Omega_1 + \Omega_2 + \omega_c)$$

$$\times \Omega_i = \omega_i - \omega_c; \quad \Omega = \omega - \omega_c. \tag{5.86}$$

The frequency ω_c can conveniently be chosen as, for example the lowest frequency on which $\tilde{A}(z, \omega)$ is expanded or the centre frequency of the interval where $\tilde{A}(z, \omega)$ is assumed finite. Thus, we can consider to do our Fourier transformations on a grid starting at, or centred at, zero frequency. In the following, the frequency shift ω_c will be suppressed from the notation, that is we will write

$$\tilde{A}(z, \Omega + \omega_c) \to \tilde{A}(z, \omega); \quad \Omega \to \omega, \tag{5.87}$$

etc.

We now consider a function $f(t)$ which is periodic in time with period T, so that $f(t + T) = f(t)$ for all t. The basis of the Fourier analysis is that such a function can be expanded in complex exponential functions with this periodicity, that is as

$$f(t) = \sum_{m=1}^{\infty} f_m e^{i\omega_m t}; \quad \omega_m = (m-1)\frac{2\pi}{T}. \tag{5.88}$$

The frequency summation starts with $\omega = 0$, which gives the average value of f, since all exponentials with $m > 1$ average to zero over a full time period. The summation in principle runs to infinitely high frequencies. In practice, we must restrict ourselves to consider a finite range of frequencies, which is also physically justified, as discussed earlier. So the expansion actually used would look like this:

$$f(t) = \sum_{m=1}^{N} f_m e^{i\omega_m t}; \quad \omega_m = (m-1)\frac{2\pi}{T} \equiv m\Delta\omega, \tag{5.89}$$

where the total number of frequencies ω_m, including $\omega = 0$, is N. The use of a finite frequency cut-off implies a finite resolution of the temporal grid on which $f(t)$ is represented. This can be seen from the relations

$$\Delta t = \frac{2\pi}{\omega_{N+1}} = \frac{T}{N} \Rightarrow \omega_N \Delta t = 2\pi \Rightarrow e^{i\omega_{N+1}\Delta t} = 1. \tag{5.90}$$

So, on the temporal grid given by Equation (5.90), the frequency ω_{N+1}, which we left out of the expansion, is indistinguishable from $\omega = 0$. However, if we would use a temporal grid with a shorter value of Δt, we would need ω_{N+1}, and possibly higher values of ω as well, to expand an arbitrary function which could be expressed on such a grid. We see that the number of gridpoints is the same in the time and frequency domains, which was to be expected, because the Fourier transform is invertible, and so the total amount of information in $f(t)$ and $f(\omega)$ should be the same.

Instead of considering the frequencies to run from zero to $(N-1)\Delta\omega$, we could equally well think of them as running from $-((N/2)-1)\Delta\omega$ to $(N/2)\Delta\omega$. This can be seen from the relation

$$e^{i\omega_m t_n} = e^{i(\omega_m t_n - 2(n-1)\pi)} = e^{i(\omega_m t_n - \omega_{N+1}t_n)} = e^{i(\omega_m - \omega_{N+1})t_n}; \quad t_n = (n-1)\Delta t. \tag{5.91}$$

This implies that any frequency on the grid can be shifted by ω_{N+1}, or an integer multiple of ω_{N+1}, without affecting the expansion. We may choose to shift all frequencies higher than $(N/2)\Delta\omega$ down by ω_{N+1}, which leads to the frequency range indicated earlier. The ordering of different frequency ranges on the grid used by the MATLAB FFT routines is illustrated in Figure 5.4.

To evaluate the RHS of Equation (5.42) by the FFT procedure outlined in Section 5.5.1, we must represent the modal amplitudes $A(z, \omega)$ as well as the Raman response function on the Fourier frequency grid. For the Raman response function, one can either implement the time-domain form in Equation (5.68) and do a numerical Fourier transform or establish the frequency-domain form by analytic Fourier transformation. If the latter approach is pursued, one must keep in mind that the equivalence between positive and negative frequencies illustrated in Figure 5.4 does *not* hold for the

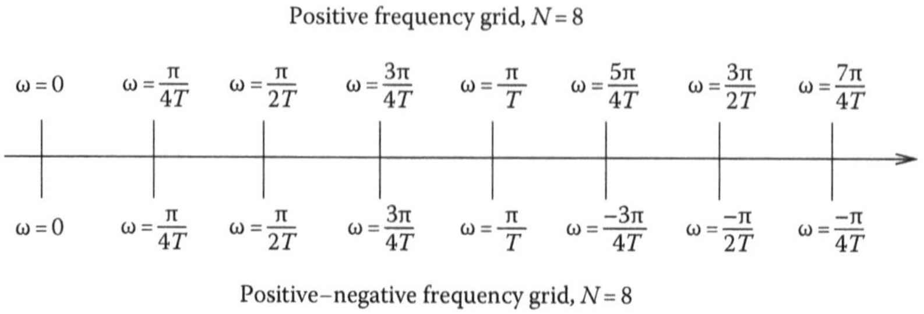

Positive frequency grid, $N = 8$

$\omega = 0$	$\omega = \dfrac{\pi}{4T}$	$\omega = \dfrac{\pi}{2T}$	$\omega = \dfrac{3\pi}{4T}$	$\omega = \dfrac{\pi}{T}$	$\omega = \dfrac{5\pi}{4T}$	$\omega = \dfrac{3\pi}{2T}$	$\omega = \dfrac{7\pi}{4T}$

$\omega = 0$	$\omega = \dfrac{\pi}{4T}$	$\omega = \dfrac{\pi}{2T}$	$\omega = \dfrac{3\pi}{4T}$	$\omega = \dfrac{\pi}{T}$	$\omega = \dfrac{-3\pi}{4T}$	$\omega = \dfrac{-\pi}{2T}$	$\omega = \dfrac{-\pi}{4T}$

Positive–negative frequency grid, $N = 8$

FIGURE 5.4 Illustration of the FFT grid in frequency, with the frequencies being either all positive or centred on $\omega = 0$.

Fourier transforms of continuous functions. It is necessary to insert the frequency values which are centred on $\omega = 0$ into the analytical result in order to capture the full low-frequency dynamics of the Raman response function.

5.5.4 IMPLEMENTATION IN MATLAB

To implement the GNLS equation in MATLAB, we may take advantage of the built-in functions for doing FFTs and solving ODEs. Specifically, we will use the *ode45* solver, along with the functions *fft* and *ifft* to perform the discrete direct and inverse Fourier transforms.

The relations between the Fourier integrals used so far and the discrete Fourier transforms provided by MATLAB can be seen by approximating the Fourier integral as a sum over gridpoints:

$$\frac{1}{\sqrt{2\pi}} \int dt\, e^{-i\omega t} F(t) \approx \frac{\Delta t}{\sqrt{2\pi}} \sum_n e^{-i\omega t_n} F(t_n) = \frac{\Delta t}{\sqrt{2\pi}} \sum_n e^{-i\omega(n-1)\Delta t} F(t_n), \quad (5.92)$$

where Δt is the spacing between points on the (uniform) temporal grid. Since the frequencies are also represented on a finite grid, given by $\omega_m = (m-1)\Delta \omega$, we need only evaluate the Fourier transforms for frequencies on the grid, so we have

$$\frac{1}{\sqrt{2\pi}} \int dt\, e^{-i\omega_m t} F(t) \approx \frac{\Delta t}{\sqrt{2\pi}} \sum_n e^{-i(m-1)\Delta\omega(n-1)\Delta t} F(t_n)$$

$$= \frac{\Delta t}{\sqrt{2\pi}} \sum_n e^{-i(n-1)(m-1)\Delta\omega\Delta t} F(t_n). \quad (5.93)$$

The discrete time and frequency grids are related by

$$\Delta t \Delta \omega = \frac{2\pi}{N}, \quad (5.94)$$

so we can finally write

$$\frac{1}{\sqrt{2\pi}} \int dt e^{-i\omega_m t} F(t) \approx \frac{\Delta t}{\sqrt{2\pi}} \sum_n e^{-i(n-1)(m-1)\frac{2\pi}{N}} F(t_n).$$ (5.95)

Except for the prefactor of $\Delta t/\sqrt{2\pi}$, this is the sum performed by the MATLAB function $fft(\tilde{F})$.

Similarly, the inverse Fourier transform can be written as a sum over the frequency grid:

$$\frac{1}{\sqrt{2\pi}} \int d\omega e^{i\omega t_n} G(\omega) \approx \frac{\Delta\omega}{\sqrt{2\pi}} \sum_m e^{i(n-1)(m-1)\frac{2\pi}{N}} G(\omega_m).$$ (5.96)

The sum over m in this equation is done by the MATLAB function $ifft(\tilde{G})$. The functions fft and $ifft$ handle both real and complex FFTs.

The discrete version of the recipe for evaluating the derivative of $A(z, \omega)$, Equations (5.75) through (5.79), can thus be summarized as

---

Given an input $A(z, \omega_m)$, calculate $\tilde{A}(z, t_n)$ using

$$\tilde{A}(z, \omega_m) = e^{-i\beta(\omega_m)z} A(z, \omega_m); \quad \tilde{A}(z, t_n) = \frac{\Delta\omega}{\sqrt{2\pi}} ifft(\tilde{A}(z, \omega_m)).$$ (5.97)

Evaluate the function $F(\omega_m)$ given by

$$F(\omega_m) = \frac{\Delta t}{\sqrt{2\pi}} fft(|\tilde{A}(z, t_n)|^2).$$ (5.98)

Multiply by the Raman response function and go back into the time domain:

$$G(\omega_m) = [(1 - f_R) + f_R R(\omega_m)] F(\omega_m); \quad G(t_n) = \frac{\Delta\omega}{\sqrt{2\pi}} ifft(G(\omega_m)).$$ (5.99)

Multiply by $\tilde{A}(z, t_n)$ and transform back into the frequency domain:

$$I(t_n) = G(t_n)\tilde{A}(z, t_n); \quad I(\omega_m) = \frac{\Delta t}{\sqrt{2\pi}} fft(I(t_n)).$$ (5.100)

Evaluate the $A(z, \omega_m)$ derivative as

$$\partial_z A(z, \omega_m) = -\frac{i\omega_m n_2 e^{i\beta(\omega_m)z}}{c A_{eff}\sqrt{2\pi}} I(\omega_m).$$ (5.101)

EXERCISES

To solve the exercises in this section, you should start by setting up a code to propagate Equation (5.42) as outlined in the previous section, using the ODE stepper *ode45* provided by MATLAB. The exercises can largely be regarded as a series of tests that will allow you to check whether the routine works correctly. To this end, the various analytical and semi-analytical results in Section 5.3 will be utilized.

In the following exercises, use values of $n_2 = 2.2 \cdot 10^{-8}\,\mu\text{m}^2/\text{W}$, as measured for unpolarized light at 1550 nm [20,21].

5.1 We start out by considering the NLS equation without dispersion, so you should set $\beta_2 = f_R = 0$ and also neglect self-steepening, that is replace ω by ω_0 in the frontfactor of Equation (5.42). Let ω_0 correspond to a wavelength of 1 μm. Set the input pulse to be a Gaussian, with a temporal FWHM, t_{FWHM} of 100 fs, and a peak power of 1 kW. Set $A_{eff} = 30\,\mu\text{m}^2$, a typical value for a standard fibre which is single moded at 1 μm. Verify that the temporal power profile, $\tilde{A}(z,t)$, is unchanged as the pulse propagates. Compare the numerically calculated spectrum with the one predicted by Fourier transforming the analytical solution in Equation (5.59).

5.2 Add self-steepening into the equation, that is use ω instead of ω_0 in the prefactor on the RHS of Equation (5.42). Repeat the calculations from Exercise 5.1, and verify the self-steepening behaviour of the temporal pulse profile. How does the spectrum evolve as you approach the shock front?

5.3 We now go back to the NLS equation, and consider an optical soliton in a standard telecommunications fibre. Let ω_0 correspond to a wavelength of 1.55 μm, and set $\beta_2 = -2.2 \cdot 10^{-8}\,\text{ps}^2/\mu\text{m}$, $A_{eff} = 80\,\mu\text{m}^2$. These are typical values for telecom transmission fibres such as the SMF28. Set up an input pulse given by Equation (5.62) with $t_s = 0.1$ ps. Verify that the soliton propagates without changing its temporal or spectral power density. Reintroduce self-steepening into the equation; what is the effect of this? Try also a simulation including only dispersion, without any nonlinear effects; what do you see now?

5.4 Turn on the Raman scattering, parametrized according to Equation (5.68), and with $f_R = 0.2$. Leave out self-steepening for this exercise. Let the wavelength and fibre parameters be as in Exercise 5.3. Consider first a soliton with $t_s = 200$ fs. Use Equation (5.67) to estimate the length for which the frequency shift of the soliton is equal to its own bandwidth (FWHM). Check the prediction numerically. Repeat this exercise for other t_s values, and estimate the range of validity of Equation (5.67).

The ODE solver will be more time consuming as the pulse grows shorter and more powerful. To avoid excessively long calculations for short pulses, it may be an advantage to scale the fibre length with t_s^4, so that the redshift magnitude will stay constant as long as Equation (5.67) holds.

5.5 Add third-order dispersion to the propagation equation, writing

$$\beta(\omega) = \frac{\beta_2}{2}(\omega - \omega_0)^2 + \frac{\beta_3}{6}(\omega - \omega_0)^3. \qquad (5.102)$$

Use $\beta_3 = 10^{-10}$ ps^3/µm as a reasonable value for the dispersion slope of a standard telecom fibre. Calculate the redshift of a $t_s = 50$ fs soliton, for a propagation distance of 20 m with and without β_3, and explain the results.

Check also that the propagation conserves pulse energy without self-steepening and photon number when self-steepening is turned on.

REFERENCES

1. R. W. Boyd, *Nonlinear Optics*, 3rd edn. San Diego, CA: Academic Press, 2008.
2. G. P. Agrawal, *Nonlinear Fiber Optics*. San Diego, CA: Academic Press, 2007.
3. P. S. Pershan, Nonlinear optical properties of solids: Energy considerations, *Phys. Rev.*, 130, 919–929 (1963).
4. N. Bloembergen, R. K. Chang, S. S. Jha, and C. H. Lee, Optical second-harmonic generation in reflection from media with inversion symmetry, *Phys. Rev.*, 174, 813–822 (1968).
5. J. A. Armstrong, N. Bloembergen, J. Ducuing, and P. S. Pershan, Interactions between light waves in a nonlinear dielectric, *Phys. Rev.*, 127, 1918–1939 (1962).
6. E. Garmire, F. Pandarese, and C. H. Townes, Coherently driven molecular vibrations and light modulation, *Phys. Rev. Lett.*, 11, 160–163 (1963).
7. R. Y. Chiao, C. H. Townes, and B. P. Stoicheff, Stimulated Brillouin scattering and coherent generation of intense hypersonic waves, *Phys. Rev. Lett.*, 12, 592–595 (1964).
8. W. Kaiser and C. G. B. Garrett, Two-photon excitation in caf$_2$: eu^{2+}, *Phys. Rev. Lett.*, 7, 229–231 (1961).
9. J. Armstrong, Saturable optical absorption in phthalocyanine dyes, *J. Appl. Phys.*, 36, 471–473 (1965).
10. J. W. Arkwright, P. Elango, G. R. Atkins, T. Whitbread, and M. J. F. Digonnet, Experimental and theoretical analysis of the resonant nonlinearity in ytterbium-doped fiber, *J. Lightwave Technol.*, 16, 798 (1998).
11. S. Guha, E. W. V. Stryland, and M. J. Soileau, Self-defocusing in CdSe induced by charge carriers created by two-photon absorption, *Opt. Lett.*, 10, 285–287 (1985).
12. W. Koechner, Thermal lensing in a Nd:YAG laser rod, *Appl. Opt.*, 9, 2548–2553 (1970).
13. G. Martin and R. Hellwarth, Infrared-to-optical image conversion by Bragg reflection from thermally induced index gratings, *Appl. Phys. Lett.*, 34, 371–373 (1979).
14. S. J. Bentley, R. W. Boyd, W. E. Butler, and A. C. Melissinos, Measurement of the thermal contribution to the nonlinear refractive index of air at 1064 nm, *Opt. Lett.*, 25, 1192–1194 (2000).
15. A. V. Smith and J. J. Smith, Mode instability in high power fiber amplifiers, *Opt. Express*, 19, 10180–10192 (2011).
16. R. W. Hellwarth, 3rd-order optical susceptibilities of liquids and solids, *Prog. Quantum Electron.*, 5, 1–68 (1977).
17. J. Lægsgaard, Mode profile dispersion in the generalised nonlinear Schrödinger equation, *Opt. Express*, 15, 16110–16123 (2007).
18. K. J. Blow and D. Wood, Theoretical description of transient stimulated Raman scattering in optical fibers, *IEEE J. Quantum Electron.*, 25, 2665–2673 (1989).
19. D. Milam, Review and assessment of measured values of the nonlinear refractive-index coefficient of fused silica, *Appl. Opt.*, 37, 546–550 (1998).
20. A. Boskovic, L. Gruner-Nielsen, O. A. Levring, S. V. Chernikov, and J. R. Taylor, Direct continuous-wave measurement of n2 in various types of telecommunication fiber at 1.55 µm, *Opt. Lett.*, 21, 1966–1968 (1996).

21. A. Melloni, M. Martinelli, and A. Fellegara, Frequency characterization of the nonlinear refractive index in optical fiber, *Fiber Integr. Opt.*, 18, 1–13 (1999).

22. V. E. Zakharov and A. B. Shabat, Exact theory of 2-dimensional self-focusing and one-dimensional self-modulation of waves in nonlinear media, *Sov. Phys. JETP*, 34, 62–69 (1972).

23. J. P. Gordon, Theory of the soliton self-frequency shift, *Opt. Lett.*, 11, 662–664 (1986).

24. R. Joenk and R. Landauer, Laser pulse distortion in a nonlinear dielectric, *Phys. Lett. A*, 24, 228–229 (1967).

25. F. DeMartini, C. H. Townes, T. K. Gustafson, and P. L. Kelley, Self-steepening of light pulses, *Phys. Rev.*, 164, 312–323 (1967).

26. J. M. Dudley, G. Genty, and S. Coen, Supercontinuum generation in photonic crystal fiber, *Rev. Mod. Phys.*, 78, 1135–1184 (2006).

6 The Modal Method

6.1 INTRODUCTION

The modal method (MM) is a family of mode expansion techniques for the frequency domain developed over the past three decades. These techniques were initially developed as a grating theory under the names of rigorously coupled waveguide analysis (RCWA) [1] and Fourier modal method (FMM) [2] to study diffraction of periodic structures. Subsequently, a mode expansion technique was used to analyse rotationally symmetric structures including vertical-cavity surface-emitting lasers (VCSELs) and micropillar cavities under the name eigenmode expansion technique (EET) [3]. In recent years, the method has been applied to simulate open geometries [4] without rotational or translational symmetries such as the photonic crystal membrane under the name of aperiodic Fourier modal method (a-FMM) or simply FMM. The conceptual differences between the individual methods are small and are related to the symmetries and the boundary conditions of the geometry under study as well as to the technique used to compute the modes.

In a mode expansion technique, the geometry is divided into layers of uniform relative permittivity profile along a propagation axis, usually chosen as the z axis. For each layer, we consider the lateral permittivity profile with uniformity along the z axis, and we determine the corresponding transverse eigenmodes to the wave equation. In the following, we will simply refer to these as *eigenmodes*. As the name implies, the optical field in each layer is expanded on the corresponding eigenmodes, and the fields at each side of the interface between adjacent layers are connected using mode matching at the interface. The difference between the spatial discretization used in finite-difference techniques and that of a mode expansion technique is illustrated in Figure 6.1. Most methods introduce a subwavelength spatial discretization of the computational domain as illustrated in Figure 6.1a. However, in the MM, the geometry is only divided into layers of uniform relative permittivity profile along the z axis as sketched in Figure 6.1b. Such a subdivision is computationally advantageous in structures which consist predominantly of uniform or periodic sections. As we will see, the MM gives direct access to interesting physical parameters such as mode profiles, effective indices, and scattering coefficients.

In spite of several decades of intense development, commercial software packages implementing the method remain scarce [5], and only few open-source software packages [6] with limited functionality exist. Fortunately, the method is sufficiently simple so that a student with basic programming skills can implement the method. This chapter presents the basics of the method and outlines advanced extensions such as absorbing boundary conditions and the Bloch mode formalism.

We first introduce the concept of expansion on eigenmodes in Section 6.2. Then in Section 6.3 we study the most simple case, the 1D geometry, where the concept of subdividing the geometry into uniform layers along a propagation axis will be

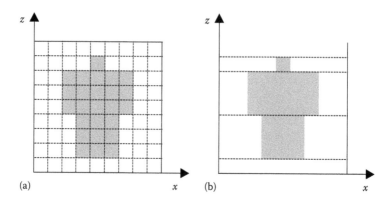

FIGURE 6.1 (a) A computational domain with subwavelength spatial discretization. (b) The corresponding layer division for the MM.

introduced. Subsequently, in Section 6.4 we will extend the formalism to 2D structures with expansion on multiple lateral eigenmodes. In Section 6.5, we discuss formalisms for handling the particular case of a periodic structure. The presence of current sources is treated in Section 6.6. Finally, in Section 6.7 we will discuss the extension of the theory to full 3D structures.

In this chapter, we assume non-magnetic materials and employ for the time dependence the convention $e^{-i\omega t}$ predominant in the literature on the MMs.

6.2 EIGENMODES

The starting point for the MM is the determination of eigenmodes that serve as a basis set for the expansion of the optical field. To define these, the relative permittivity profile is sliced into layers having translational symmetry along the z axis. In Figure 6.2a,

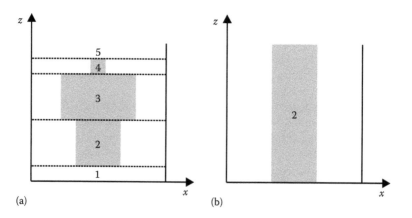

FIGURE 6.2 (a) The division of an example geometry into five layers. (b) The profile for layer 2 with uniformity along the z axis.

the permittivity profile of a random glass sample is illustrated. The sample itself can be separated into three layers, 2–4, uniform along the z axis. Also including the surrounding air layers 1 and 5, the structure consists of five layers in total.

For each layer, we now consider the geometry with permittivity profile in the (x, y) plane corresponding to the original geometry, but with uniformity along the z axis. As an example in Figure 6.2b, the (x, z) profile corresponding to layer 2 in Figure 6.2a is depicted. When we refer to the eigenmodes of particular layer in the following, it is understood that they are the solutions to the corresponding geometry with uniformity along the z axis as shown in Figure 6.2b. In the full geometry, the layers of course have finite thickness; however, this non-uniformity will be handled separately using recursive matrix formalism.

The eigenmodes are solutions to the eigenvalue problem equation (2.58) for the electric field repeated here for convenience:

$$\nabla_\perp^2 \mathbf{e}_{\perp m} + \nabla_\perp(\mathbf{e}_{\perp m} \cdot \nabla \ln \varepsilon(\mathbf{r}_\perp)) + \varepsilon(\mathbf{r}_\perp) k_0^2 \mathbf{e}_{\perp m} = \beta_m^2 \mathbf{e}_{\perp m}. \tag{6.1}$$

Equation (6.1) is a central equation for the MMs. By solving it, we obtain eigenmodes $\mathbf{e}_{\perp m}(\mathbf{r}_\perp)$ and eigenvalues that are the squares of the propagation constants β_m. In some literature, the mode propagation is described instead using an *effective index* $n_{\text{eff},m}$ related to the propagation constant as $n_{\text{eff},m} = \beta_m/k_0$.

These eigenmodes constitute the basis set upon which the optical field in the layer is expanded. In order to represent any optical field using eigenmode expansion, the basis set must be complete. For geometries featuring a real-valued relative permittivity profile and uniformity along the y-axis, it has been proven [7] that the eigenmodes obtained by solving Equation (6.1) form a complete set. It is unclear whether the completeness extends to general geometries with complex permittivity profiles, but so far no evidence has indicated that it is not the case and the completeness of the eigenmodes is thus generally assumed in spite of the lack of formal proof.

We can then describe an arbitrary optical field using

$$\mathbf{E}_\perp(\mathbf{r}) = \sum_{m=1}^{\infty} a_m \mathbf{e}_{\perp m}(\mathbf{r}_\perp) e^{i\beta_m z} + \sum_{m=1}^{\infty} b_m \mathbf{e}_{\perp m}(\mathbf{r}_\perp) e^{-i\beta_m z}, \tag{6.2}$$

where the expansion coefficients a_m and b_m refer to the forward and backward travelling eigenmodes, respectively, which have been classified according to the scheme summarized in Table 2.1. Ideally, the summations are over the infinite set of eigenmodes; however, for practical calculations, a truncation is necessary leading to a summation over a finite number N of modes.

Equations (6.1) and (6.2) only involve the lateral electric field components $\mathbf{e}_\perp$; however, the remaining components of the six-component electromagnetic field can be derived from $\mathbf{e}_\perp$ as discussed in Section 2.7.1. Instead of solving the eigenvalue problem (Equation 6.1), another possible approach is to consider the eigenvalue problem (Equation 2.64) for the magnetic field, compute the lateral magnetic field $\mathbf{h}_\perp$, and finally derive the remaining field components from the $\mathbf{h}_\perp$ components. These two approaches are equivalent, and both are found in the literature.

6.3 1D GEOMETRY

Before studying 2D and 3D structures, we will introduce the recursive matrix formalisms describing the propagation of the field along the z axis in multilayer structures in the simple 1D geometry. Following the 1D transfer matrix literature, we will employ the refractive index $n = \sqrt{\varepsilon\mu} = \sqrt{\varepsilon}$ for a non-magnetic material instead of the relative permittivity to describe the structure. In the 1D geometry, the index $n(z)$ varies along the propagation z axis but is independent of the lateral x and y coordinates. One-dimensional theory is simple and efficiently describes light propagation in fibre Bragg gratings and VCSELs in the effective index approximation. An example of such a refractive-index profile is illustrated in Figure 6.3a. Following the MM, we proceed by subdividing the refractive-index profile $n(z)$ into layers q of constant index n_q ($q = 1, 2, 3, \ldots$) as shown in Figure 6.3a, and we will employ a recursive matrix formalism as described later to relate the fields in the different layers.

We now consider the geometry of a uniform layer q as illustrated in Figure 6.3b. This is the simplest possible geometry, where the refractive index is independent of position. In the 1D case, the eigenmode problem (Equation 6.1) for the layer q reduces to

$$\nabla_\perp^2 \mathbf{e}_q + n_q^2 k_0^2 \mathbf{e}_q = \beta^2 \mathbf{e}_q, \tag{6.3}$$

We will solve this equation for a $\mathbf{r}_\perp$-independent field travelling along the z axis. In this case, we have only one lateral eigenmode $\mathbf{e}_q$ in the layer q and our expansion coefficient sets in Equation (6.2) for the forward and backward travelling eigenmodes thus simply reduce to the scalars a_q and b_q. Since the lateral field components in the 1D case do not couple, the lateral eigenmode profile can be written simply as $\mathbf{e}_q(\mathbf{r}_\perp) = C\hat{\mathbf{u}}$, where C is a constant and $\hat{\mathbf{u}}$ is a unit vector in the (x, y) plane.

From Equation (6.3), we obtain the simple dispersion relation

$$n_q^2 k_0^2 = \beta_q^2, \tag{6.4}$$

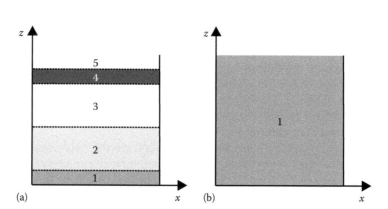

FIGURE 6.3 (a) A geometry consisting of layers with refractive-index profile $n(z)$, represented by different colours, depending only on the z coordinate. (b) The single-layer geometry.

and the scalar field $E_q(z)$ inside the layer q is thus given by

$$E_q(z) = a_q e^{i\beta_q z} + b_q e^{-i\beta_q z}, \tag{6.5}$$

where $\beta_q = n_q k_0$ and the full vectorial field is $\mathbf{E}_q(z) = E_q(z)\hat{\mathbf{u}}$.

Equation (6.5) is a sum of a forward and a backward travelling field, where a_q and b_q are constants describing the phases and amplitudes of the two contributions. It represents a complete analytical solution meaning that once the coefficients a_q and b_q are known, the field in all spatial positions z can be directly deduced. We thus only need to determine the coefficients a_q and b_q, and then we can subsequently exploit the analytical dependence on z to compute the field wherever we need to evaluate it.

For future reference, we note that the *intensity* I defined as the power flow per unit area for a 1D wave $a_q e^{i n_q k_0 z}$ is given by $I = n_q |a_q|^2/(2\mu_0 c)$.

6.3.1 RECURSIVE MATRIX FORMALISM

In geometries consisting of multiple layers, we employ a recursive matrix formalism to relate the fields in the different layers. In the following, we will consider two such formalisms, namely the transfer or T matrix formalism [8] and the S matrix formalism [9].

Let us consider the 1D N-layer geometry depicted in Figure 6.4, which is illuminated with two incoming fields, a forward travelling field a_1 in layer 1 and a backward travelling field b_N in layer N. At this point, the structure in between is a black box, and the details of light propagation inside the structure are unknown. However, scattering of the incoming fields leads to outgoing fields b_1 and a_N, and this scattering is described by the reflection and transmission coefficients defined by

$$r_{1N} = \frac{b_1}{a_1} \quad (b_N = 0) \tag{6.6}$$

$$t_{1N} = \frac{a_N}{a_1} \quad (b_N = 0) \tag{6.7}$$

$$r_{N1} = \frac{a_N}{b_N} \quad (a_1 = 0) \tag{6.8}$$

$$t_{N1} = \frac{b_1}{b_N} \quad (a_1 = 0) \tag{6.9}$$

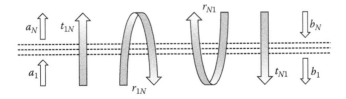

FIGURE 6.4 Forward and backward travelling light in layers 1 and N in a N-layer geometry. The fields are related by the reflection and transmission coefficients r_{1N}, t_{1N}, r_{N1}, and t_{N1}.

such that the outgoing fields are given as

$$a_N = t_{1N}a_1 + r_{N1}b_N. \tag{6.10}$$

$$b_1 = r_{1N}a_1 + t_{N1}b_N \tag{6.11}$$

The S matrix formalism relates the outgoing scattered fields for a structure with the incoming fields. We describe this relation using the matrix $\overline{\overline{S}}$ as

$$\begin{bmatrix} a_N \\ b_1 \end{bmatrix} = \overline{\overline{S}} \begin{bmatrix} a_1 \\ b_N \end{bmatrix}, \tag{6.12}$$

where the $\overline{\overline{S}}$ matrix elements are given directly from Equations (6.10) and (6.11) as

$$\overline{\overline{S}} = \begin{bmatrix} S_{11} & S_{12} \\ S_{21} & S_{22} \end{bmatrix} = \begin{bmatrix} t_{1N} & r_{N1} \\ r_{1N} & t_{N1} \end{bmatrix}. \tag{6.13}$$

However, it is also convenient to be able to relate the forward and backward travelling fields at one point in the geometry with those at another point. This is done in the T matrix formalism, where the fields at each side of the structure are related as

$$\begin{bmatrix} a_N \\ b_N \end{bmatrix} = \overline{\overline{T}} \begin{bmatrix} a_1 \\ b_1 \end{bmatrix}. \tag{6.14}$$

The $\overline{\overline{T}}$ matrix elements can be derived from Equations (6.10) and (6.11) by rearranging terms to write a_N and b_N as function of a_1 and b_1, respectively. They are given by

$$\overline{\overline{T}} = \begin{bmatrix} T_{11} & T_{12} \\ T_{21} & T_{22} \end{bmatrix} = \frac{1}{t_{N1}} \begin{bmatrix} t_{1N}t_{N1} - r_{1N}r_{N1} & r_{N1} \\ -r_{1N} & 1 \end{bmatrix}. \tag{6.15}$$

In Equations (6.13) and (6.15), the S and T matrices are defined from knowledge of the reflection and transmission coefficients. Inversely, having computed the S or the T matrix for an arbitrary geometry, we can derive its reflection and transmission properties from the S and T matrices. In the S matrix formalism, the coefficients r_{1N}, t_{1N}, r_{N1}, and t_{N1} are extracted directly from Equation (6.13). The coefficients can be derived in the T matrix formalism starting from Equation (6.14) and using the defining Equations (6.6) through (6.9) leading to

$$r_{1N} = -\frac{T_{21}}{T_{22}} \tag{6.16}$$

$$t_{1N} = T_{11} - \frac{T_{12}T_{21}}{T_{22}} \tag{6.17}$$

$$r_{N1} = \frac{T_{12}}{T_{22}} \tag{6.18}$$

$$t_{N1} = \frac{1}{T_{22}}. \tag{6.19}$$

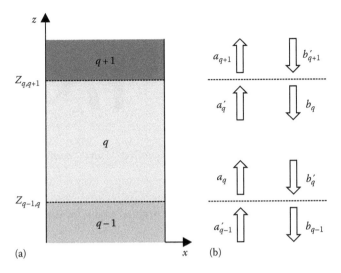

FIGURE 6.5   (a) A multilayer geometry. (b) The corresponding reference coordinate positions of the field expansion coefficients.

We will see in the following that the T matrix formalism is well adapted to handle 1D geometries, whereas the S matrix formalism is required in the 2D and 3D cases.

So far, the expansion coefficients a_q and b_q for the layer q are defined relative to $z = 0$ such that $E_q(0) = a_q + b_q$. However, when we consider geometries with many layers, it will be convenient to define expansion coefficients a_q^z and b_q^z relative to various positions z inside the layer q. In particular, we will introduce the notation outlined in Figure 6.5 for expansion coefficients defined relative to the positions of the interfaces. From now on, we thus define the coefficients a_q and b_q' relative to the z coordinate $z_{q-1,q}$ of the interface between layers $q-1$ and q, and similarly, we define the coefficients a_q' and b_q relative to the z coordinate $z_{q,q+1}$ of the interface between layers q and $q+1$. The coefficients a_q and b_q can be thought of as *having their back against the interface*, where the coefficients a_q' and b_q' are *facing the interface*.

6.3.2   1D INTERFACE

Equipped with the solution to a single layer, we now consider the two-layer geometry illustrated in Figure 6.6 consisting of layers 1 and 2 of different refractive indices n_1 and n_2 separated by an interface located at $z = z_{1,2}$. The refractive-index profile is thus piecewise constant and is given by

$$n(z) = n_1 \quad (z < z_{12})$$
$$n(z) = n_2 \quad (z > z_{21}).$$

We first consider the scenario where we illuminate the interface from layer 1 only as illustrated in Figure 6.6b. Due to the index contrast, the forward travelling incoming mode a_1' will be scattered at the interface and will be partly reflected to the

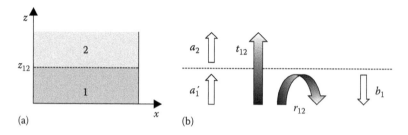

FIGURE 6.6 (a) A two-layer geometry. (b) The reflection and transmission of light at the interface.

backward travelling mode b_1 and partly transmitted to the forward travelling mode a_2. The field is thus given by

$$E(z) = a_1' e^{in_1 k_0 (z - z_{12})} + b_1 e^{-in_1 k_0 (z - z_{12})} \quad (z \leq z_{12}) \tag{6.20}$$

$$E(z) = a_2 e^{in_2 k_0 (z - z_{12})} \quad (z \geq z_{12}), \tag{6.21}$$

where we have expanded the field in each layer of the two-layer geometry on the solutions for each separate layer.

To determine the field in this geometry, we need to identify the free parameters a_1', b_1, and a_2. The parameter a_1' is chosen by us as our illumination condition, and the parameters b_1 and a_2 are subsequently computed from a_1 using the Fresnel reflection and transmission coefficients r_{12} and t_{12} such that $b_1 = r_{12} a_1'$ and $a_2 = t_{12} a_1'$. The Fresnel coefficients are defined by the requirements of continuity of the tangential components of the electric and magnetic fields across the interface at the position z_{12}, and for modes travelling along the z axis, they are given by

$$r_{12} = \frac{n_1 - n_2}{n_1 + n_2} \tag{6.22}$$

$$t_{12} = \frac{2n_1}{n_1 + n_2} \tag{6.23}$$

$$r_{21} = \frac{n_2 - n_1}{n_1 + n_2} \tag{6.24}$$

$$t_{21} = \frac{2n_2}{n_1 + n_2}. \tag{6.25}$$

From the Fresnel coefficients (Equations 6.22 through 6.25), we can define the *interface matrices* $\overline{\overline{S}}_{12}$ for the S matrix formalism and $\overline{\overline{T}}_{12}$ for the T matrix formalism describing the interface using Equations (6.13) and (6.15).

6.3.3 MULTILAYER STRUCTURE

Using the S or the T matrix formalism, we can describe the optical field in a two-layer geometry. Now, we will extend the T matrix formalism to treat a 1D geometry

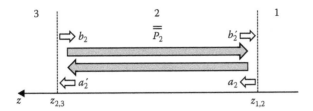

FIGURE 6.7 Propagation of light inside a 1D uniform layer.

with an arbitrary number of layers. The corresponding extension for the S matrix formalism is discussed in Section 6.4.4.

First, we define the *propagation matrix* $\overline{\overline{P}}_q$ in the T matrix formalism for a uniform layer q of length $L_q = z_{q,q+1} - z_{q-1,q}$ and refractive index n_q sketched in Figure 6.7. The choice for the z axis orientation will be justified later. The propagation matrix $\overline{\overline{P}}_q$ relates the fields at each side of a uniform layer as

$$\begin{bmatrix} a'_q \\ b_q \end{bmatrix} = \overline{\overline{P}}_q \begin{bmatrix} a_q \\ b'_q \end{bmatrix}. \tag{6.26}$$

and its matrix elements are given by

$$\overline{\overline{P}}_q = \begin{bmatrix} e^{in_q k_0 L_q} & 0 \\ 0 & e^{-in_q k_0 L_q} \end{bmatrix}. \tag{6.27}$$

More generally, the propagation matrix $\overline{\overline{P}}_q(\Delta z)$ can be used inside the layer q to relate field coefficients a_q^z and b_q^z defined at z with the coefficients $a_q^{z'}$ and $b_q^{z'}$ defined at z' by replacing the layer length L_q in Equation (6.27) with the distance $\Delta z = z' - z$. However, for the propagation matrix $\overline{\overline{P}}_q(L_q)$ over the full layer length L_q, we will omit the argument (L_q) to keep the notation short and instead simply use $\overline{\overline{P}}_q$ as given in Equation (6.27) to represent this matrix.

Having identified the interface matrix $\overline{\overline{T}}_{q,q+1}$ relating the fields at each side of an interface between layers q and $q+1$ and the propagation matrix $\overline{\overline{P}}_q$ relating the fields at different positions inside the uniform layer q, we are now ready to treat a structure with an arbitrary number of layers such as the one illustrated in Figure 6.8. In this

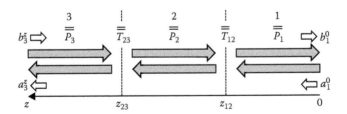

FIGURE 6.8 Propagation of light described by $\overline{\overline{T}}$ and $\overline{\overline{P}}$ matrices for a three-layer geometry.

example geometry, we have two interfaces described by the interface matrices $\overline{\overline{T}}_{12}$ and $\overline{\overline{T}}_{23}$ and three uniform layers described by the propagation matrices $\overline{\overline{P}}_1$, $\overline{\overline{P}}_2$, and $\overline{\overline{P}}_3$.

The relation between the expansion coefficients can then be obtained by simply cascading the propagation and interface matrices, such that

$$
\begin{bmatrix} a_3^z \\ b_3^z \end{bmatrix} = \overline{\overline{T}}_t \begin{bmatrix} a_1^0 \\ b_1^0 \end{bmatrix} = \overline{\overline{P}}_3(z - z_{23}) \, \overline{\overline{T}}_{23} \, \overline{\overline{P}}_2 \, \overline{\overline{T}}_{12} \, \overline{\overline{P}}_1(z_{12}) \begin{bmatrix} a_1^0 \\ b_1^0 \end{bmatrix},
\tag{6.28}
$$

and we see that the choice of z axis orientation in Figures 6.7 and 6.8 allows for correspondence between the order of the geometry segments and order of the T matrix multiplication in Equation (6.28).

Using this procedure of cascading $\overline{\overline{T}}$ and $\overline{\overline{P}}$ matrices, the field at any point of the geometry can be evaluated. Furthermore, the reflection and transmission properties of the entire structure can be determined from the total T matrix $\overline{\overline{T}}_t$ using Equations (6.16) through (6.19). As an example, we consider the distributed Bragg reflector (DBR) consisting of p alternating layers of GaAs and AlAs as illustrated in Figure 6.9. Taking advantage of the periodic nature of the structure, the total T matrix [8] describing this geometry is given by

$$
\overline{\overline{T}}_t = \left(\overline{\overline{T}}_{23} \, \overline{\overline{P}}_{\text{AlAs}} \, \overline{\overline{T}}_{12} \, \overline{\overline{P}}_{\text{GaAs}} \right)^p.
\tag{6.29}
$$

We will illuminate the geometry from the left side by setting $a_1^0 = 1$. In the T matrix formalism, we need both a_1^0 and b_1^0 coefficients to compute the fields in the rest of the geometry. The reflection coefficient r_{1N} is extracted from $\overline{\overline{T}}_t$ using Equation (6.16) leading to $b_1^0 = r_{1N}$, and from knowledge of the field coefficients at $z = 0$, the field at any point z' is computed using the T matrix relating the positions $z = z'$ and $z = 0$.

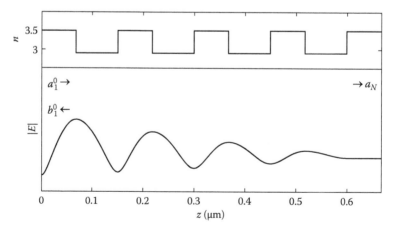

FIGURE 6.9 The refractive-index profile and the field profile for a 1D Bragg reflector consisting of $p = 4$ GaAs/AlAs layer pairs. $\lambda = 950$ nm, $n_{\text{GaAs}} = 3.5$, $n_{\text{AlAs}} = 2.9$, and the layer thicknesses are chosen as $\lambda/(4n)$.

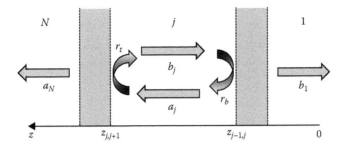

FIGURE 6.10   A cavity layer j surrounded by mirrors of reflectivity r_b and r_t.

6.3.4   1D CAVITY

A general 1D cavity is illustrated in Figure 6.10. It consists of a uniform layer j of length $L_j = z_{j,j+1} - z_{j-1,j}$ surrounded by partially reflecting elements described by reflection coefficients r_b and r_t. Whereas we can illuminate a geometry from outside using any wavelength, the cavity supports a discrete set of cavity modes at wavelengths corresponding to the cavity resonances, and we must first determine these resonance wavelengths before we can plot the field.

The boundary condition for a cavity mode is that the incoming fields a_1 and b_N are zero, corresponding to a situation where light is leaking out of the cavity. Since light is dissipating from the system, the cavity resonance frequency will take a complex value $\omega = \omega_R - i\omega_I$ with $\omega_R = \mathrm{Re}(\omega)$ and $\omega_I = \mathrm{Im}(\omega) > 0$, leading to a time dependence given as

$$E(z, t) = E(z)e^{-i\omega t} = E(z)e^{-i\omega_R t}e^{-\omega_I t}, \tag{6.30}$$

where the damping of the field is explicit.

By setting $a_1 = 0$ in Equation (6.14), we see that the requirement $b_N = 0$ implies that $T_{t;22} = 0$. We can thus determine the cavity mode wavelengths by looking for poles in the complex frequency plane for the function $T_{t;22}(\omega)$. Once such a pole is found, the field is computed from the conditions $a_1^0 = 0$ and $b_1^0 = 1$. An example calculation of the cavity mode profile in a 1D λ-cavity is shown in Figure 6.11. Inside the cavity, we observe a standing wave pattern resulting from the interference of forward and backward travelling fields, and outside the cavity, only outgoing fields are present.

The cavity *quality (Q) factor* is generally employed to describe the rate of leakage of light from the cavity. When the resonance frequency ω is identified, the Q factor can be computed as

$$Q = \frac{\omega_R}{2\omega_I}. \tag{6.31}$$

Note: If one chooses to illuminate the structure from outside at a complex frequency where $T_{t;22}(\omega) = 0$, a quick inspection of Equations (6.16) through (6.19) reveals that the reflected and transmitted light becomes infinite. However, here we

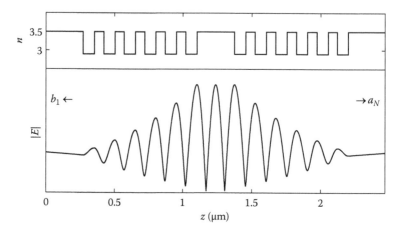

FIGURE 6.11   A λ cavity of thickness λ/n surrounded by two DBRs with $p = 6$ layer pairs. The parameters are identical to those in Figure 6.9.

are violating the initial boundary condition that the incoming fields $a_1 = b_N = 0$. The cavity mode is determined from the strict requirement that only outgoing fields are present.

6.4   2D GEOMETRY

We now consider geometries where the permittivity profile depends not only on z but also on the lateral x coordinate. As in the 1D case, our strategy will be to split the geometry into layers q uniform along the z axis as illustrated in Figure 6.2a. We then compute the eigenmodes for each layer assuming a z independent permittivity profile. The optical field will be expanded on these solutions, and the z dependence will be handled using recursive matrix formalism. The presence of evanescent field components in the expansion (Equation 6.2) leads to numerical instability of the T matrix formalism in the 2D case, and we will thus only employ the S matrix formalism in the following.

Several methods are available for computing the eigenmodes, each with their advantages and drawbacks. The finite-difference method was described in detail in Chapter 4, and in the following, we will also describe the *plane-wave expansion* method predominantly used in the FMM. When the lateral profile possesses translational symmetry as in the case of the 2D geometry, exact analytical descriptions of the eigenmodes are available, and these can be determined using a *semi-analytical* (SA) approach.

In this section, we compute the eigenmodes in 2D by solving the eigenvalue problems (Equation 2.68) for TE polarization and (Equation 2.70) for TM polarization. We will use the forms

$$\partial_x^2 e_y + \varepsilon(x)k_0^2 e_y = \beta^2 e_y \quad \text{(TE)} \tag{6.32}$$

$$\varepsilon(x)\partial_x \frac{1}{\varepsilon(x)} \partial_x h_y + \varepsilon(x)k_0^2 h_y = \beta^2 h_y \quad \text{(TM)} \tag{6.33}$$

where we have slightly modified the equation for TM polarization to make the boundary conditions more transparent in the following discussion.

For use in the subsequent sections, we consider the appropriate continuity conditions for the fields, which can be derived either from physical or mathematical arguments. Physically, we require that the field components tangential to the ε variations along the x-axis be continuous. In the TE case, these are the e_y and h_z components. From Maxwell's equation (2.55), we obtain for TE polarization $h_z = \frac{1}{i\omega\mu_0}\partial_x e_y$, and we thus require the continuity of e_y and $\partial_x e_y$ in the TE case. For TM polarization, the tangential components are h_y and e_z, and in this case, Maxwell's equation (2.56) gives $e_z = \frac{i}{\omega\varepsilon_0\varepsilon}\partial_x h_y$, leading to the requirement in the TM case of continuity of h_y and $\frac{1}{\varepsilon}\partial_x h_y$. Mathematically, the same conditions are obtained from requiring that terms after a derivate in Equations (6.32) and (6.33) be continuous for the derivatives to be well defined.

6.4.1 PLANE-WAVE EXPANSION

We will first discuss the approach to solving Equations (6.32) and (6.33) based on plane-wave expansion used in the RCWA and FMM methods. Here, the field and the dielectric constant are expanded in plane-wave series. These methods were originally used to study periodic structures, and in the following, we consider a periodic geometry of period L_x as illustrated in Figure 2.2b such that $\varepsilon(x+L_x) = \varepsilon(x)$.

6.4.1.1 Li's Factorization Rules

Before studying the eigenvalue problem, we will first study a few basic properties of plane-wave expansion needed in the following. Let us consider two period functions $f(x)$ and $g(x)$ with period L_x and their corresponding expansions

$$f(x) = \sum_{m=-\infty}^{m=\infty} f_m e^{imKx} \tag{6.34}$$

$$g(x) = \sum_{n=-\infty}^{n=\infty} g_n e^{inKx} \tag{6.35}$$

where $K = 2\pi/L_x$. The expansions (Equations 6.34 and 6.35) are also referred to as the *Fourier series*, hence the name FMM. Let us also define the product $h(x)$ of the two functions:

$$h(x) = f(x)g(x) \tag{6.36}$$

with Fourier series

$$h(x) = \sum_{m=-\infty}^{m=\infty} h_m e^{imKx}. \tag{6.37}$$

The expansion coefficients can be identified by inserting the expansions (Equations 6.34 and 6.35) into Equation (6.36). We obtain

$$h(x) = \sum_{m=-\infty}^{m=\infty} f_m e^{imKx} \sum_{n=-\infty}^{n=\infty} g_n e^{inKx} = \sum_{m=-\infty}^{m=\infty} \sum_{n=-\infty}^{n=\infty} f_{m-n} g_n e^{imKx}, \tag{6.38}$$

such that

$$h_m = \sum_{n=-\infty}^{n=\infty} f_{m-n} g_n. \tag{6.39}$$

The functions f, g, and h are now represented by their Fourier series given by vectors $\bar{f}$, $\bar{g}$, and $\bar{h}$ containing the expansion coefficients f_m, g_m, and h_m, and in the following, we employ this bar notation for vectors representing expansion coefficients. These expansions are infinite series; however, to fit our expansions in a computer, we need to truncate the summations such that our Fourier series for, for example, the function f is given by the vector $\bar{f} = [f_{-M}; \ldots; f_M]$ of finite length $N = 2M + 1$.

Now, from Equation (6.39), we observe that the expansion vector $\bar{h}$ for the function $h(x)$ can be computed from the Fourier series $\bar{f}$ and $\bar{g}$ as

$$\bar{h} = \overline{\overline{F}}\,\bar{g}, \tag{6.40}$$

if we define the *Toeplitz matrix* $\overline{\overline{F}}$ representing the Fourier series $\bar{f}$ as the $N \times N$ matrix with elements $F_{m,n}$ equal to the coefficients f_{m-n} such that

$$\overline{\overline{F}} = \begin{bmatrix} f_0 & f_{-1} & f_{-2} & f_{-3} & \cdots \\ f_1 & f_0 & f_{-1} & f_{-2} & \cdots \\ f_2 & f_1 & f_0 & f_{-1} & \cdots \\ \cdots & \cdots & \cdots & \cdots \end{bmatrix}. \tag{6.41}$$

The matrix product (Equation 6.40) is referred to as *Laurent's rule* or the *direct rule* for the product $h = fg$.

Let us now consider the function $f^{\dagger}(x) = 1/f(x)$ with series expansion:

$$f^{\dagger}(x) = \sum_{m=-\infty}^{m=\infty} f_m^{\dagger} e^{imKx} \tag{6.42}$$

and corresponding Toeplitz $\overline{\overline{F^{\dagger}}}$. From the expression $h = \left(f^{\dagger}\right)^{-1} g$, it is clear that we can also compute the Fourier series for h as

$$\bar{h} = \left(\overline{\overline{F^{\dagger}}}\right)^{-1} \bar{g}. \tag{6.43}$$

This matrix product (Equation 6.43) is referred to as the *inverse rule*.

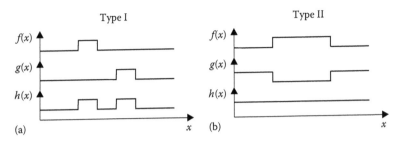

FIGURE 6.12 Functions $f(x)$ and $g(x)$ without (a) and with (b) complementary jump discontinuities. The discontinuity of $h(x) = f(x)g(x)$ in (b) is removable.

When $N \to \infty$, the two matrix products (Equations 6.40 and 6.43) converge towards the same limit; however, it turns out that the convergence properties of the two products are very different. Depending on the properties of the functions f and g, either the direct rule (Equation 6.40) or the inverse rule (Equation 6.43) will provide the fastest convergence. The rules for choosing the correct product are known as *Li's factorization rules* [10].

To understand these rules, let us first consider the discontinuous functions shown in Figure 6.12. The nature of the discontinuities of the functions f and g govern which matrix product should be chosen. We refer to the positions of the discontinuities as *jump points*. In Figure 6.12a, the functions f and g both feature jump points but at different positions. The function h thus features jump points corresponding to the positions of the jump points of f and g. This scenario is referred to as type I. Another scenario is depicted in Figure 6.12b. Here, the functions f and g feature concurrent jump points. In the case where the functions f and g are discontinuous but their product h is continuous, we say that the discontinuity in h is *removable*, and this case is referred to as type II. The factorization rules are then given as follows:

1. If the discontinuities are of type I, the fastest convergence is obtained by choosing the direct rule given by the matrix product (Equation 6.40).
2. If the discontinuities are of type II, the fastest convergence is given by the inverse rule of Equation (6.43).

A further discussion of these rules can be found in Ref. [10].

6.4.1.2 Eigenvalue Problem

We are now ready to solve Equations (6.32) and (6.33) using plane-wave expansion. We employ periodic boundary conditions such that $e_y(x + L_x) = e_y(x)e^{i\alpha}$, and the parameter α is chosen freely from the outset. We start by considering TE polarization, where the eigenproblem is given by

$$\partial_x^2 e_y(x) + k_0^2 \varepsilon(x) e_y(x) = \beta^2 e_y(x), \tag{6.44}$$

and then we write the mode profile $e_y(x)$ and the dielectric constant $\varepsilon(x) = n^2(x)$ as the plane-wave expansions:

$$e_y(x) = \sum_{j=-\infty}^{j=\infty} u_j e^{i(jK + \alpha/L_x)x} \tag{6.45}$$

$$\varepsilon(x) = \sum_{j=-\infty}^{j=\infty} \varepsilon_j e^{ijKx}. \tag{6.46}$$

As discussed earlier, the electric field e_y profile in TE polarization is a continuous function. The product εe_y is thus of type I, and we should use the direct rule for this product. Equation (6.44) can then be written in matrix form as

$$\left(\overline{\overline{K}}^2 + k_0^2 \overline{\overline{E}} \right) \overline{u} = \beta^2 \overline{u}, \tag{6.47}$$

where $\overline{\overline{K}}$ is the $N \times N$ diagonal matrix with diagonal elements $(iK[-M; \ldots; M] + i\alpha/L_x)$ and $\overline{\overline{E}}$ is the Toeplitz matrix representing the expansion (6.46). The elements ε_j are computed using the integral

$$\varepsilon_j = \frac{1}{L_x} \int_0^{L_x} \varepsilon(x) e^{-ijKx} \, dx, \tag{6.48}$$

which for simple permittivity profiles can be evaluated analytically. For more complicated geometries, a fast Fourier transform (FFT) algorithm is usually employed.

Equation (6.47) is a standard eigenvalue problem of the form

$$\overline{\overline{O}} \overline{u} = \lambda \overline{u}, \tag{6.49}$$

which we solve numerically using standard eigenproblem solvers such as the MATLAB® EIG command to obtain eigenvectors $\overline{u}_m$ and eigenvalues $\beta_m^2 = \lambda_m$ describing the electric field profiles e_{ym} of index m.

In the particular case of a uniform geometry with a dielectric constant $\varepsilon(x) = \varepsilon$, the Toeplitz matrix $\overline{\overline{E}}$ is diagonal. If we numerate the eigenmodes m from $-M$ to M, the solution e_{ym} to the eigenproblem (Equation 6.47) for the uniform geometry is then simply a single uncoupled plane wave with eigenvector coefficients $u_{m,j} = \delta_{mj}$ and propagation constant $\beta^2 = \varepsilon k_0^2 - (mK + \alpha/L_x)^2$.

Once the e_{ym} profile has been computed, we can determine the expansions for the magnetic fields:

$$h_{xm}(x) = \sum_{j=-M}^{j=M} v_{m,j} e^{i(jK + \alpha/L_x)x} \tag{6.50}$$

$$h_{zm}(x) = \sum_{j=-M}^{j=M} w_{m,j} e^{i(jK + \alpha/L_x)x} \tag{6.51}$$

using the Maxwell's equation (2.55) leading to

$$h_x = -\frac{\beta}{\omega\mu_0}e_y \tag{6.52}$$

$$h_z = \frac{1}{i\omega\mu_0}\partial_x e_y. \tag{6.53}$$

Using (6.52) and (6.53), the Fourier series for the magnetic fields of the mode with index m are determined from electric e_{ym} profile as

$$\overline{v_m} = -\frac{\beta_m}{\omega\mu_0}\overline{u_m} \tag{6.54}$$

$$\overline{w_m} = \frac{1}{i\omega\mu_0}\overline{K}\,\overline{u_m}. \tag{6.55}$$

The case of TM polarization can be handled in a similar manner. We rewrite Equation (6.33) as

$$\varepsilon(x)\left(\partial_x\frac{1}{\varepsilon(x)}\partial_x + k_0^2\right)h_y = \beta^2 h_y, \tag{6.56}$$

and we introduce the Fourier series

$$h_{ym}(x) = \sum_{j=-M}^{j=M} v_{m,j}e^{i(jK+\alpha/L_x)x} \tag{6.57}$$

$$e_{xm}(x) = \sum_{j=-M}^{j=M} u_{m,j}e^{i(jK+\alpha/L_x)x} \tag{6.58}$$

$$e_{zm}(x) = \sum_{j=-M}^{j=M} w'_{m,j}e^{i(jK+\alpha/L_x)x} \tag{6.59}$$

for the field components.

Whereas the functions $1/\varepsilon$ and $\partial_x h_y$ are each discontinuous functions, the product $1/\varepsilon(\partial_x h_y)$ must be continuous as discussed in the beginning of this section and is thus of type II. We should thus use the inverse rule for this product. Furthermore, we should consider the product of ε and the parenthesis times h_y on the LHS of Equation (6.56). These factors are discontinuous functions; however, their product equals $\beta^2 h_y$, a continuous function, and this product is thus also of type II. To handle this last product, we define the Fourier series for $1/\varepsilon$ as

$$\frac{1}{\varepsilon(x)} = \sum_{j=-\infty}^{j=\infty} a_j e^{ijKx} \tag{6.60}$$

$$a_j = \frac{1}{L_x}\int_0^{L_x}\frac{1}{\varepsilon(x)}e^{-ijKx}dx, \tag{6.61}$$

represented by the Toeplitz matrix $\overline{\overline{A}}$. We should thus employ the inverse product rule twice, and the matrix form of the eigenvalue problem becomes

$$\overline{\overline{A}}^{-1}\left(\overline{\overline{K}}\,\overline{\overline{E}}^{-1}\overline{\overline{K}} + k_0^2\overline{\overline{I}}\right)\overline{v} = \beta^2\overline{v}. \tag{6.62}$$

When the magnetic field h_{ym} of the mode m is known, we can similarly to the TE case use Maxwell's equation (2.56) to derive the electric field components as

$$e_x = \frac{\beta}{\omega\varepsilon_0\varepsilon(x)}h_y \tag{6.63}$$

$$e_z = \frac{i}{\omega\varepsilon_0\varepsilon(x)}\partial_x h_y. \tag{6.64}$$

Inspection of Equation (6.63) reveals that the requirement of continuity of the magnetic field h_y component leads to the requirement of continuity of the product $e_x\varepsilon$. For TM polarization, we can thus expect discontinuities in the electric field e_x component at discontinuity points in the permittivity profile.

The electric field expansion vectors become

$$\overline{u_m} = \frac{\beta_m}{\omega\varepsilon_0}\overline{\overline{A}}\,\overline{v_m} \tag{6.65}$$

$$\overline{w'_m} = \frac{i}{\omega\varepsilon_0}\overline{\overline{E}}^{-1}\overline{\overline{K}}\overline{v_m}, \tag{6.66}$$

where we have used the direct rule in Equation (6.65) and the inverse rule in Equation (6.66).

Note: We relate the forward and backward travelling eigenmodes $(\mathbf{e}_m^{\pm}, \mathbf{h}_m^{\pm})$ using the conventions of Equations (2.62 and 2.63), where the transverse magnetic field depends on the sign of β but the transverse electric field does not. If the forward travelling eigenmode is characterized by a lateral field $(\overline{u_m}, \overline{v_m})$, the corresponding backward travelling eigenmode is characterized by $(\overline{u_m}, -\overline{v_m})$. This corresponds to using Equation (6.65) for the backward travelling field and subsequently flipping the signs of both vectors to respect the convention.

The refractive-index profile and the fundamental mode of a simple GaAs waveguide geometry computed using the plane-wave expansion technique for TE and TM polarizations are depicted in Figure 6.13. The corresponding modes computed using the SA approach described later, which can be considered exact to machine precision, are also illustrated. For TE polarization, we observe that already for $M = 5$, the mode agrees very well with the exact profile and for $M = 25$, the two profiles appear indistinguishable. As discussed earlier, the mode profile for TM polarization features discontinuities at the waveguide boundaries. The description of the strong field discontinuities for the GaAs waveguide using plane-wave expansion requires many modes, and we observe that even for $M = 25$, the mode profile still deviates substantially from the exact profile. Generally, for increasing M, we can expect

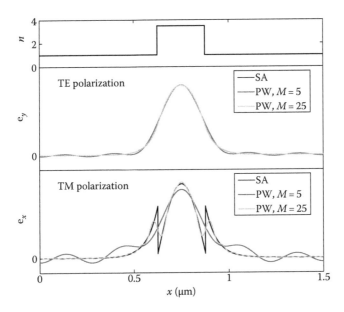

FIGURE 6.13 (**See colour insert.**) The refractive-index geometry (top) of a GaAs waveguide of diameter d surrounded by air. The corresponding fundamental mode computed using plane-wave expansion (PW) with M equaling 5 or 25 and using the SA approach in TE (centre) and TM polarizations (bottom). The geometrical parameters are $d = 250$ nm, $n_{GaAs} = 3.5$, $\lambda = 1$ μm, and $L_x = 1.5$ μm, and periodic boundary conditions with $\mathbf{e}(L_x) = \mathbf{e}(0)$ are employed.

calculations performed for TM polarization to converge more slowly than those for TE polarizations due to this phenomenon.

The solution of the matrix eigenmode problems (Equations 6.47 and 6.62) will automatically yield all N eigenmodes for the geometry under study. The fact that we do not need to worry about overlooking a solution represents a major advantage of the plane-wave expansion technique as compared to the SA approach. Furthermore, the plane-wave technique can be used to find modes in arbitrary geometries including those without analytical solutions. However, many modes are needed to properly describe the field discontinuities occurring for TM polarization. Also, since we need to store N expansion coefficients for each mode, the plane-wave expansion is generally more memory demanding than the SA description of the eigenmodes presented in the following.

6.4.2 SEMI-ANALYTICAL APPROACH

In this section, we will employ a different approach to solve the eigenmode problem in 2D based on analytical expressions for the mode profiles. This procedure can be used when the lateral geometry consists of regions of piecewise constant permittivity as illustrated in Figure 6.14a. This particular geometry consists of three lateral regions, but the theory can be extended in a straightforward manner to handle an arbitrary number of lateral regions.

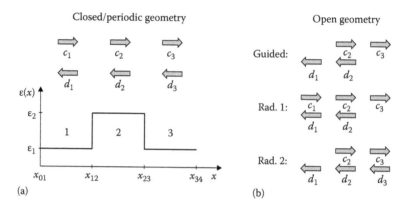

FIGURE 6.14 (a) A three-region lateral permittivity profile with the field coefficients employed for closed/periodic boundary conditions. (b) The corresponding coefficients for the open boundary condition.

We first consider TE polarization and employ Equation (6.32) to compute the eigenmode profiles $e_{ym}(x)$ of mode index m for the example geometry in Figure 6.14a. The eigenmode problem for the lateral region j of constant permittivity ε_j is

$$\partial_x^2 e_y(x) + \varepsilon_j k_0^2 e_y(x) = \beta^2 e_y(x), \tag{6.67}$$

where we have skipped the mode index m for simplicity. Its general solution is given by

$$e_y(x) = c_j e^{ik_j x} + d_j e^{-ik_j x}, \tag{6.68}$$

where k_j and β are related as

$$k_j^2 + \beta^2 = \varepsilon_j k_0^2. \tag{6.69}$$

In each of the three regions of the geometry in Figure 6.14a, the eigenmode can be described using the solution (Equation 6.69) of the corresponding region, such that the mode is given by

$$
\begin{aligned}
e_y(x) &= c_1 e^{ik_1 x} + d_1 e^{-ik_1 x} & (x_{01} \leq x \leq x_{12}) \\
e_y(x) &= c_2 e^{ik_2 x} + d_2 e^{-ik_2 x} & (x_{12} \leq x \leq x_{23}) \\
e_y(x) &= c_3 e^{ik_3 x} + d_3 e^{-ik_3 x} & (x_{23} \leq x \leq x_{34}).
\end{aligned}
\tag{6.70}
$$

The parameters c_j, d_j, and β thus fully characterize the mode profile. The coefficients c_j and d_j are determined by requiring that boundary conditions are fulfilled at the boundaries between the regions. The problem of determining the mode profile along the x-axis resembles the calculation of the 1D field along the z axis in Section 6.3, and it can be tempting to introduce a T matrix formalism to compute the mode profile. However, the possibility of having complex-valued k_j in Equation (6.70) leads to numerical instability of the T matrix formalism. We will thus instead derive a set of linear equations in c_j and d_j for the boundary conditions between the various regions j. In the TE case, we demand the continuity of e_y and $\partial_x e_y$ leading to the conditions

$$c_1 e^{ik_1 x_{12}} + d_1 e^{-ik_1 x_{12}} - c_2 e^{ik_2 x_{12}} - d_2 e^{-ik_2 x_{12}} = 0 \tag{6.71}$$

$$c_1 k_1 e^{ik_1 x_{12}} - d_1 k_1 e^{-ik_1 x_{12}} - c_2 k_2 e^{ik_2 x_{12}} + d_2 k_2 e^{-ik_2 x_{12}} = 0 \tag{6.72}$$

$$c_2 e^{ik_2 x_{23}} + d_2 e^{-ik_2 x_{23}} - c_3 e^{ik_3 x_{23}} - d_3 e^{-ik_3 x_{23}} = 0 \tag{6.73}$$

$$c_2 k_2 e^{ik_2 x_{23}} - d_2 k_2 e^{-ik_2 x_{23}} - c_3 k_3 e^{ik_3 x_{23}} + d_3 k_3 e^{-ik_3 x_{23}} = 0, \tag{6.74}$$

describing the boundary conditions at the positions x_{12} and x_{23}.

The boundary conditions at the outer boundaries x_{01} and x_{34} depend on the type of geometry. For the closed geometry, we require that the e_y field vanish at the boundaries such that

$$\begin{aligned} c_1 e^{ik_1 x_{01}} + d_1 e^{-ik_1 x_{01}} &= 0 \\ c_3 e^{ik_3 x_{34}} + d_3 e^{-ik_3 x_{34}} &= 0, \end{aligned} \tag{6.75}$$

whereas for the periodic geometry the field and its derivative at the outer boundary should reproduce itself multiplied by a phase term $e^{i\alpha}$ such that

$$\begin{aligned} c_1 e^{ik_1 x_{01}} e^{i\alpha} + d_1 e^{-ik_1 x_{01}} e^{i\alpha} - c_3 e^{ik_3 x_{34}} - d_3 e^{-ik_3 x_{34}} &= 0 \\ c_1 k_1 e^{ik_1 x_{01}} e^{i\alpha} - d_1 k_1 e^{-ik_1 x_{01}} e^{i\alpha} - c_3 k_3 e^{ik_3 x_{34}} + d_3 k_3 e^{-ik_3 x_{34}} &= 0. \end{aligned} \tag{6.76}$$

The boundary condition for the open geometry illustrated in Figure 2.2a is not directly compatible with the MM due to the requirement of working with a discrete set of expansion modes. However, the SA approach still allows for a calculation of the guided modes and the continuous set of radiation modes of the open geometry. The relevant field coefficients for the open boundary condition are illustrated in Figure 6.14b. As the open geometry extends to $\pm\infty$ along the x-axis, there is no boundary condition at the points x_{01} and x_{34}. For guided modes, k_1 and k_3 will be imaginary leading to exponentially increasing outward travelling d_1 and c_3 fields, and to ensure that the fields outside the waveguide are decaying, we impose the conditions $d_1 = c_3 = 0$ for the guided modes. For the radiation modes, k_1 and k_3 are real, and all field components can be non-zero. The four conditions (Equations 6.71 through 6.74) for the six field coefficients lead to two linearly independent solutions for any $\beta^2 < \varepsilon_1 k_0^2$. These two solutions can be separated into radiation mode sets 1 and 2 fulfilling either $d_3 = 0$ or $c_1 = 0$ corresponding to modes defined by illumination of the structure from the left or the right, respectively, as illustrated in Figure 6.14b.

When determining the discrete set of modes for any category of boundary conditions, the system of linear equations given by the boundary conditions can be written as

$$\overline{\overline{B}} \, \overline{c} = \overline{0}, \tag{6.77}$$

where $\overline{\overline{B}}$ is the matrix representing the boundary conditions and $\overline{c}$ is the field coefficient vector $[c_1; d_1; c_2; d_2; c_3; d_3]$. The parameters k_j can be computed from β^2 using Equation (6.69), and the elements of $\overline{\overline{B}}$ can thus be considered functions of β^2. The system of linear equations allows for a non-trivial solution when

$$F(\beta^2) \equiv \det(\overline{\overline{B}}(\beta^2)) = 0, \tag{6.78}$$

and in order to compute our eigenmodes, we thus need to search the complex β^2 plane and determine the nodes of the function $F(\beta^2)$. At each node, we have a non-trivial solution to Equation (6.77) corresponding to a mode obeying our boundary conditions. The mode coefficients are then determined by computing the null space of the matrix $\overline{\overline{B}}$.

For a real-valued permittivity profile, the nodes usually lie on the real axis as illustrated in Figure 2.2e. To find the solutions, we thus start from $\varepsilon_2 k_0^2$ and scan the real axis of the β^2 plane towards $-\infty$ looking for nodes of $F(\beta^2)$.

Note: When determining the null space of $\overline{\overline{B}}$ for guided modes with $\text{Im}(k_j) \neq 0$, the magnitudes of the coefficients c_j and d_j can be very different leading to significant numerical errors. These errors can be avoided by introducing the coefficients c_j' and d_j' given by

$$\begin{aligned} c_j' = c_j |e^{ik_j x_{j-1,j}}| \quad d_j' = c_j |e^{ik_j x_{j,j+1}}| \quad (\text{Im}(k_j > 0)) \\ c_j' = c_j |e^{ik_j x_{j,j+1}}| \quad d_j' = c_j |e^{ik_j x_{j-1,j}}| \quad (\text{Im}(k_j < 0)) \end{aligned} \tag{6.79}$$

and setting up the system of equations for the coefficients c_j' and d_j' instead. Once the corresponding null space has been computed, the eigenvector coefficients c_j and d_j can then be recovered from c_j' and d_j' using Equation (6.79).

Note: When $\beta^2 = \varepsilon_j k_0^2$, we have $k_j = 0$ and $F(\beta^2) = 0$ leading to a mathematical but non-physical solution to the system of linear equations. This non-physical solution can be avoided by replacing the boundary conditions (Equations 6.72 and 6.74) with

$$c_1 e^{ik_1 x_{12}} - d_1 e^{-ik_1 x_{12}} - c_2 \frac{k_2}{k_1} e^{ik_2 x_{12}} + d_2 \frac{k_2}{k_1} e^{-ik_2 x_{12}} = 0 \tag{6.80}$$

$$c_2 e^{ik_2 x_{23}} - d_2 e^{-ik_2 x_{23}} - c_3 \frac{k_3}{k_2} e^{ik_3 x_{23}} + d_3 \frac{k_3}{k_2} e^{-ik_3 x_{23}} = 0. \tag{6.81}$$

Note: Even when the permittivity profile is strictly real valued, nodes away from the real axis with complex eigenvalues β^2 as discussed in Section 2.7.3 may in rare cases appear. For these geometries, the off-axis nodes may be determined using an algorithm similar to the one for complex permittivity profiles discussed later.

In the particular case of a uniform single-region permittivity profile with dielectric constant $\varepsilon(x) = \varepsilon_1$ and periodic boundary conditions $e_y(x + L_x) = e_y(x)e^{i\alpha}$ with $\alpha = 0$, the eigenmodes are plane waves of the form of Equation (6.68) with $j = 1$ and $k_1 = m2\pi/L_x$, where m is an integer. For each $k_1 \neq 0$, the eigenmodes appear in degenerate pairs with expansion coefficients (c_1, d_1) which can be written as $(1, 0)$ and $(1, 0)$. However, the field can equally well be expanded on any linear

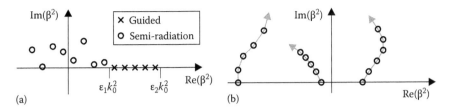

FIGURE 6.15 (a) Distribution of propagation constants squared β^2 in the complex plane with guided modes (crosses) and semi-radiation modes (circles) for a geometry with complex permittivity profile or with absorbing boundary condition. (b) Example trajectories of β^2 positions as the imaginary part of the permittivity profile is increased.

combination of degenerate eigenmodes, and to maintain compatibility with the scalar product (Equation 6.98), it is preferable to instead employ eigenmodes in the uniform geometry with expansion coefficients (c_1, d_1) given by $(1, -1)$ and $(1, 1)$. This issue is discussed further in the following section. However, when $\alpha \neq 0$, all the eigenmodes are non-degenerate, and in this case, the linear combinations $(1, 0)$ and $(1, 0)$ must be used.

When computing modes for a geometry featuring a complex permittivity profile $\varepsilon(x) = \varepsilon_R(x) + i\varepsilon_I(x)$ or an absorbing boundary condition as discussed further in Section 6.4.5, the nodes no longer appear on the real axis but move into the complex plane as illustrated in Figure 6.15a. Here, we face the challenge of determining nodes of an unknown function in a 2D domain. One method that we may adopt is the following two-step routine:

Algorithm for Identification of Nodes

The propagation constants β_m^2 representing nodes of Equation (6.78) in the complex plane in the case of a complex permittivity profile can be identified using the following steps:

1. We initially consider the real-valued profile $\varepsilon_0(x) = \varepsilon_R(x)$. The nodes will lie on the real axis, and by scanning along it, we determine an initial set of propagation constants corresponding to the profile $\varepsilon_0(x)$.
2. We then consider the permittivity profile $\varepsilon_\gamma(x) = \varepsilon_R(x) + \gamma\varepsilon_I(x)i$. For $\gamma = 0$, this profile is identical to $\varepsilon_0(x)$. We then gradually increase γ towards 1 and carefully follow each of the nodes. As γ increases, the nodes move away from the real axis and into the complex β^2 plane as illustrated in Figure 6.15b. In this iterative scheme, we determine the node position of a mode m for a specific value γ_j, and then we look for its new position for the value $\gamma_{j+1} = \gamma_j + \Delta\gamma$. The increment $\Delta\gamma$ should be chosen such that

$$|\beta_m^2(\gamma_{j+1}) - \beta_m^2(\gamma_j)| \ll \min_{n \neq m}(|\beta_m^2(\gamma_j) - \beta_n^2(\gamma_j)|). \qquad (6.82)$$

> Using this criterion, the incremental displacement of the node position β_m^2 is small compared to the distance to the closest neighbouring nodal position, which ensures that mistaking the nodes $\beta_m^2(\gamma_{j+1})$ and $\beta_n^2(\gamma_{j+1})$ is avoided.
>
> The node positions of the original profile $\varepsilon(x)$ are then obtained for $\gamma = 1$.

Once the electric field profiles for the eigenmodes are determined, we can compute the magnetic fields using Equations (6.52) and (6.53).

The mode profiles $e_y m(x)$ and the propagation constants β_m obtained using the SA technique are exact to machine precision. However, we need to manually search the β^2 plane for nodes. To date, no truly efficient method for identifying these nodes has been proposed, and this issue represents a drawback of the SA approach.

Finally, we outline the corresponding procedure for computing modes in TM polarization. The eigenproblem (Equation 6.33) for a piecewise constant lateral region is given by

$$\partial_x^2 h_y(x) + \varepsilon_j k_0^2 h_y(x) = \beta^2 h_y(x), \tag{6.83}$$

and we obtain the TM formalism simply by performing the substitution $e_y \rightarrow h_y$ in Equations (6.68) through (6.71) and (6.73). However, we now instead require continuity of the terms h_y and $(1/\varepsilon)\partial_x h_y$, and we should thus replace Equations (6.80) and (6.81) with

$$c_1 \frac{1}{\varepsilon_1} e^{ik_1 x_{12}} - d_1 \frac{1}{\varepsilon_1} e^{-ik_1 x_{12}} - c_2 \frac{k_2}{\varepsilon_2 k_1} e^{ik_2 x_{12}} + d_2 \frac{k_2}{\varepsilon_2 k_1} e^{-ik_2 x_{12}} = 0 \tag{6.84}$$

$$c_2 \frac{1}{\varepsilon_2} e^{ik_2 x_{23}} - d_2 \frac{1}{\varepsilon_2} e^{-ik_2 x_{23}} - c_3 \frac{k_3}{\varepsilon_3 k_2} e^{ik_3 x_{23}} + d_3 \frac{k_3}{\varepsilon_3 k_2} e^{-ik_3 x_{23}} = 0, \tag{6.85}$$

for the boundary conditions at the positions x_{12} and x_{23} in TM polarization. The electric field can then be derived using Equations (6.63) and (6.64).

Fundamental mode profiles for the GaAs waveguide computed using the SA approach in TE and TM polarizations are shown in Figure 6.13. The field profiles are exact to machine precision, and, unlike the plane-wave expansion technique, the discontinuities present for TM polarization pose no difficulty. This represents an advantage of the SA approach.

6.4.3 INTERFACE

Having determined eigenmodes for layers uniform along the z axis, we again turn to the two-layer structure. In each of the two layers, the field is expanded on eigenmodes as in Equation (6.2). We now consider an incoming field $\overline{a}_1$ consisting of a single mode m such that the vector elements are $a_{1j} = \delta_{jm}$. The field is travelling forward in layer 1 as sketched in Figure 6.16b; it reaches the layer interface and is partly reflected giving rise to a backward travelling field in layer 1. The incoming field,

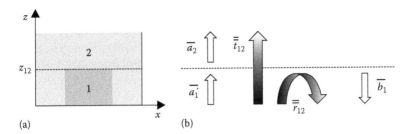

FIGURE 6.16 (a) A two-layer geometry. (b) The reflection and transmission of light at the interface.

however, is also partly transmitted, and this results in a forward travelling field in layer 2. We then write the electric and magnetic fields in the two layers as

$$\mathbf{E}_1(\mathbf{r}) = a'_{1m}\mathbf{e}^+_{1m}(\mathbf{r}_\perp)e^{i\beta_{1m}(z-z_{12})} + \sum_j^N b_{1j}\mathbf{e}^-_{1j}(\mathbf{r}_\perp)e^{-i\beta_{1j}(z-z_{12})} \tag{6.86}$$

$$\mathbf{E}_2(\mathbf{r}) = \sum_j^N a_{2j}\mathbf{e}^+_{2j}(\mathbf{r}_\perp)e^{i\beta_{2j}(z-z_{12})} \tag{6.87}$$

$$\mathbf{H}_1(\mathbf{r}) = a'_{1m}\mathbf{h}^+_{1m}(\mathbf{r}_\perp)e^{i\beta_{1m}(z-z_{12})} + \sum_j^N b_{1j}\mathbf{h}^-_{1j}(\mathbf{r}_\perp)e^{-i\beta_{1j}(z-z_{12})} \tag{6.88}$$

$$\mathbf{H}_2(\mathbf{r}) = \sum_j^N a_{2j}\mathbf{h}^+_{2j}(\mathbf{r}_\perp)e^{i\beta_{2j}(z-z_{12})}. \tag{6.89}$$

The fields are then given by $(\mathbf{E}, \mathbf{H}) = (\mathbf{E}_1, \mathbf{H}_1)$ when $z \leq z_{12}$ and by $(\mathbf{E}, \mathbf{H}) = (\mathbf{E}_2, \mathbf{H}_2)$ when $z \geq z_{12}$. We will now fix $a'_{1m} = 1$; however, the coefficients b_{1j} and a_{2j} are yet unknown.

We now introduce the reflection and transmission matrices $\overline{\overline{r}}_{12}$ and $\overline{\overline{t}}_{12}$ defined by $b_{1j} = r_{12;jm}a'_{1m}$ and $a_{2j} = t_{12;jm}a'_{1m}$. The coefficients of the matrices $\overline{\overline{r}}_{12}$ and $\overline{\overline{t}}_{12}$ are obtained by requiring that the boundary conditions are fulfilled at the interface between layers 1 and 2, meaning that the tangential components of the electric and magnetic fields should be continuous. From Equations (6.86) and (6.89), we obtain

$$\mathbf{e}_{\perp 1m}(\mathbf{r}_\perp) + \sum_j^N \mathbf{e}_{\perp 1j}(\mathbf{r}_\perp)r_{12;jm} = \sum_j^N \mathbf{e}_{\perp 2j}(\mathbf{r}_\perp)t_{12;jm} \tag{6.90}$$

$$\mathbf{h}_{\perp 1m}(\mathbf{r}_\perp) - \sum_j^N \mathbf{h}_{\perp 1j}(\mathbf{r}_\perp)r_{12;jm} = \sum_j^N \mathbf{h}_{\perp 2j}(\mathbf{r}_\perp)t_{12;jm}, \tag{6.91}$$

where a minus sign appears on the LHS of Equation (6.91) in front of the backward travelling magnetic field lateral components as discussed in Section 2.7.

The procedure to determine the reflection and transmission matrices from Equations (6.90) and (6.91) now depends on the method chosen to describe the eigenmodes. When working with eigenmodes described using plane-wave expansion, the lateral electric $\mathbf{e}_\perp$ and magnetic $\mathbf{h}_\perp$ fields can simply be described using their Fourier series $\bar{u}$ and $\bar{v}$, respectively, such that

$$\overline{u_{1m}} + \sum_j^N \overline{u_{1j}}\, r_{12;jm} = \sum_j^N \overline{u_{2j}}\, t_{12;jm} \tag{6.92}$$

$$\overline{v_{1m}} - \sum_j^N \overline{v_{1j}}\, r_{12;jm} = \sum_j^N \overline{v_{2j}}\, t_{12;jm}. \tag{6.93}$$

The matrix forms of these equations are

$$\overline{\overline{u_1}} + \overline{\overline{u_1}}\,\overline{\overline{r_{12}}} = \overline{\overline{u_2}}\,\overline{\overline{t_{12}}} \tag{6.94}$$

$$\overline{\overline{v_1}} - \overline{\overline{v_1}}\,\overline{\overline{r_{12}}} = \overline{\overline{v_2}}\,\overline{\overline{t_{12}}}, \tag{6.95}$$

where $\overline{\overline{u}}$ and $\overline{\overline{v}}$ are matrices with columns m given by the Fourier series $\overline{u_m}$ and $\overline{v_m}$ for the eigenmode m. Multiplying Equations (6.94) and (6.95) with $\overline{\overline{u_1}}^{-1}$ and $\overline{\overline{v_1}}^{-1}$, respectively, and taking their sums and differences leads to

$$\overline{\overline{t_{12}}} = 2\left(\overline{\overline{u_1}} \backslash \overline{\overline{u_2}} + \overline{\overline{v_1}} \backslash \overline{\overline{v_2}}\right)^{-1} \tag{6.96}$$

$$\overline{\overline{r_{12}}} = \frac{1}{2}\left(\overline{\overline{u_1}} \backslash \overline{\overline{u_2}} - \overline{\overline{v_1}} \backslash \overline{\overline{v_2}}\right)\overline{\overline{t_{12}}}, \tag{6.97}$$

where $\backslash$ denotes matrix division, that is $\overline{\overline{A}} \backslash \overline{\overline{B}} = \left(\overline{\overline{A}}\right)^{-1}\overline{\overline{B}}$. The expressions (6.96) and (6.97) are valid for any periodic boundary condition $\mathbf{E}(x + L_x) = \mathbf{E}(x)e^{i\alpha}$ chosen, including the case with $e^{i\alpha} \neq 1$.

When working with eigenmodes described using the SA approach, we cannot match the field coefficients of the eigenmodes directly, and the implementation of a scalar product is required to derive the reflection and transmission matrices from Equations (6.90) and (6.91). We thus introduce the scalar product given by

$$\langle \mathbf{e}_m | \mathbf{h}_n \rangle = \int (\mathbf{e}_{\perp m} \times \mathbf{h}_{\perp n}) \cdot \hat{z} d\mathbf{r}_\perp = \int (\mathbf{e}_m \times \mathbf{h}_n) \cdot \hat{z} d\mathbf{r}_\perp, \tag{6.98}$$

and in the following, we skip the subindex $\perp$ for simplicity.

Note: The previous notation is known as *bra–ket* notation. Here, the isolated term $\langle \mathbf{e}_m |$ is called the *bra*, the isolated term $|\mathbf{h}_{in}\rangle$ is called the *ket*, and these isolated terms simply represent the functions $\mathbf{e}_m$ and $\mathbf{h}_n$, respectively. However, the product term $\langle \mathbf{e}_m | \mathbf{h}_n \rangle$ is understood to describe the integral of their cross product as defined in Equation (6.98).

It is convenient to work with eigenmodes, which have been normalized according to the scalar product. The normalization can be performed by computing the normalization constant $C_m = \langle e_m | h_m \rangle$ and performing the substitution $(e_m, h_m) \rightarrow (e_m, h_m)/\sqrt{C_m}$ such that $\langle e_m | h_m \rangle = 1$. In the following, we assume that our mode profiles have been normalized using this scalar product. We then have from the orthogonality relation of Equation (2.90) that $\langle e_{qm} | h_{qn} \rangle = \delta_{mn}$, where q is the layer index.

Note: For a uniform single-region geometry with $\varepsilon(x) = \varepsilon_1$ and periodic boundary conditions $e(x + L_x) = e(x)e^{i\alpha}$ with $\alpha = 0$, a degenerate eigenmode m with expansion coefficients (c_1, d_1) given by $(1,0)$ or $(0,1)$ has a normalization constant $C_m = \langle e_m | h_m \rangle = 0$. To perform the normalization, it is thus preferable to work with the linear combinations $(1,-1)$ and $(1,1)$ for the degenerate eigenmodes for which $C_m = \langle e_m | h_m \rangle \neq 0$.

Now, multiplying the bra version of Equation (6.90) with $|h_{1n}\rangle$ and the ket version of Equation (6.91) with $\langle e_{1n} |$ gives

$$\delta_{mn} + r_{12;nm} = \sum_j^N \langle e_{2j} | h_{1n} \rangle t_{12;jm} \tag{6.99}$$

$$\delta_{mn} - r_{12;nm} = \sum_j^N \langle e_{1n} | h_{2j} \rangle t_{12;jm}. \tag{6.100}$$

Adding and subtracting these two equations, we obtain

$$2\delta_{mn} = \sum_j^N (\langle e_{2j} | h_{1n} \rangle + \langle e_{1n} | h_{2j} \rangle) t_{12;jm} \tag{6.101}$$

$$r_{12;nm} = \frac{1}{2} \sum_j^N (\langle e_{2j} | h_{1n} \rangle - \langle e_{1n} | h_{2j} \rangle) t_{12;jm}. \tag{6.102}$$

In matrix form, this can be written as

$$\overline{\overline{t}}_{12} = 2 \left(\overline{\overline{\langle e_2 | h_1 \rangle}}^T + \overline{\overline{\langle e_1 | h_2 \rangle}} \right)^{-1} \tag{6.103}$$

$$\overline{\overline{r}}_{12} = \frac{1}{2} \left(\overline{\overline{\langle e_2 | h_1 \rangle}}^T - \overline{\overline{\langle e_1 | h_2 \rangle}} \right) \overline{\overline{t}}_{12}, \tag{6.104}$$

where superscript T denotes transposition, and these equations define the reflection and transmission matrices. When the reflection and transmission matrices are known, the coefficients $\overline{b}_1$ and $\overline{a}_2$ can be determined, and the field is well defined by Equations (6.86) through (6.89).

In Figure 6.16, the structure is illuminated from below. However, we could equally well illuminate the geometry from above, and in this case, we would use the matrices $\overline{\overline{r}}_{21}$ and $\overline{\overline{t}}_{21}$ to describe the field. The expressions for these matrices are obtained by simply inverting the indices 1 and 2 in Equations (6.103) and (6.104).

In the derivation of the reflection and transmission matrices for eigenmodes described using the SA approach, we have employed the orthogonality relation

(Equation 2.90), which as noted in Section 2.7.4 does not hold for periodic boundary conditions when $e^{i\alpha} \neq 1$. However, the expressions (Equations 6.103 and 6.104) can be extended to also handle the situation with non-orthogonal eigenmodes. This extension is the subject of Exercise 6.7.

6.4.4 S MATRIX THEORY

We can describe a two-layer geometry using the reflection and transmission matrices introduced earlier. Now, let us consider the three-layer geometry illustrated in Figure 6.17a.

As previously, we illuminate the structure from below with light described by the eigenmode m. The electric field is given by

$$\mathbf{E}_1(\mathbf{r}) = a'_{1m}\mathbf{e}^+_{1m}(\mathbf{r}_\perp)e^{i\beta_{1m}(z-z_{12})} + \sum_j^N b_{1j}\mathbf{e}^-_{1j}(\mathbf{r}_\perp)e^{-i\beta_{1j}(z-z_{12})} \qquad (6.105)$$

$$\mathbf{E}_2(\mathbf{r}) = \sum_j^N a_{2j}\mathbf{e}^+_{2j}(\mathbf{r}_\perp)e^{i\beta_{2j}(z-z_{12})} + \sum_j^N b_{2j}\mathbf{e}^-_{2j}(\mathbf{r}_\perp)e^{-i\beta_{2j}(z-z_{23})} \qquad (6.106)$$

$$\mathbf{E}_3(\mathbf{r}) = \sum_j^N a_{3j}\mathbf{e}^+_{3j}(\mathbf{r}_\perp)e^{i\beta_{3j}(z-z_{23})}, \qquad (6.107)$$

and similarly for the magnetic field.

For the layer 2, we define coefficients a'_{2j} and b'_{2j} by $a'_{2j} = p_{2j}a_{2j}$ and $b'_{2j} = p_{2j}b_{2j}$, where $p_{2j} = e^{i\beta_{2j}L_2}$ and $L_2 = z_{23} - z_{12}$. The coefficients a_j and b_j can be thought of as *having their back against the interface*, where the coefficients a'_j and b'_j are *facing the interface*.

Referring to Figure 6.17b, we now consider the reflection of light by the three-layer geometry. The light of mode m travels forward in layer 1. At the interface, it is

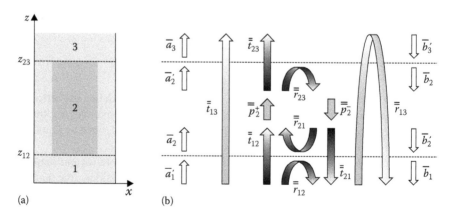

(a)   (b)

FIGURE 6.17 (a) A three-layer geometry. (b) The reflection and transmission at the two interfaces and the multiple scattering of light inside the central layer are illustrated.

partly reflected back into layer 1 with reflection described by the matrix $\overline{\overline{r_{12}}}$. However, part of the light is also transmitted into layer 2. It propagates towards the interface between layers 2 and 3 and is partly reflected. It then travels backwards towards the first interface and is partly transmitted at the interface into layer 1. This journey can be described by the matrix $\overline{\overline{t_{21}}}\,\overline{\overline{p_2^-}}\,\overline{\overline{r_{23}}}\,\overline{\overline{p_2^+}}\,\overline{\overline{t_{12}}}$, where $\overline{\overline{p_2^+}} = \overline{\overline{p_2^-}} = \overline{\overline{p_2}}$ is the propagation matrix for the layer 2, a diagonal matrix consisting of the elements p_{2j}. At this stage, the $\overline{\overline{p_2^\pm}}$ matrix does not depend on the direction $\pm$ of the propagation; however, we include the $+$ and $-$ signs for compatibility with Section 6.5. This journey consisting of one round trip inside layer 2 adds an additional reflection to that described by $\overline{\overline{r_{12}}}$. There is, however, also contributions from light making two, three, etc. round trips in layer 2. The scattering reflection matrix $\overline{\overline{r_{13}}}$ describing reflection at the interface between layers 1 and 2 taking into account multiple reflections inside layer 2 is thus given by

$$\overline{\overline{r_{13}}} = \overline{\overline{r_{12}}} + \overline{\overline{t_{21}}}\,\overline{\overline{p_2^-}}\,\overline{\overline{r_{23}}}\,\overline{\overline{p_2^+}}\,\overline{\overline{t_{12}}} + \overline{\overline{t_{21}}}\,\overline{\overline{p_2^-}}\,\overline{\overline{r_{23}}}\,\overline{\overline{p_2^+}} \left(\overline{\overline{r_{21}}}\,\overline{\overline{p_2^-}}\,\overline{\overline{r_{23}}}\,\overline{\overline{p_2^+}} \right) \overline{\overline{t_{12}}}$$

$$+ \overline{\overline{t_{21}}}\,\overline{\overline{p_2^-}}\,\overline{\overline{r_{23}}}\,\overline{\overline{p_2^+}} \left(\overline{\overline{r_{21}}}\,\overline{\overline{p_2^-}}\,\overline{\overline{r_{23}}}\,\overline{\overline{p_2^+}} \right)^2 \overline{\overline{t_{12}}} + \cdots, \tag{6.108}$$

where the product $\overline{\overline{r_{21}}}\,\overline{\overline{p_2^-}}\,\overline{\overline{r_{23}}}\,\overline{\overline{p_2^+}}$ describes a round trip in layer 2. We now employ the relation for a geometric series $\overline{\overline{I}} + \overline{\overline{a}} + \overline{\overline{a}}^2 + \cdots = (\overline{\overline{I}} - \overline{\overline{a}})^{-1}$, where $\overline{\overline{I}}$ is the identity matrix. This relation holds for matrices with absolute eigenvalues below unity. The defining relation becomes

$$\overline{\overline{r_{13}}} = \overline{\overline{r_{12}}} + \overline{\overline{t_{21}}}\,\overline{\overline{p_2^-}}\,\overline{\overline{r_{23}}}\,\overline{\overline{p_2^+}} \left(\overline{\overline{I}} - \overline{\overline{r_{21}}}\,\overline{\overline{p_2^-}}\,\overline{\overline{r_{23}}}\,\overline{\overline{p_2^+}} \right)^{-1} \overline{\overline{t_{12}}}. \tag{6.109}$$

Again, considering the multiple reflections in layer 2, one can in a similar manner derive the scattering transmission matrix $\overline{\overline{t_{13}}}$ describing transmission of light from layer 1 to layer 3 and the matrices $\overline{\overline{r_{31}}}$ and $\overline{\overline{t_{31}}}$ corresponding to reflection and transmission of a backward travelling field in layer 3. They are given by

$$\overline{\overline{t_{13}}} = \overline{\overline{t_{23}}}\,\overline{\overline{p_2^+}} \left(\overline{\overline{I}} - \overline{\overline{r_{21}}}\,\overline{\overline{p_2^-}}\,\overline{\overline{r_{23}}}\,\overline{\overline{p_2^+}} \right)^{-1} \overline{\overline{t_{12}}}, \tag{6.110}$$

$$\overline{\overline{r_{31}}} = \overline{\overline{r_{32}}} + \overline{\overline{t_{23}}}\,\overline{\overline{p_2^+}} \left(\overline{\overline{I}} - \overline{\overline{r_{21}}}\,\overline{\overline{p_2^-}}\,\overline{\overline{r_{23}}}\,\overline{\overline{p_2^+}} \right)^{-1} \overline{\overline{r_{21}}}\,\overline{\overline{p_2^-}}\,\overline{\overline{t_{32}}} \tag{6.111}$$

$$\overline{\overline{t_{31}}} = \overline{\overline{t_{21}}}\,\overline{\overline{p_2^-}}\,\overline{\overline{t_{32}}} + \overline{\overline{t_{21}}}\,\overline{\overline{p_2^-}}\,\overline{\overline{r_{23}}}\,\overline{\overline{p_2^+}} \left(\overline{\overline{I}} - \overline{\overline{r_{21}}}\,\overline{\overline{p_2^-}}\,\overline{\overline{r_{23}}}\,\overline{\overline{p_2^+}} \right)^{-1} \overline{\overline{r_{21}}}\,\overline{\overline{p_2^-}}\,\overline{\overline{t_{32}}}. \tag{6.112}$$

Slightly shortened expressions for $\overline{\overline{r_{31}}}$ and $\overline{\overline{t_{31}}}$ can be obtained by implementing the inverse matrix $\left(\overline{\overline{I}} - \overline{\overline{r_{23}}}\,\overline{\overline{p_2^+}}\,\overline{\overline{r_{21}}}\,\overline{\overline{p_2^-}} \right)^{-1}$ instead; however, it may be numerically preferable to only have to invert one matrix.

The relations $\overline{b_1} = \overline{\overline{r_{13}}}\,\overline{a_1'}$ and $\overline{a_3} = \overline{\overline{t_{13}}}\,\overline{a_1'}$ give the fields in layers 1 and 3. In the central layer 2, the field is determined from the expansion vectors $\overline{a_2}$ and $\overline{b_2}$

describing the forward and backward travelling parts of the field. The vector $\overline{a_2}$ consists of the contributions $\overline{\overline{t_{12}}}\,\overline{a_1'}$ and $\overline{\overline{r_{21}}}\,\overline{p_2^-}\,\overline{\overline{t_{32}}}\,\overline{b_3'}$ due to illumination from below and above, respectively. Its complete expression becomes

$$\overline{a_2} = \left(\overline{\overline{I}} - \overline{\overline{r_{21}}}\,\overline{p_2^-}\,\overline{\overline{r_{23}}}\,\overline{p_2^+} \right)^{-1} \left(\overline{\overline{t_{12}}}\,\overline{a_1'} + \overline{\overline{r_{21}}}\,\overline{p_2^-}\,\overline{\overline{t_{32}}}\,\overline{b_3'} \right), \tag{6.113}$$

where the inverse matrix in a similar way as in Equation (6.109) takes into account the effects of multiple round trips of light inside layer 2. The corresponding expression for $\overline{b_2}$ becomes

$$\overline{b_2} = \left(\overline{\overline{I}} - \overline{\overline{r_{23}}}\,\overline{p_2^+}\,\overline{\overline{r_{21}}}\,\overline{p_2^-} \right)^{-1} \left(\overline{\overline{r_{23}}}\,\overline{p_2^+}\,\overline{\overline{t_{12}}}\,\overline{a_1'} + \overline{\overline{t_{32}}}\,\overline{b_3'} \right), \tag{6.114}$$

where the multiple reflections are again taken into account. The scattering matrices equations (6.109) through (6.112) and Equations (6.113) and (6.114) for the fields in the central layer completely describe the field in the entire three-layer geometry.

We can introduce a convenient shorthand notation for the procedure in Equations (6.109) through (6.112) used to compute full S matrix linking layers 1 and 3. We first define the scattering matrix $\overline{\overline{S_{pq}}}$ between layers p and q as

$$\begin{bmatrix} a_q \\ b_p' \end{bmatrix} = \overline{\overline{S_{pq}}} \begin{bmatrix} a_p \\ b_q' \end{bmatrix}, \tag{6.115}$$

where we are taking into account the thickness of the layer p but not the presence of neighbouring layers below the layer p and above the layer q. The shorthand notation is then given as

$$S_{13} = S_{12} \otimes S_{23}, \tag{6.116}$$

where the subindices here are understood to refer to the layer numbers and the double bars have been skipped for simplicity.

When scattering reflection and transmission matrices are determined in a three-layer structure, we can compute the field in a structure of a given number of layers recursively. The recursive procedure can be understood by inspecting the four-layer geometry of Figure 6.18. We first calculate the scattering matrices $\overline{\overline{r_{13}}}$, $\overline{\overline{t_{13}}}$, $\overline{\overline{r_{31}}}$, and $\overline{\overline{t_{31}}}$ using the procedure outlined earlier. We can then conceptually treat the four-layer structure as a three-layer geometry, where reflection and transmission between layers 1 and 3 are described using the scattering matrices listed earlier. The scattering reflection matrix $\overline{\overline{r_{14}}}$ is then given by an expression similar to Equation (6.109) where the matrices $\overline{\overline{r_{12}}}$, $\overline{\overline{t_{12}}}$, etc. are simply replaced by $\overline{\overline{r_{13}}}$, $\overline{\overline{t_{13}}}$, etc.

More generally, once the reflection and transmission matrices $\overline{\overline{r_{1q}}}$, $\overline{\overline{t_{1q}}}$, $\overline{\overline{r_{q1}}}$, and $\overline{\overline{t_{q1}}}$ for the q'th layer geometry are known, we can compute the matrices $\overline{\overline{r_{1,q+1}}}$, $\overline{\overline{t_{1,q+1}}}$, $\overline{\overline{r_{q+1,1}}}$, and $\overline{\overline{t_{q+1,1}}}$ by conceptually reducing the structure to a three-layer structure and describing the bottom reflection and transmission using the matrices $\overline{\overline{r_{1q}}}$, $\overline{\overline{t_{1q}}}$, $\overline{\overline{r_{q1}}}$,

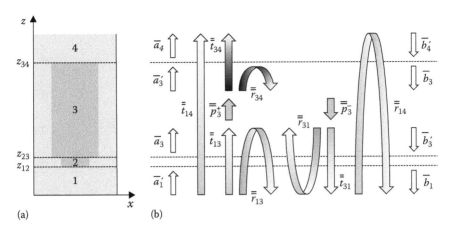

FIGURE 6.18 (a) A four-layer geometry. (b) The reflection and transmission of light can be described by conceptually reducing the structure to a three-layer geometry.

and $\overline{\overline{t_{q1}}}$. This recursion formula connects the fields in the first with the q'th layer and can be summarized in shorthand notation as

$$S_{1,q+1} = S_{1q} \otimes S_{q,q+1}. \tag{6.117}$$

In a structure with n total layers, we will also need to compute reflection and transmission matrices $\overline{\overline{r_{qn}}}$, $\overline{\overline{t_{qn}}}$, $\overline{\overline{r_{nq}}}$, and $\overline{\overline{t_{nq}}}$ connecting the fields between the q'th and the final n'th layers. The recursive procedure is the same, except that here we start by solving the three-layer geometry consisting of the layers $n-2$, $n-1$, and n. Having obtained the $\overline{\overline{r_{n-2,n}}}$, $\overline{\overline{t_{n-2,n}}}$, $\overline{\overline{r_{n,n-2}}}$, and $\overline{\overline{t_{n,n-2}}}$, we proceed recursively by conceptually reducing the four-layer structure consisting of layers $n-3$ to n to a three-layer geometry by introducing the previously calculated matrices $\overline{\overline{r_{n-2,n}}}$, $\overline{\overline{t_{n-2,n}}}$, $\overline{\overline{r_{n,n-2}}}$, and $\overline{\overline{t_{n,n-2}}}$ to describe reflection and transmission at the top. In this recursive manner, we compute the matrices $\overline{\overline{r_{q-1,n}}}$, $\overline{\overline{t_{q-1,n}}}$, $\overline{\overline{r_{n,q-1}}}$, and $\overline{\overline{t_{n,q-1}}}$ from knowledge of $\overline{\overline{r_{qn}}}$, $\overline{\overline{t_{qn}}}$, $\overline{\overline{r_{nq}}}$, and $\overline{\overline{t_{nq}}}$, and the iterative procedure can be written in shorthand notation as

$$S_{q-1,N} = S_{q-1,q} \otimes S_{qn}. \tag{6.118}$$

The defining recursive relations for all the scattering matrices used in the S matrix formalism are

$$\overline{\overline{r_{1,q+1}}} = \overline{\overline{r_{1q}}} + \overline{\overline{t_{q1}}}\,\overline{\overline{p_q^-}}\,\overline{\overline{r_{q,q+1}}}\,\overline{\overline{p_q^+}} \left(\overline{\overline{I}} - \overline{\overline{r_{q1}}}\,\overline{\overline{p_q^-}}\,\overline{\overline{r_{q,q+1}}}\,\overline{\overline{p_q^+}}\right)^{-1}\overline{\overline{t_{1q}}} \tag{6.119}$$

$$\overline{\overline{t_{1,q+1}}} = \overline{\overline{t_{q,q+1}}}\,\overline{\overline{p_q^+}} \left(\overline{\overline{I}} - \overline{\overline{r_{q1}}}\,\overline{\overline{p_q^-}}\,\overline{\overline{r_{q,q+1}}}\,\overline{\overline{p_q^+}}\right)^{-1}\overline{\overline{t_{1q}}} \tag{6.120}$$

$$\overline{\overline{r_{q+1,1}}} = \overline{\overline{r_{q+1,q}}} + \overline{\overline{t_{q,q+1}}}\,\overline{\overline{p_q^+}} \left(\overline{\overline{I}} - \overline{\overline{r_{q1}}}\,\overline{\overline{p_q^-}}\,\overline{\overline{r_{q,q+1}}}\,\overline{\overline{p_q^+}}\right)^{-1}\overline{\overline{r_{q1}}}\,\overline{\overline{p_q^-}}\,\overline{\overline{t_{q+1,q}}} \tag{6.121}$$

$$\overline{\overline{t_{q+1,1}}} = \overline{\overline{t_{q1}}}\,\overline{\overline{P_q^-}}\,\overline{\overline{t_{q+1,q}}} + \overline{\overline{t_{q1}}}\,\overline{\overline{P_q^-}}\,\overline{\overline{r_{q,q+1}}}\,\overline{\overline{P_q^+}}\left(\overline{\overline{I}} - \overline{\overline{r_{q1}}}\,\overline{\overline{P_q^-}}\,\overline{\overline{r_{q,q+1}}}\,\overline{\overline{P_q^+}}\right)^{-1}\overline{\overline{r_{q1}}}\,\overline{\overline{P_q^-}}\,\overline{\overline{t_{q+1,q}}} \tag{6.122}$$

$$\overline{\overline{r_{q-1,n}}} = \overline{\overline{r_{q-1,q}}} + \overline{\overline{t_{q,q-1}}}\,\overline{\overline{P_q^-}}\,\overline{\overline{r_{qn}}}\,\overline{\overline{P_q^+}}\left(\overline{\overline{I}} - \overline{\overline{r_{q,q-1}}}\,\overline{\overline{P_q^-}}\,\overline{\overline{r_{qn}}}\,\overline{\overline{P_q^+}}\right)^{-1}\overline{\overline{t_{q-1,q}}} \tag{6.123}$$

$$\overline{\overline{t_{q-1,n}}} = \overline{\overline{t_{qn}}}\,\overline{\overline{P_q^+}}\left(\overline{\overline{I}} - \overline{\overline{r_{q,q-1}}}\,\overline{\overline{P_q^-}}\,\overline{\overline{r_{qn}}}\,\overline{\overline{P_q^+}}\right)^{-1}\overline{\overline{t_{q-1,q}}} \tag{6.124}$$

$$\overline{\overline{r_{n,q-1}}} = \overline{\overline{r_{nq}}} + \overline{\overline{t_{qn}}}\,\overline{\overline{P_q^+}}\left(\overline{\overline{I}} - \overline{\overline{r_{q,q-1}}}\,\overline{\overline{P_q^-}}\,\overline{\overline{r_{qn}}}\,\overline{\overline{P_q^+}}\right)^{-1}\overline{\overline{r_{q,q-1}}}\,\overline{\overline{P_q^-}}\,\overline{\overline{t_{nq}}} \tag{6.125}$$

$$\overline{\overline{t_{n,q-1}}} = \overline{\overline{t_{q,q-1}}}\,\overline{\overline{P_q^-}}\,\overline{\overline{t_{nq}}} + \overline{\overline{t_{q,q-1}}}\,\overline{\overline{P_q^-}}\,\overline{\overline{r_{qn}}}\,\overline{\overline{P_q^+}}\left(\overline{\overline{I}} - \overline{\overline{r_{q,q-1}}}\,\overline{\overline{P_q^-}}\,\overline{\overline{r_{qn}}}\,\overline{\overline{P_q^+}}\right)^{-1}\overline{\overline{r_{q,q-1}}}\,\overline{\overline{P_q^-}}\,\overline{\overline{t_{nq}}}. \tag{6.126}$$

Finally the field expansion vectors $\overline{a_q}$ and $\overline{b_q}$ describing the forward and backward travelling parts of the field in a layer q in a structure with a total of n layers are given by the expressions

$$\overline{a_q} = \left(\overline{\overline{I}} - \overline{\overline{r_{q1}}}\,\overline{\overline{P_q^-}}\,\overline{\overline{r_{qn}}}\,\overline{\overline{P_q^+}}\right)^{-1}\left(\overline{\overline{t_{1q}}}\,\overline{a_1'} + \overline{\overline{r_{q1}}}\,\overline{\overline{P_q^-}}\,\overline{\overline{t_{nq}}}\,\overline{b_n'}\right) \tag{6.127}$$

$$\overline{b_q} = \left(\overline{\overline{I}} - \overline{\overline{r_{qn}}}\,\overline{\overline{P_q^+}}\,\overline{\overline{r_{q1}}}\,\overline{\overline{P_q^-}}\right)^{-1}\left(\overline{\overline{r_{qn}}}\,\overline{\overline{P_q^+}}\,\overline{\overline{t_{1q}}}\,\overline{a_1'} + \overline{\overline{t_{nq}}}\,\overline{b_n'}\right). \tag{6.128}$$

From these expansion vectors, the electric and magnetic fields in the q'th layer are then given by

$$\mathbf{E}_q(\mathbf{r}) = \sum_{j}^{N}\left(a_{qj}e^{i\beta_{qj}(z-z_{q-1,q})}\mathbf{e}_{qj}^+(\mathbf{r}_\perp) + b_{qj}e^{-i\beta_{qj}(z-z_{q,q+1})}\mathbf{e}_{qj}^-(\mathbf{r}_\perp)\right) \tag{6.129}$$

$$\mathbf{H}_q(\mathbf{r}) = \sum_{j}^{N}\left(a_{qj}e^{i\beta_{qj}(z-z_{q-1,q})}\mathbf{h}_{qj}^+(\mathbf{r}_\perp) + b_{qj}e^{-i\beta_{qj}(z-z_{q,q+1})}\mathbf{h}_{qj}^-(\mathbf{r}_\perp)\right). \tag{6.130}$$

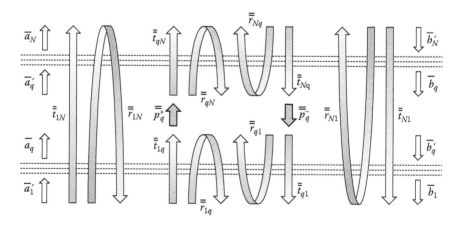

FIGURE 6.19 The reflection and transmission of light in the layer q are described by conceptually reducing the complete N-layer structure to a three-layer geometry.

Using the scattering matrix formalism, we can illuminate the structure from below or from above (or both simultaneously), and from these boundary conditions, we can then completely determine the optical field in the entire structure.

Algorithm for Field Computation

We determine the fields in arbitrary layer q of the N-layer geometry sketched in Figure 6.19 using the following steps:

1. Compute the eigenmodes in layer from 1 to N using one of the techniques discussed in Sections 6.4.1 and 6.4.2.
2. Compute the interface reflection and transmission matrices $\overline{\overline{r}}_{q,q+1}$, $\overline{\overline{t}}_{q,q+1}$, $\overline{\overline{r}}_{q+1,q}$, and $\overline{\overline{t}}_{q+1,q}$ using Equations (6.103) and (6.104) for all interfaces for q ranging from 1 to $n-1$.
3. Starting from $q = 1$, compute recursively $\overline{\overline{r}}_{1,q+1}$, $\overline{\overline{t}}_{1,q+1}$, $\overline{\overline{r}}_{q+1,1}$, and $\overline{\overline{t}}_{q+1,1}$ from $\overline{\overline{r}}_{1q}$, $\overline{\overline{t}}_{1q}$, $\overline{\overline{r}}_{q1}$, and $\overline{\overline{t}}_{q1}$ using Equations (6.119) through (6.122).
4. Starting from $q = n-1$, compute recursively $\overline{\overline{r}}_{q-1,n}$, $\overline{\overline{t}}_{q-1,n}$, $\overline{\overline{r}}_{n,q-1}$, and $\overline{\overline{t}}_{n,q-1}$ from $\overline{\overline{r}}_{qn}$, $\overline{\overline{t}}_{qn}$, $\overline{\overline{r}}_{nq}$, and $\overline{\overline{t}}_{nq}$ using Equations (6.123) through (6.126).
5. Compute the field expansion vectors $\overline{a}_q$ and $\overline{b}_q$ for the q'th layer using Equations (6.127) and (6.128).
6. Compute the field in the layer q using Equations (6.129) and (6.130).

6.4.5 Absorbing Boundary Conditions

In the MMs, the field is expanded on a discretized set of eigenmodes. Structures with closed or periodic boundary conditions, in the following referred to as non-open geometries, naturally feature discretized sets of guided and semi-radiation modes, as discussed previously in Section 2.7.3. These geometries are thus naturally compatible with the MMs.

However, when performing a simulation for an open geometry, the field should ideally be expanded on a discrete set of guided modes and on the continuous set of radiation modes. This continuous set represents an infinite number of modes, which cannot be stored in memory of a PC. We are thus forced to introduce a discretization of the continuous set of radiation modes. The effect of implementing this discretization is to restore non-open boundary conditions. Since the optical field is dependent on the choice of boundary conditions, the fields will generally differ for open and non-open boundary conditions. The phenomena are the same as that discussed for the finite-difference time-domain method in Section 3.4.

Let us consider the example of the truncated waveguide illustrated in Figure 6.20a. We want to compute the light emission, illustrated using orange arrows, for this waveguide in open space. However, if we use non-open boundary conditions, the

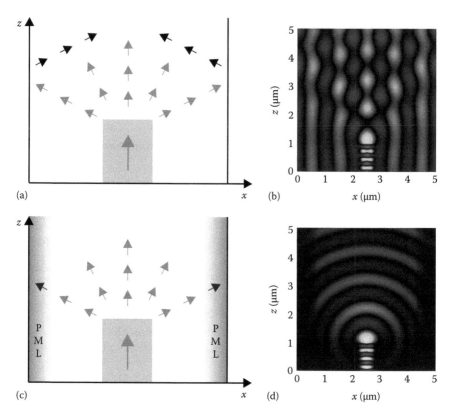

FIGURE 6.20 (**See colour insert.**) (a) Truncated waveguide geometry in the absence of PML. The red arrows indicate parasitic reflections. (b) Electric field profile without PML. (c) Same geometry with PML coatings and (d) corresponding electric field profile.

emitted light will either be reflected by the metal walls of the closed geometry or interfere with the optical field of the neighbouring waveguide in the case of periodic boundary conditions. These parasitic reflections are illustrated in Figure 6.20a using red arrows. The presence of these reflections due to the non-open boundary conditions leads to advanced interference patterns for the corresponding electric field profile illustrated in Figure 6.20b. These patterns do not occur for the open geometry and represent an artefact of the non-open boundary conditions.

When working with open geometries in the MMs, we thus need a strategy to suppress the effects of the parasitic reflections, arising due to the discretization of the continuous set of radiation modes, on the optical field profile.

One simple strategy is to use non-open boundary conditions and then make the computational domain large. As its size increases, the effects of parasitic reflections are reduced, especially in structures when light is predominantly propagating along the propagation z axis. However, when there is non-negligible light propagating in the transverse direction, as illustrated in Figure 6.20b, a very large computational domain is required. In a calculation, the number of modes required to obtain proper convergence scales with the size of the computational domain. Increasing its size may thus quickly lead to increases in computation time and memory requirement beyond what is acceptable for a practical calculation.

Another strategy consists of introducing absorbing layers, also known as *perfectly matched layers* (PMLs) [11]. The PML, also discussed in the context of the finite-difference time-domain technique in Section 3.4 and of the finite element technique in Sections 8.2.2 and 8.3.4, is placed at the boundaries of the computational domain as illustrated in Figure 6.20c. Here, light is still reflected at the outermost boundaries; however, it is also attenuated in the absorbing layers, and the parasitic reflections are thus suppressed. This leads to the light emission profile without interference patterns shown in Figure 6.20d as we would expect for a waveguide emitting into open space.

Let us consider the propagation of a plane wave in a medium of thickness L and refractive index n. The propagation is described analytically by

$$\mathbf{E}(L) = e^{ik_0 nL}\mathbf{E}(0), \tag{6.131}$$

where n and L are real-valued parameters.

One way of introducing absorption is to implement a complex refractive index $n' = n_R + in_I$ in Equation (6.131), which will lead to a damping of the field amplitude given by $e^{-n_I k_0 L}$. By increasing the imaginary part n_I, the damping can be made arbitrarily strong. However, in this approach, an index contrast is introduced between the absorbing layer of refractive index $n' = n_R + in_I$ and the inner region with index $n = n_R$. From the Fresnel reflection coefficient in Equation (6.22), we know that this index contrast will introduce new reflections between the inner region and the absorbing layer, and we are thus back where we started.

As discussed in detail in Section 3.4, these reflections can indeed be fully suppressed by modifying both the permittivity $\overline{\overline{\varepsilon}}(x)$ and permeability $\overline{\overline{\mu}}(x)$ tensors. However, another very elegant approach based on a *coordinate transformation* exists to perform the same task. Inspection of Equation (6.131) reveals that it is possible to implement absorption by introducing a complex thickness $L' = L_R + iL_I$. This will lead to a damping of the field given by $e^{-nk_0 L_I}$, which again can be made arbitrarily strong by increasing L_I. The complex thickness can be implemented in the context of a generalized coordinate transformation $x \rightarrow X(x)$ [4] transforming the real-valued x coordinate space into the complex coordinate space X. Such a transformation can in 2D be written as

$$X = X(x) \tag{6.132}$$

$$Z = z, \tag{6.133}$$

where X now represents a complex coordinate function of the real-valued parameter x.

The permittivity profile of a lateral waveguide geometry featuring a PML at the outer boundary is shown in Figure 6.21a, and the coordinate transformation $x \rightarrow X(x)$ allowing for a complex thickness of the PML is sketched in Figure 6.21b. The relative permittivities in the inner and in the PML region are identical, and absorption results solely from the complex coordinate transformation. No reflection of light occurs

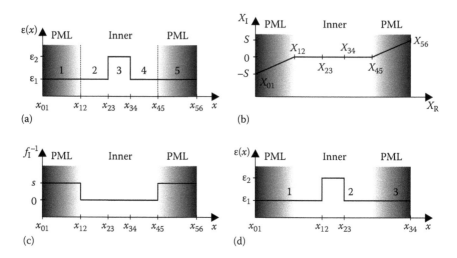

FIGURE 6.21 (a) Relative permittivity profile of the geometry implementing PMLs. (b) Parametrized coordinate trajectory in complex X coordinate space. (c) Imaginary part f_I^{-1} of the function $f^{-1} = dX/dx$ employed in the plane-wave expansion technique. The real part is simply $f_R^{-1} = 1$. (d) Relative permittivity profile with boundary coordinates relevant for the SA approach.

at the interface between the inner and the PML regions, and for this reason, the absorption region is referred to as a *perfectly matched* layer.

The complex coordinate transformation is mathematically represented by a parametrized curve mapping the real-valued x parameter to a coordinate trajectory $X(x)$ in complex X space. The most simple transformation $X(x)$ is given by

$$
\begin{aligned}
X &= x + is(x - x_{12}) && (x \leq x_{12}) \\
X &= x && (x_{12} \leq x \leq x_{45}) \\
X &= x + is(x - x_{45}) && (x_{45} \leq x),
\end{aligned}
\tag{6.134}
$$

and is shown in Figure 6.21b with $S = s(x_{12} - x_{01}) = s(x_{56} - x_{45})$. The permittivity profile in X space is simply given by $\varepsilon(X(x)) = \varepsilon(x)$. The coordinate transformation (Equation 6.134) represents the simplest transformation leading to absorption. For a discussion of more advanced transformations for the plane-wave expansion technique, we refer to Ref. [4].

To implement the PML, we exploit that the general form of Maxwell's equations remains unchanged [12] under a general coordinate transformation. This means that the eigenproblem for TE polarization in the complex coordinate space (X, Z) given by

$$
\frac{\partial^2}{\partial X^2} e_y(X) + \varepsilon(X) k_0^2 e_y(X) = \beta^2 e_y(X)
\tag{6.135}
$$

is equivalent to an eigenproblem in the usual coordinate system (x, z) with modified permittivity $\overline{\overline{\varepsilon}}'(x)$ and permeability $\overline{\overline{\mu}}'(x)$ tensors. These modified tensors obey the condition for impedance matching discussed in Section 3.4 leading to absorption in the PML without any reflection at the boundary between the inner region and the absorbing layer. The PML formulation based on the coordinate transformation and an unperturbed permittivity profile in the PML region is thus fully equivalent to the PML implementation in real space with modified permittivity and permeability tensors.

Solving the eigenproblem along the complex path $X(x)$ ensures that the field is subject to absorption in the PML. However, we should keep in mind that the complex coordinate X is purely a mathematical construct, and a complex length does not have a direct physical meaning. We should thus understand the field $e_y(X(x))$ and the permittivity profile $\varepsilon(X(x))$ simply as functions $e_y(x)$ and $\varepsilon(x)$ of the physical real-valued parameter x.

In order to calculate modes using the plane-wave expansion technique described in Section 6.4.1, we transform Equation (6.135) into

$$\frac{\partial x}{\partial X}\frac{\partial}{\partial x}\frac{\partial x}{\partial X}\frac{\partial}{\partial x}e_y(x) + \varepsilon(x)k_0^2 e_y(x) = \beta^2 e_y(x). \tag{6.136}$$

By comparing Equation (6.136) to the original eigenproblem (2.68), we observe that the effect of the coordinate transformation is now simply contained in the derivative function $f(x) \equiv \partial x/\partial X$. The imaginary part of the inverse profile $f^{-1}(x) = \partial X/\partial x$ is shown in Figure 6.21c for the transformation (Equation 6.134). In the plane-wave expansion technique, it is necessary to specify the trajectory $X(x)$ from the outset. Similarly to the procedure in Section 6.4.1, we write the Fourier series of the function $f(x)$ as

$$f(x) = \sum_{j=-\infty}^{j=\infty} f_j e^{ijKx} \tag{6.137}$$

$$f_j = \frac{1}{L_x}\int_0^{L_x} f(x)e^{-ijKx}\,dx, \tag{6.138}$$

and define the associated Toeplitz matrix $\overline{\overline{F}}$ with elements $F_{m,n}$ given by the expansion coefficients f_{m-n}. The eigenproblem (Equation 6.136) for TE polarization can then be written as

$$\left(\overline{\overline{F}}\,\overline{\overline{K}}\,\overline{\overline{F}}\,\overline{\overline{K}} + k_0^2\overline{\overline{E}}\right)\bar{u} = \beta^2\bar{u}. \tag{6.139}$$

We observe that the implementation of the absorbing boundary condition in the plane-wave expansion technique simply requires the inclusion of one additional Toeplitz matrix $\overline{\overline{F}}$ in the eigenproblem formulation.

In a similar way, the coordinate transformation can be implemented in the eigenmode problem for TM polarization. The eigenproblem (Equation 6.33) for TM polarization in complex coordinate space becomes

$$\varepsilon(x)\frac{\partial x}{\partial X}\frac{\partial}{\partial x}\frac{1}{\varepsilon(x)}\frac{\partial x}{\partial X}\frac{\partial}{\partial x}h_y(x) + \varepsilon(x)k_0^2 h_y(x) = \beta^2 h_y(x), \tag{6.140}$$

and the corresponding matrix eigenproblem is given by

$$\overline{\overline{A}}^{-1}\left(\overline{\overline{F}}\,\overline{\overline{K}}\,\overline{\overline{E}}^{-1}\,\overline{\overline{F}}\,\overline{\overline{K}} + k_0^2\overline{\overline{I}}\right)\bar{u} = \beta^2\bar{u}, \tag{6.141}$$

where we again have used the appropriate Li factorization rule.

When computing eigenmodes for a geometry implementing a coordinate transformation using the SA approach described in Section 6.4.2, we simply have to make the replacement $x_{ij} \rightarrow X_{ij}$ in our system of equations (Equations 6.71 through 6.76) for the boundary conditions. Either closed or periodic boundary conditions can be chosen for the outer boundary; as the field is absorbed, this choice matters little.

In Figure 6.21a, the positions of the boundaries between the inner and the PML regions are given by the coordinates x_{12} and x_{45}. However, since the permittivities are identical in regions 1 and 2 and in regions 4 and 5, it is not necessary to specify equations for the boundary conditions at these interfaces. In the SA approach, we simply need to consider the boundary conditions for the coordinates shown in Figure 6.21d, and the detailed nature of the trajectory $X(x)$ does not enter the system of equations for the boundary conditions. Only when plotting the field profile, a trajectory $X(x)$ needs to be chosen. Thus, referring to Figure 6.21a and b, boundary condition equations only need to be specified at the coordinates X_{01}, X_{23}, X_{34}, and X_{56}.

Similarly to the case of a complex permittivity profile, the introduction of the complex coordinates X_{01} and X_{56} into the boundary conditions leads to the appearance of nodes of the determinant function $F(\beta^2)$ in the complex plane. To identify these nodes, we can first introduce the coordinates $X_{01,\gamma} = X_{R,01} + i\gamma X_{I,01}$ and $X_{56,\gamma} = X_{R,56} + i\gamma X_{I,56}$. We then employ the same algorithm described in Section 6.4.2, where the nodes are first located on the real axis for $\gamma = 0$ and then γ is gradually increased towards 1 to locate the nodes for the geometry implementing the full coordinate transformation.

The coordinate transformation for TM polarization is implemented similarly, by first setting up the system of equations for the boundary conditions for TM polarization and then making the replacement $x_{ij} \rightarrow X_{ij}$.

We remark that the implementation of the coordinate transformation for the SA approach results in artificial field divergences at the interfaces when $X_{I,01}$ and $X_{I,56}$ are increased and a large number of modes are included. The field divergences can be removed by implementing a mode-dependent coordinate transformation [13].

6.5 PERIODIC STRUCTURES

Many interesting geometries such as photonic crystals and DBRs feature sections with a finite number of repetitions of the same structure. An example geometry is the DBR in a pillar depicted in Figure 6.22a. It features a waveguide layer followed by $p = 4$ periodic elements consisting of pairs of alternating material composition and finally an air layer. It is possible to treat this structure directly using the iterative scheme presented in Section 6.4.4. Let us write the S matrix for one period as S. The total S matrix S_p for the four periods is given by

$$S_p = S \otimes S \otimes S \otimes S, \tag{6.142}$$

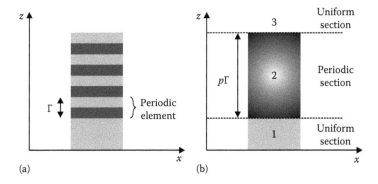

FIGURE 6.22 (a) Geometry featuring $p = 4$ periodic elements with period Γ. (b) The corresponding geometry where the periodic section has been reduced to a uniform section of length $p\Gamma$.

and the computation time in this iterative scheme is linearly proportional to the number of periods.

However, we can improve the computation time by exploiting that the S matrices for the periodic elements are identical. Instead of computing S_p in a linear manner, we now write the same matrix as

$$S_p = (S \otimes S) \otimes (S \otimes S), \tag{6.143}$$

where we have regrouped the S matrices such that the content of the two parenthesis is identical. By reusing the calculation of one of these parenthesis, we have effectively reduced the number of iterative steps from three using Equation (6.142) to two using (6.143). This procedure amounts to computing the quantity a^4 as $(a^2)^2$ instead of $aaaa$ and can be summarized as follows:

$$S_{2j} = S_j \otimes S_j \tag{6.144}$$

$$S_{2j+1} = S_{2j} \otimes S. \tag{6.145}$$

We observe that the number of concatenations in this scheme has a logarithmic rather than a linear dependence on the number of elements, and this logarithmic dependence represents a huge improvement for large p.

6.5.1 BLOCH MODES

However, an even more elegant formalism exists to handle periodic structures based on the Bloch theorem. In this formalism, we introduce an expansion on the modes describing the field propagation in a periodic structure. Let us consider a general periodic structure along the z axis with period Γ such that $\varepsilon(z+\Gamma) = \varepsilon(z)$. According to the Bloch theorem, the solutions to the wave equation in this periodic geometry can be written as a product of a periodic function along z and a phase term. We refer to these solutions as the *Bloch modes* $\Psi_{qm} = (\Psi_{qm}^E, \Psi_{qm}^H)$ of the periodic element q, where the superindices E and H indicate the electric and magnetic field components respectively and m is the mode index. The Bloch modes obey the relation

$$\boldsymbol{\Psi}_{qm}(z + \Gamma) = \boldsymbol{\Psi}_{qm}(z)\Lambda_{qm}, \tag{6.146}$$

where Λ_{qm} is the phase term. We will often skip the section subindex q in the following for simplicity.

To establish analogy between the eigenmodes of the uniform layer and the Bloch modes of the periodic element, we will define a propagation constant β_m for the Bloch mode such that $e^{i\beta_m \Gamma} = \Lambda_m$. In the same way propagation of an eigenmode with propagation constant β_j over a length L is given by $e^{i\beta_j L}$, the propagation of a Bloch mode over p periods of periodicity Γ is described by $e^{i\beta_m p\Gamma}$. We distinguish between forward $\boldsymbol{\Psi}_m^+$ and backward $\boldsymbol{\Psi}_m^-$ travelling Bloch modes, the propagation of which is described by

$$\boldsymbol{\Psi}_m^+(z + p\Gamma) = e^{i\beta_m^+ p\Gamma}\boldsymbol{\Psi}_m^+(z) \tag{6.147}$$

$$\boldsymbol{\Psi}_m^-(z - p\Gamma) = e^{i\beta_m^- p\Gamma}\boldsymbol{\Psi}_m^-(z). \tag{6.148}$$

Previously, we expanded the field in a particular layer of finite length on the eigenmodes of the corresponding geometry computed by assuming uniformity along the z axis. In the same way, we now expand the field inside a periodic section of finite total length on the Bloch modes determined by assuming an infinite periodic structure. Thus, by expanding the field inside the periodic section on the Bloch modes, we can conceptually reduce the advanced structure depicted in Figure 6.22a to the three-section geometry shown in Figure 6.22b. For conceptually reduced geometries, we refer to the various regions along the propagation axis as *sections*. Such a section can be either uniform, where we expand the field on eigenmodes as in Equations (6.129) and (6.130), or it can be periodic and consists of p periodic elements, where the field is expanded on the Bloch modes as

$$\mathbf{E}(\mathbf{r}) = \sum_m^N \left(a_m \boldsymbol{\Psi}_m^{E+}(\mathbf{r}) + b_m \boldsymbol{\Psi}_m^{E-}(\mathbf{r}) \right). \tag{6.149}$$

$$\mathbf{H}(\mathbf{r}) = \sum_m^N \left(a_m \boldsymbol{\Psi}_m^{H+}(\mathbf{r}) + b_m \boldsymbol{\Psi}_m^{H-}(\mathbf{r}) \right). \tag{6.150}$$

In all sections, we have an analytical expression for the z dependence of the modes, and the computation time thus becomes independent of the number of periods in the DBR.

To take advantage of this formalism, we first need to compute the Bloch modes of the periodic element. To this purpose, we write the electric field of the m'th Bloch mode of the periodic element as a linear combination of forward and backward travelling eigenmodes relative to a position z' inside the period element as

$$\boldsymbol{\Psi}_m^{E\pm}(\mathbf{r}_\perp, z) = \sum_j^N \left[c_{mj}^\pm \mathbf{e}_{qj}^+(\mathbf{r}_\perp)e^{i\beta_j(z-z')} + d_{mj}^\pm \mathbf{e}_{qj}^-(\mathbf{r}_\perp)e^{-i\beta_j(z-z')} \right], \tag{6.151}$$

where $\overline{c}_m^{\pm}$ and $\overline{d}_m^{\pm}$ are the expansion vectors for the m'th Bloch mode at the position z'. The corresponding magnetic field is given by

$$\mathbf{\Psi}_m^{H\pm}(\mathbf{r}_\perp, z) = \sum_j^N \left[c_{mj}^{\pm} \mathbf{h}_{qj}^+(\mathbf{r}_\perp) e^{i\beta_j(z-z')} + d_{mj}^{\pm} \mathbf{h}_{qj}^-(\mathbf{r}_\perp) e^{-i\beta_j(z-z')} \right]. \tag{6.152}$$

Here, we stress that we expand the Bloch mode at the position z' on the eigenmodes of a chosen layer q, and Equations (6.151) and (6.152) only hold for z values inside this layer q. Determination of the Bloch mode profile inside the entire periodic element will be discussed in the end of this section.

The Bloch modes are universal and do not depend on the choice of z' or the set of eigenmodes. However, to compute the Bloch modes in the MM, it is necessary to choose a z' and a set of eigenmodes. It is usually convenient to choose the z' as the position of a layer interface and the set of eigenmodes as the modes of one of the two layers adjacent to this interface. As an example, let us consider the geometry of the single period illustrated in Figure 6.23a. In this example, we consider a periodic element consisting of only two layers for simplicity; however, the formalism is valid for an arbitrary number of layers. Figure 6.23b illustrates the scattering of light when we choose $z' = z_{01}$ and expand the field on the eigenmodes of layer 1. To compute the Bloch modes, we now relate the field at the interfaces z_{01} and z_{23}. The relation between the fields at the interfaces z_{01} and z_{23} can be expressed in the T matrix formalism as

$$\begin{bmatrix} \overline{c}_3 \\ \overline{d'}_3 \end{bmatrix} = \overline{\overline{T}} \begin{bmatrix} \overline{c}_1 \\ \overline{d'}_1 \end{bmatrix} = \begin{bmatrix} \overline{\overline{T}}_{11} & \overline{\overline{T}}_{12} \\ \overline{\overline{T}}_{21} & \overline{\overline{T}}_{22} \end{bmatrix} \begin{bmatrix} \overline{c}_1 \\ \overline{d'}_1 \end{bmatrix} \tag{6.153}$$

where we have reintroduced the *facing the interface* formalism for the backward travelling d' coefficients. The corresponding relation in the S matrix formalism is

$$\begin{bmatrix} \overline{c}_3 \\ \overline{d'}_1 \end{bmatrix} = \overline{\overline{S}} \begin{bmatrix} \overline{c}_1 \\ \overline{d'}_3 \end{bmatrix} = \begin{bmatrix} \overline{\overline{S}}_{11} & \overline{\overline{S}}_{12} \\ \overline{\overline{S}}_{21} & \overline{\overline{S}}_{22} \end{bmatrix} \begin{bmatrix} \overline{c}_1 \\ \overline{d'}_3 \end{bmatrix}. \tag{6.154}$$

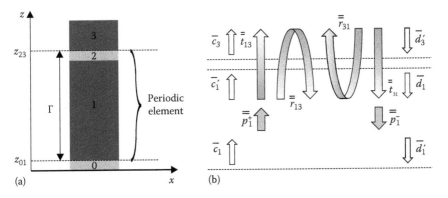

FIGURE 6.23 (a) The geometry of a single periodic element. (b) The reflection and transmission of light of this periodic element.

In Equations (6.153) and (6.154), the $\overline{\overline{T}}$ and $\overline{\overline{S}}$ matrices describe the relation of the fields at each side of the periodic element, and their subindices here refer not to layer interfaces but to subblocks of the matrices according to Equations (6.13) and (6.15). Furthermore, we have skipped the Bloch mode index j for simplicity. For the example geometry in Figure 6.23, the $\overline{\overline{S}}$ matrix elements are given by

$$\begin{bmatrix} \overline{\overline{S}}_{11} & \overline{\overline{S}}_{12} \\ \overline{\overline{S}}_{21} & \overline{\overline{S}}_{22} \end{bmatrix} = \begin{bmatrix} \overline{\overline{t}}_{13}\,\overline{\overline{p}}_1 & \overline{\overline{r}}_{31} \\ \overline{\overline{p}}_1\,\overline{\overline{r}}_{13}\,\overline{\overline{p}}_1 & \overline{\overline{p}}_1\,\overline{\overline{t}}_{31} \end{bmatrix}, \tag{6.155}$$

where the $\overline{\overline{p}}_1$ matrix is included to take into account the propagation in the first layer not included in the $\overline{\overline{r}}_{13}, \overline{\overline{t}}_{13}, \overline{\overline{r}}_{31},$ and $\overline{\overline{t}}_{31}$ matrices as illustrated in Figure 6.23.

Now, the Bloch theorem states that

$$\begin{bmatrix} \overline{c}_3 \\ \overline{d}'_3 \end{bmatrix} = \Lambda \begin{bmatrix} \overline{c}_1 \\ \overline{d}'_1 \end{bmatrix}, \tag{6.156}$$

and insertion of Equation (6.156) into Equations (6.153) and (6.154) gives

$$\begin{bmatrix} \overline{\overline{T}}_{11} & \overline{\overline{T}}_{12} \\ \overline{\overline{T}}_{21} & \overline{\overline{T}}_{22} \end{bmatrix} \begin{bmatrix} \overline{c}_1 \\ \overline{d}'_1 \end{bmatrix} = \Lambda \begin{bmatrix} \overline{c}_1 \\ \overline{d}'_1 \end{bmatrix} \tag{6.157}$$

$$\begin{bmatrix} \overline{\overline{S}}_{11} & \overline{\overline{0}} \\ \overline{\overline{S}}_{21} & -\overline{\overline{I}} \end{bmatrix} \begin{bmatrix} \overline{c}_1 \\ \overline{d}'_1 \end{bmatrix} = \Lambda \begin{bmatrix} \overline{\overline{I}} & -\overline{\overline{S}}_{12} \\ \overline{\overline{0}} & -\overline{\overline{S}}_{22} \end{bmatrix} \begin{bmatrix} \overline{c}_1 \\ \overline{d}'_1 \end{bmatrix}, \tag{6.158}$$

where $\overline{\overline{0}}$ and $\overline{\overline{I}}$ are the empty matrix and the identity matrix, respectively. Whereas Equation (6.157) is a standard eigenvalue problem, Equation (6.158) represents a generalized eigenvalue problem. Due to the numerical instability of the T matrix algorithm, it is preferable to solve the generalized eigenproblem (6.158) to determine the Bloch modes. This can be done in MATLAB using the EIG command and will lead to eigenvectors $\{\overline{c}_{1m}, \overline{d}'_{1m}\}$ and eigenvalues $\Lambda_m = e^{i\beta_m\Gamma}$ for the Bloch modes.

6.5.2 CLASSIFICATION

A calculation including N eigenmodes will lead to $2N$ Bloch modes of which N are forward and N are backward travelling. The numerical routine for solving the eigenproblem (6.158) does not automatically classify the two sets of modes, and it is necessary to perform this classification manually.

In passive structures, the Bloch modes can be categorized into three sets according to their eigenvalues Λ_m as illustrated in Table 6.1. The Bloch modes in class I are exponentially decaying as we move along the positive z axis, and these belong to the set of forward travelling modes such that

$$\mathbf{\Psi}_m^+(z + \Gamma) = \Lambda_m \mathbf{\Psi}_m^+(z) = e^{i\beta_m^+\Gamma} \mathbf{\Psi}_m^+(z). \tag{6.159}$$

TABLE 6.1

Eigenvalue-Based Bloch Mode Classification Scheme

Class	Eigenvalue	$e^{i\beta\Gamma}$	Direction
I	$\|\Lambda_m\| < 1$	Λ_m	+
II	$\|\Lambda_m\| > 1$	Λ_m^{-1}	−
III	$\|\Lambda_m\| = 1$		

The class II Bloch modes exponentially increase along the positive z axis. Since there is no gain in the structure, these modes must necessarily be backward travelling modes with

$$\Psi_m^-(z - \Gamma) = \Lambda_m^{-1}\Psi_m^-(z) = e^{i\beta_m^-\Gamma}\Psi_m^-(z). \tag{6.160}$$

The final class III consists of the Bloch modes with $|\Lambda| = 1$, and here, the eigenvalue alone cannot be used to classify the modes. Instead, we should evaluate the direction of the power flow P_m of each Bloch mode m by integrating the z component of its time-averaged Poynting vector as

$$P_m = \frac{1}{2} \int \mathrm{Re}(\Psi_m^E \times \Psi_m^{H*}) \cdot \hat{z}d\mathbf{r}_\perp, \tag{6.161}$$

where Re and $*$ denote the real part and complex conjugation, respectively. The classification of the class III Bloch modes is then based on the sign of P_m as shown in Table 6.2.

When the structure contains loss, the eigenvalue-based classification scheme presented in Table 6.1 still holds. However, when the geometry includes gain, a forward travelling mode can now be exponentially increasing such that $|\Lambda_m| > 1$. In this case, the class III should be extended to include eigenvalues $1 - \gamma < |\Lambda_m| < 1/(1 - \gamma)$, where γ is a number chosen to ensure that exponentially increasing modes are correctly classified. Even for passive structures, it is usually a good idea to implement a small γ to take into account numerical inaccuracy.

After the classification, we are equipped with N forward and N backward travelling Bloch modes Ψ_m^+ and Ψ_m^- with corresponding expansion vectors given by $(\overline{c_m^+, d_m'^+})$ and $(\overline{c_m^-, d_m'^-})$, respectively.

TABLE 6.2

Power-Flow-Based Bloch Mode Classification Scheme

Power flow	$e^{i\beta\Gamma}$	Direction
$P_m > 0$	Λ	+
$P_m < 0$	Λ^{-1}	−

6.5.3 INTERFACE

We will now compute reflection and transmission between neighbouring sections which can be either uniform or periodic. Let us consider the advanced micropillar structure illustrated in Figure 6.24a. By conceptually reducing the original geometry to uniform and periodic sections, we obtain the geometry illustrated in Figure 6.24b, where the field is expanded on the Bloch modes in the periodic sections 2 and 4 and on regular eigenmodes in uniform sections 1, 3, and 5.

With the purpose of generalizing the boundary conditions of Equations (6.90) and (6.91), we will now treat sections 1, 3, and 5 as pseudo-periodic sections, where the field is expanded on *pseudo-Bloch modes* $\mathbf{\Psi}_m$ defined by the expansion coefficients

$$\begin{aligned} \mathbf{\Psi}_m^+ : \quad & c_{mj}^+ = \delta_{mj} \quad d_{mj}'^+ = 0 \\ \mathbf{\Psi}_m^- : \quad & c_{mj}^- = 0 \quad d_{mj}'^- = \delta_{mj}, \end{aligned} \tag{6.162}$$

such that the pseudo-Bloch modes of the uniform section are simply identical to its eigenmodes. The propagation constant of the pseudo-Bloch mode $\mathbf{\Psi}_m$ of the uniform structure is then given simply by the eigenmode propagation constant β_m, and for practical purposes, we do not need to distinguish between the Bloch modes of true periodic sections and the pseudo-Bloch modes of the uniform sections. Using this definition, we can formally consider the field expanded on the Bloch modes in all sections of the conceptually reduced structure as depicted in Figure 6.24c, and we can generalize Equations (6.90) and (6.91) for the boundary conditions at the interface as

$$\mathbf{\Psi}_{\perp 1m}^{E+}(\mathbf{r}_\perp) + \sum_j^N \mathbf{\Psi}_{\perp 1j}^{E-}(\mathbf{r}_\perp) r_{12;jm} = \sum_j^N \mathbf{\Psi}_{\perp 2j}^{E+}(\mathbf{r}_\perp) t_{12;jm} \tag{6.163}$$

$$\mathbf{\Psi}_{\perp 1m}^{H+}(\mathbf{r}_\perp) + \sum_j^N \mathbf{\Psi}_{\perp 1j}^{H-}(\mathbf{r}_\perp) r_{12;jm} = \sum_j^N \mathbf{\Psi}_{\perp 2j}^{H+}(\mathbf{r}_\perp) t_{12;jm}. \tag{6.164}$$

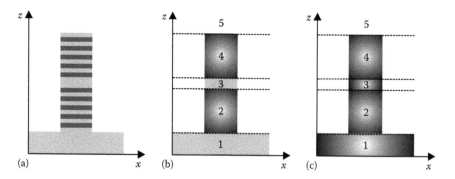

FIGURE 6.24 (a) Micropillar geometry featuring both uniform and periodic sections. (b) Conceptually reduced structure where the field inside the periodic regions is expanded on Bloch modes. (c) The structure where the field is formally expanded on the Bloch modes in all sections.

Insertion of the expressions (Equations 6.151 and 6.152) into Equations (6.163) and (6.164) leads to

$$\sum_i^N \mathbf{e}_{\perp 1i}(\mathbf{r}_\perp) \left(\left(c_{1mi}^+ + d_{1mi}'^+ \right) + \sum_j^N \left(c_{1ji}^- + d_{1ji}'^- \right) r_{12;jm} \right)$$

$$= \sum_{i,j}^N \mathbf{e}_{\perp 2i}(\mathbf{r}_\perp) \left(c_{2ji}^+ + d_{2ji}'^+ \right) t_{12;jm} \tag{6.165}$$

$$\sum_i^N \mathbf{h}_{\perp 1i}(\mathbf{r}_\perp) \left(\left(c_{1mi}^+ - d_{1mi}'^+ \right) + \sum_j^N \left(c_{1ji}^- - d_{1ji}'^- \right) r_{12;jm} \right)$$

$$= \sum_{i,j}^N \mathbf{h}_{\perp 2i}(\mathbf{r}_\perp) \left(c_{2ji}^+ - d_{2ji}'^+ \right) t_{12;jm}, \tag{6.166}$$

where we have used $\mathbf{e}_\perp^- = \mathbf{e}_\perp^+ = \mathbf{e}_\perp$ and $\mathbf{h}_\perp^- = -\mathbf{h}_\perp^+ = -\mathbf{h}_\perp$.

Similarly to the derivation in Section 6.4.3, multiplication of the bra version of Equation (6.165) with $|\mathbf{h}_{1n}\rangle$ and the ket version of Equation (6.166) with $\langle \mathbf{e}_{1n}|$, introduction of matrix notation and rearrangement lead to the following generalized expressions for the reflection and transmission matrices:

$$\overline{\overline{t}}_{12} = \left[\left(\overline{\overline{c_1^-}} + \overline{\overline{d_1'^-}} \right)^{-1} \overline{\overline{\langle \mathbf{e}_2|\mathbf{h}_1\rangle}}^T \left(\overline{\overline{c_2^+}} + \overline{\overline{d_2'^+}} \right) \right.$$
$$\left. - \left(\overline{\overline{c_1^-}} - \overline{\overline{d_1'^-}} \right)^{-1} \overline{\overline{\langle \mathbf{e}_1|\mathbf{h}_2\rangle}} \left(\overline{\overline{c_2^+}} - \overline{\overline{d_2'^+}} \right) \right]^{-1}$$
$$\times \left[\left(\overline{\overline{c_1^-}} + \overline{\overline{d_1'^-}} \right)^{-1} \left(\overline{\overline{c_1^+}} + \overline{\overline{d_1'^+}} \right) - \left(\overline{\overline{c_1^-}} - \overline{\overline{d_1'^-}} \right)^{-1} \left(\overline{\overline{c_1^+}} - \overline{\overline{d_1'^+}} \right) \right] \tag{6.167}$$

$$\overline{\overline{r}}_{12} = \left(\overline{\overline{c_1^-}} + \overline{\overline{d_1'^-}} \right)^{-1} \overline{\overline{\langle \mathbf{e}_2|\mathbf{h}_1\rangle}}^T \left(\overline{\overline{c_2^+}} + \overline{\overline{d_2'^+}} \right) \overline{\overline{t}}_{12} - \left(\overline{\overline{c_1^-}} + \overline{\overline{d_1'^-}} \right)^{-1} \left(\overline{\overline{c_1^+}} + \overline{\overline{d_1'^+}} \right). \tag{6.168}$$

The corresponding expressions for the $\overline{\overline{r}}_{21}$ and $\overline{\overline{t}}_{21}$ matrices are obtained by interchanging the indices 1 and 2 and by inverting the $+$ and $-$ signs for the $\overline{\overline{c}}$ and $\overline{\overline{d'}}$ matrices.

The propagation of the Bloch modes inside the periodic sections is described by Equations (6.147) and (6.148), and the reflection and transmission between adjacent sections are computed using Equations (6.167) and (6.168). The S matrix formalism describing the scattering of light on the conceptually reduced structure is then identical to that presented in Section 6.4.4 except for two changes: the reflection and transmission matrices for the single interface of the conceptually reduced structure are now given by Equations (6.167) and (6.168) and the propagation matrices $\overline{\overline{p_q^+}}$ and $\overline{\overline{p_q^-}}$ are no longer identical, since the eigenvalues Λ_m^+ and Λ_m^- generally differ for the forward and backward travelling Bloch modes $\mathbf{\Psi}_m^+$ and $\mathbf{\Psi}_m^-$. The exception is when the periodic section features inversion symmetry along the z axis. We can thus compute scattering of light on conceptually reduced structures featuring three or more sections and correctly take into account multiple reflections inside periodic sections using the iterative equations (6.119) through (6.126).

6.5.4 FIELD PROFILE IN A PERIODIC ELEMENT

Finally, we will consider the determination of the full field profile inside a periodic section of a geometry as shown in Figure 6.25a. Using Equations (6.167) and (6.168), Equations (6.147) and (6.148), and the S matrix formalism of Section 6.4.4, we can determine the field profile at an arbitrary interface between two periodic sections. Referring to Figure 6.25a, let us assume that the field expansions on the Bloch modes at the interfaces z_A and z_B have been computed previously and are given by

$$\mathbf{E}(z_A) = \sum_m^N \left(a_{Am} \mathbf{\Psi}_m^{E+} + b_{Am} \mathbf{\Psi}_m^{E-} \right) \tag{6.169}$$

$$\mathbf{E}(z_B) = \sum_m^N \left(a_{Bm} \mathbf{\Psi}_m^{E+} + b_{Bm} \mathbf{\Psi}_m^{E-} \right) \tag{6.170}$$

with corresponding expressions for the magnetic field. To compute the field between the interfaces, we need to calculate the corresponding field expansions directly on eigenmodes. By inserting (6.151) into Equations (6.169) and (6.170), we obtain

$$\mathbf{E}(z_A) = \sum_j^N \left(a_{1j}^\dagger \mathbf{e}_j^+ + b_{1j}^\dagger \mathbf{e}_j^- \right) \tag{6.171}$$

$$\mathbf{E}(z_B) = \sum_j^N \left(a_{3j}^\dagger \mathbf{e}_j^+ + b_{3j}^\dagger \mathbf{e}_j^- \right), \tag{6.172}$$

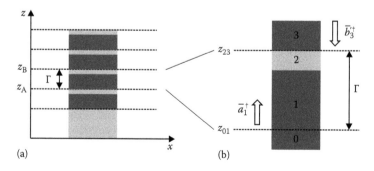

FIGURE 6.25 (a) The periodic structure. (b) The geometry of a single period with defining incoming fields $\overline{a_1^\dagger}$ and $\overline{b_3^\dagger}$ used to compute periodic element field profile.

where the coefficients $a_{1j}^\dagger$, $b_{1j}^{\prime\dagger}$, $a_{3j}^\dagger$, and $b_{3j}^{\prime\dagger}$ are given by

$$a_{1j}^\dagger = \sum_m^N \left(a_{Am} c_{mj}^+ + b_{Am} c_{mj}^- \right) \tag{6.173}$$

$$b_{1j}^{\prime\dagger} = \sum_m^N \left(a_{Am} d_{mj}^+ + b_{Am} d_{mj}^- \right) \tag{6.174}$$

$$a_{3j}^\dagger = \sum_m^N \left(a_{Bm} c_{mj}^+ + b_{Bm} c_{mj}^- \right) \tag{6.175}$$

$$b_{3j}^{\prime\dagger} = \sum_m^N \left(a_{Bm} d_{mj}^+ + b_{Bm} d_{mj}^- \right). \tag{6.176}$$

We now consider the geometry of the single periodic element shown in Figure 6.25b, where we have defined $z_{01} = z_A$ and $z_{23} = z_B$. The fields at the positions z_{01} and z_{23} are given by Equations (6.171) and (6.172), where the vectors $\overline{a_1^\dagger}$ and $\overline{b_3^{\prime\dagger}}$ describe the incoming fields and $\overline{b_1^{\prime\dagger}}$ and $\overline{a_3^\dagger}$ are the outgoing fields.

We can now easily compute the field inside layers 1 and 2 of the single periodic element simply by illuminating the geometry with the fields $\overline{a_1^\dagger}$ and $\overline{b_3^{\prime\dagger}}$ as illustrated in Figure 6.25b and by computing the scattering properties of the geometry using the standard S matrix formalism. In the full structure, the layer 0 is identical to layer 2; however, for the field calculation of the single periodic element, the layer 0 is chosen to be identical to layer 1 to mark that no scattering at the interface between layers 0 and 1 should be included in the field profile calculation. On the other hand, since the fields are expanded on the eigenmodes of layers 1 and 3, we should take into account reflection and transmission at the interface between layers 2 and 3.

Using these illumination conditions, we can compute the field profile inside the complete periodic element. As for any geometry, the scattering of light will lead to outgoing fields. Referring to Figure 6.23b, the Bloch modes are defined such that illumination with the fields described by the vectors $\overline{c_1}$ and $\overline{d_3}$ results in outgoing fields given by $\overline{d_1}$ and $\overline{c_3}$. Since the incoming fields $\overline{a_1^\dagger}$ and $\overline{b_3^{\prime\dagger}}$ are simply linear combinations of incoming fields of the total set of the Bloch modes, the same property holds, and the outgoing fields obtained from a regular field calculation for the geometry in Figure 6.23b will thus automatically equal the scattered outgoing fields $\overline{b_1^{\prime\dagger}}$ and $\overline{a_3^\dagger}$ from Equations (6.171) and (6.172).

6.6 CURRENT SOURCES

So far, we have only considered passive structures without any current density in this chapter. However, we now consider the case of a non-zero source current density distribution $\mathbf{J}(\mathbf{r})e^{-i\omega t}$. The current distribution could represent, for example the important situation of a dipole emitter placed inside a micro-cavity. The source distribution will generate an electromagnetic field $(\mathbf{E}(\mathbf{r})e^{-i\omega t}, \mathbf{H}(\mathbf{r})e^{-i\omega t})$, which interacts with the photonic environment.

6.6.1 Uniform Layer

We will first compute the field generated by a current distribution of the form $\mathbf{J}(x,y)\delta(z-z_\mathrm{J})$ placed inside single-layer geometry with uniformity along the propagation axis. As usual, we suppress the time dependence for simplicity. The field can be computed in an elegant manner using the reciprocity theorem (Equation 2.78) given by

$$\int_S (\mathbf{E}_1 \times \mathbf{H}_2 - \mathbf{E}_2 \times \mathbf{H}_1) \cdot \hat{\mathbf{n}} dS = \int_V (\mathbf{J}_1 \cdot \mathbf{E}_2 - \mathbf{J}_2 \cdot \mathbf{E}_1) dV. \tag{6.177}$$

The field 1 described by $(\mathbf{E}_1, \mathbf{H}_1, \mathbf{J}_1)$ with $\mathbf{J}_1 \equiv \mathbf{J}$ is the field we want to compute. As usual, we will write the generated field as an eigenmode expansion in the form

$$\mathbf{E}_1(\mathbf{r}) = \sum_j a_j^\mathrm{J} \mathbf{e}_j^+(\mathbf{r}_\perp) e^{i\beta_m(z-z_\mathrm{J})} \quad (z > z_\mathrm{J}) \tag{6.178}$$

$$\mathbf{H}_1(\mathbf{r}) = \sum_j a_j^\mathrm{J} \mathbf{h}_j^+(\mathbf{r}_\perp) e^{i\beta_m(z-z_\mathrm{J})} \quad (z > z_\mathrm{J}) \tag{6.179}$$

$$\mathbf{E}_1(\mathbf{r}) = \sum_j b_j^\mathrm{J} \mathbf{e}_j^-(\mathbf{r}_\perp) e^{-i\beta_m(z-z_\mathrm{J})} \quad (z < z_\mathrm{J}) \tag{6.180}$$

$$\mathbf{H}_1(\mathbf{r}) = \sum_j b_j^\mathrm{J} \mathbf{h}_j^-(\mathbf{r}_\perp) e^{-i\beta_m(z-z_\mathrm{J})} \quad (z < z_\mathrm{J}) \tag{6.181}$$

and we thus need to determine the expansion vectors $\overline{a^\mathrm{J}}$ and $\overline{b^\mathrm{J}}$ describing the forward and backward travelling fields defined relative to the position $z = z_\mathrm{J}$ of the current distribution.

To compute these expansion coefficients, we choose a test field 2 given by $\mathbf{J}_2 = 0$ and $(\mathbf{E}_2, \mathbf{H}_2) = (\mathbf{e}_m^+, \mathbf{h}_m^+) e^{i\beta_m(z-z_\mathrm{J})}$ corresponding to the eigenmode of index m travelling in the forward direction. Let us now consider the current distribution $\mathbf{J}(x,y)\delta(z-z_\mathrm{J})$ inside the closed volume V as illustrated in Figure 6.26. Similarly,

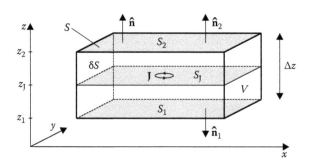

FIGURE 6.26 Current distribution $\mathbf{J}(x,y)\delta(z-z_\mathrm{J})$ inside the volume V. The bottom S_1, top S_2, and side δS surfaces cover the volume V with normal unit vector $\hat{\mathbf{n}}$. The current distribution is limited to the surface S_J defined by $z = z_\mathrm{J}$.

to the discussion in Section 2.7.4, the surface integral in Equation (6.177) over the boundary δS disappears. Furthermore, the delta function for the spatial z dependence converts the volume integral on the RHS into a surface integral, and we obtain

$$-\int_{S_1} \left(\mathbf{E}_1 \times \mathbf{h}_m^+ - \mathbf{e}_m^+ \times \mathbf{H}_1\right) \cdot \hat{\mathbf{z}} e^{i\beta_m(z_1 - z_J)} dS$$

$$+\int_{S_2} \left(\mathbf{E}_1 \times \mathbf{h}_m^+ - \mathbf{e}_m^+ \times \mathbf{H}_1\right) \cdot \hat{\mathbf{z}} e^{i\beta_m(z_2 - z_J)} dS = \int_{S_J} \mathbf{J} \cdot \mathbf{e}_m^+ dS, \tag{6.182}$$

where S_J is the surface in the plane $z = z_J$ of the current distribution, and we have used $\hat{\mathbf{n}}_2 = -\hat{\mathbf{n}}_1 = \hat{\mathbf{z}}$. Since the z components of the fields do not contribute to the surface integrals, we may replace the eigenmode fields with their lateral components $(\mathbf{e}_\perp, \mathbf{h}_\perp)$. Inserting the expansions (Equations 6.178 through 6.181), we obtain

$$-b_m^J \int_{S_1} \left(\mathbf{e}_{\perp m}^- \times \mathbf{h}_{\perp m}^+ - \mathbf{e}_{\perp m}^+ \times \mathbf{h}_{\perp m}^-\right) \cdot \hat{\mathbf{z}} dS$$

$$+a_m^J \int_{S_2} \left(\mathbf{e}_{\perp m}^+ \times \mathbf{h}_{\perp m}^+ - \mathbf{e}_{\perp m}^+ \times \mathbf{h}_{\perp m}^+\right) \cdot \hat{\mathbf{z}} e^{i2\beta_m(z_2 - z_J)} dS = \int_{S_J} \mathbf{J} \cdot \mathbf{e}_m^+ dS, \tag{6.183}$$

where we have used the orthonormality relation $\int (\mathbf{e}_m \times \mathbf{h}_n) \cdot \hat{\mathbf{z}} dS = \delta_{mn}$ to eliminate the expansion terms in Equations (6.178) through (6.181) for which $j \neq m$. As discussed in Section 2.7.1, the forward and backward travelling lateral field components are related as $\mathbf{e}_{\perp m}^- = \mathbf{e}_{\perp m}^+ = \mathbf{e}_{\perp m}$ and $\mathbf{h}_{\perp m}^- = -\mathbf{h}_{\perp m}^+ = -\mathbf{h}_{\perp m}$. Using these relations, Equation (6.183) reduces to

$$b_m^J = -\frac{1}{2} \int_{S_J} \mathbf{J} \cdot \mathbf{e}_m^+ dS. \tag{6.184}$$

The corresponding coefficient a_m^J can be derived in a similar manner by choosing a test field 2 given by $\mathbf{J}_2 = 0$ and $(\mathbf{E}_2, \mathbf{H}_2) = (\mathbf{e}_m^-, \mathbf{h}_m^-) e^{-i\beta_m(z - z_J)}$ corresponding to the backward travelling eigenmode and is given by

$$a_m^J = -\frac{1}{2} \int_{S_J} \mathbf{J} \cdot \mathbf{e}_m^- dS. \tag{6.185}$$

When the current density vector is oriented in the lateral plane such that $\mathbf{J} = \mathbf{J}_\perp$, the expansion vectors $\overline{a^J}$ and $\overline{b^J}$ will be identical. However, if the current density has a non-zero J_z contribution, the inversion symmetry is broken, and we have $\overline{a^J} \neq \overline{b^J}$.

Finally, we consider the important example where the current distribution represents a point dipole at the position $\mathbf{r}_{PD}$ such that $\mathbf{J} = \mathbf{J}_{PD} \delta(\mathbf{r} - \mathbf{r}_{PD})$. In this case, the expansion coefficients become

$$a_m^J = -\frac{\mathbf{J}_{PD} \cdot \mathbf{e}_m^-(\mathbf{r}_{PD})}{2} \tag{6.186}$$

$$b_m^J = -\frac{\mathbf{J}_{PD} \cdot \mathbf{e}_m^+(\mathbf{r}_{PD})}{2}. \tag{6.187}$$

From these expressions, we observe that the coupling strength of a point dipole at the position $\mathbf{r}_{PD}$ to a particular eigenmode is proportional to the field amplitude of the electric field profile of the eigenmode at the position $\mathbf{r}_{PD}$.

6.6.2 MULTILAYER GEOMETRY

In the previous section, we considered a dipole inside a layer uniform along the z axis. We now generalize the formalism to take into account advanced geometries with multiple layers along the propagation axis. Let us now consider the example of a dipole emitter represented by a current distribution $\mathbf{J}\delta(z - z_J)$ placed inside the micropillar geometry illustrated in Figure 6.27a. The structure consists of a total of N layers, and the current distribution appears at the coordinate $z = z_J$ inside the layer q.

Again, a field will be generated by the current distribution inside the layer q. Initially, not taking into account the presence of the surrounding structure, the expansion vectors for the field are $\overline{a^j}$ and $\overline{b^j}$ given by Equations (6.185) and (6.184). However, the presence of structuring above and below the layer q will lead to light reflection at the interfaces $z_{q-1,q}$ and $z_{q,q+1}$. Using the S matrix formalism developed in Section 6.4.4, the reflections at these interfaces are fully described by the reflection matrices $\overline{\overline{r_{q1}}}$ and $\overline{\overline{r_{qN}}}$, and we can thus conceptually reduce the full structure to the three-level geometry illustrated in Figure 6.27b.

We then write the electric field inside the layer q as a sum of forward and backward travelling contributions arising from the reflections at the interfaces:

$$\mathbf{E}_q(\mathbf{r}) = \sum_j \left(a_j^{z_J} \mathbf{e}_{qj}^+(\mathbf{r}_\perp) e^{i\beta_m(z-z_J)} + b_{qj} \mathbf{e}_{qj}^-(\mathbf{r}_\perp) e^{-i\beta_m(z-z_{q,q+1})} \right) \quad (z > z_J) \quad (6.188)$$

$$\mathbf{E}_q(\mathbf{r}) = \sum_j \left(a_{qj} \mathbf{e}_{qj}^+(\mathbf{r}_\perp) e^{i\beta_m(z-z_{q-1,q})} + b_j^{z_J} \mathbf{e}_{qj}^-(\mathbf{r}_\perp) e^{-i\beta_m(z-z_J)} \right) \quad (z < z_J) \quad (6.189)$$

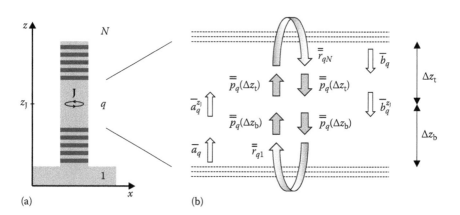

FIGURE 6.27   (a) The micropillar geometry containing a current source in the cavity layer. (b) The reflection of light generated by the current distribution $\mathbf{J}(x, y)\delta(z - z_J)$ in layer q by the parts of the geometry above and below the layer q.

To determine the field inside the layer q, we thus need to compute the expansion vectors $\overline{a^{z_{\mathrm{J}}}}$, $\overline{b^{z_{\mathrm{J}}}}$, $\overline{a}_q$, and $\overline{b}_q$. The expansion vector $\overline{a^{z_{\mathrm{J}}}}$ can be determined by taking into account the contributions from multiple round trips inside the layer q. Similarly, to the discussion in Section 6.4.4, the initial contribution $\overline{a}_{\mathrm{J}}$ will perform a round trip described by $\overline{\overline{p}}_q(\Delta z_{\mathrm{b}}) \overline{\overline{r}}_{q1} \overline{\overline{p}}_q \overline{\overline{r}}_{qN} \overline{\overline{p}}_q(\Delta z_{\mathrm{t}})$, where $\overline{\overline{p}}_q(\Delta z)$ is the propagation matrix for the layer q over the length Δz, and $\Delta z_{\mathrm{b}} = z_{\mathrm{J}} - z_{q-1,q}$ and $\Delta z_{\mathrm{t}} = z_{q,q+1} - z_{\mathrm{J}}$ are the distances from the current distribution to the bottom and top interfaces. This contribution together with those from 2, 3, etc. round trips will add up as

$$\overline{a}_{\mathrm{J}} + \overline{\overline{p}}_q(\Delta z_{\mathrm{b}}) \overline{\overline{r}}_{q1} \overline{\overline{p}}_q \overline{\overline{r}}_{qN} \overline{\overline{p}}_q(\Delta z_{\mathrm{t}}) \overline{a}_{\mathrm{J}} + \left(\overline{\overline{p}}_q(\Delta z_{\mathrm{b}}) \overline{\overline{r}}_{q1} \overline{\overline{p}}_q \overline{\overline{r}}_{qN} \overline{\overline{p}}_q(\Delta z_{\mathrm{t}}) \right)^2 \overline{a}_{\mathrm{J}} + \cdots$$
$$= \left(\overline{\overline{I}} - \overline{\overline{p}}_q(\Delta z_{\mathrm{b}}) \overline{\overline{r}}_{q1} \overline{\overline{p}}_q \overline{\overline{r}}_{qN} \overline{\overline{p}}_q(\Delta z_{\mathrm{t}}) \right)^{-1} \overline{a}_{\mathrm{J}},$$

where we have once again employed the relation for the geometric series $\overline{\overline{I}} + \overline{\overline{a}} + \overline{\overline{a}}^2 + \cdots = (\overline{\overline{I}} - \overline{\overline{a}})^{-1}$. Here, we should keep in mind that the downward travelling field also gives a contribution $\overline{\overline{p}}_q(\Delta z_{\mathrm{b}}) \overline{\overline{r}}_{q1} \overline{\overline{p}}_q(\Delta z_{\mathrm{b}}) \overline{b}_{\mathrm{J}}$ to the forward travelling field, which also experiences multiple reflections. The total forward travelling field $\overline{a^{z_{\mathrm{J}}}}$ thus becomes

$$\overline{a^{z_{\mathrm{J}}}} = \left(\overline{\overline{I}} - \overline{\overline{p}}_q(\Delta z_{\mathrm{b}}) \overline{\overline{r}}_{q1} \overline{\overline{p}}_q \overline{\overline{r}}_{qN} \overline{\overline{p}}_q(\Delta z_{\mathrm{t}}) \right)^{-1} \left(\overline{a}_{\mathrm{J}} + \overline{\overline{p}}_q(\Delta z_{\mathrm{b}}) \overline{\overline{r}}_{q1} \overline{\overline{p}}_q(\Delta z_{\mathrm{b}}) \overline{b}_{\mathrm{J}} \right). \tag{6.190}$$

The backward travelling field inside the layer q is determined similarly as

$$\overline{b^{z_{\mathrm{J}}}} = \left(\overline{\overline{I}} - \overline{\overline{p}}_q(\Delta z_{\mathrm{t}}) \overline{\overline{r}}_{qN} \overline{\overline{p}}_q \overline{\overline{r}}_{q1} \overline{\overline{p}}_q(\Delta z_{\mathrm{b}}) \right)^{-1} \left(\overline{b}_{\mathrm{J}} + \overline{\overline{p}}_q(\Delta z_{\mathrm{t}}) \overline{\overline{r}}_{qN} \overline{\overline{p}}_q(\Delta z_{\mathrm{t}}) \overline{a}_{\mathrm{J}} \right), \tag{6.191}$$

and the remaining expansion vectors in Equations (6.188) and (6.189) are given as $\overline{a}_q = \overline{\overline{r}}_{q1} \overline{\overline{p}}_q(\Delta z_{\mathrm{b}}) \overline{b^{z_{\mathrm{J}}}}$ and $\overline{b}_q = \overline{\overline{r}}_{qN} \overline{\overline{p}}_q(\Delta z_{\mathrm{t}}) \overline{a^{z_{\mathrm{J}}}}$.

The field inside the layer q is now fully described by Equations (6.188) through (6.191), which take into account multiple reflections from the structuring surrounding the layer q.

What remains is now to determine the field generated by the current distribution in the layers above and below the layer q. To do this, we first consider the structure consisting of the layers q to N as illustrated in Figure 6.28a. The forward travelling field inside layer q is given by $\overline{a^{z_{\mathrm{J}}}}$, and we have $\overline{a}'_q = \overline{\overline{p}}_q(\Delta z_{\mathrm{t}}) \overline{a^{z_{\mathrm{J}}}}$. We then proceed by considering the top structure in Figure 6.28a as an isolated geometry illuminated from below by the field described by the coefficient vector $\overline{a}'_q$, and we can then compute the fields in the layers $q + 1$ to N using the standard S matrix formalism developed in Section 6.4.4. Here, the influence of the bottom structure is not neglected, but it is contained in the expression for the coefficient vector (6.190).

The fields inside the layers 1 to q are computed similarly: we first consider the isolated structure in Figure 6.28b consisting of the layers 1 to q, we then illuminate the structure from above with the field described by $\overline{b}'_q = \overline{\overline{p}}_q(\Delta z_{\mathrm{b}}) \overline{b^{z_{\mathrm{J}}}}$, and finally the fields in the layers 1 to $q - 1$ are determined using the S matrix formalism.

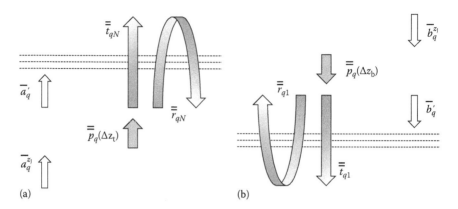

FIGURE 6.28 The illumination conditions for the calculation of the field in the isolated structures above (a) and below (b) the current distribution layer q.

6.7 3D GEOMETRIES

Finally, we briefly outline extensions of the theory to full 3D geometries. Two classes of geometries are considered. In the general case, without any simplifying symmetries, the matrix form of the eigenvalue (Equation 6.1) in Cartesian coordinates (x, y) is given as

$$\begin{bmatrix} \nabla_\perp^2 + \varepsilon k_0^2 + \partial_x \left(\partial_x \ln \varepsilon \right) & \partial_x \left(\partial_y \ln \varepsilon \right) \\ \partial_y \left(\partial_x \ln \varepsilon \right) & \nabla_\perp^2 + \varepsilon k_0^2 + \partial_y \left(\partial_y \ln \varepsilon \right) \end{bmatrix} \begin{bmatrix} e_x \\ e_y \end{bmatrix} = \beta^2 \begin{bmatrix} e_x \\ e_y \end{bmatrix}. \tag{6.192}$$

We observe that, when the relative permittivity is constant, the non-diagonal terms of the operator matrix are zero. The two field components are thus coupled by variations in the dielectric constant.

The eigenvalue problem (6.192) generally does not allow for analytic solutions, and instead, a plane-wave expansion approach should be used. Whereas the 2D problem discussed in Section 6.4.1 involves an expansion of the field on N plane waves along the x-axis, the corresponding 3D problem requires an expansion on a total of $2N^2$ plane waves in the (x, y) plane. The eigenvalue problem in general 3D geometries thus quickly becomes extremely numerically demanding, and we refer to Ref. [2] for further details of its numerical implementation.

However, if the 3D geometry features rotational symmetry as the waveguide illustrated in Figure 2.1a, the introduction of cylindrical coordinates greatly simplifies the eigenvalue problem. Using cylindrical coordinates (r, ϕ), we can write the electric field as a sum of contributions of different angular quantum numbers:

$$\mathbf{E}(\mathbf{r}) = \sum_l \mathbf{E}_l(r) e^{il\phi}, \tag{6.193}$$

where l is the angular quantum number. In a rotationally symmetric structure described by $\varepsilon(r)$, the field contributions having different angular quantum number

l are decoupled, and it is thus necessary to perform calculations for only one angular quantum number a time. This reduces the number of total effective dimensions from three to two, making exact 3D vectorial calculations feasible with only modest computing power.

When expressed in cylindrical coordinates (r, ϕ), the matrix form of the eigenvalue problem (Equation 6.1) for a rotationally symmetric permittivity profile becomes

$$
\begin{bmatrix}
\nabla_\perp^2 - \dfrac{1}{r^2} + \varepsilon k_0^2 + \partial_r(\partial_r \ln\varepsilon) & -\dfrac{2il}{r^2} \\
\dfrac{2il}{r^2} + \dfrac{il}{r}(\partial_r \ln\varepsilon) & \nabla_\perp^2 - \dfrac{1}{r^2} + \varepsilon k_0^2
\end{bmatrix}
\begin{bmatrix} e_r \\ e_\phi \end{bmatrix}
= \beta^2 \begin{bmatrix} e_r \\ e_\phi \end{bmatrix}, \qquad (6.194)
$$

where we have used the shorthand notation $\partial_r \equiv \frac{\partial}{\partial r}$. We observe that, except when the angular quantum number l is equal to zero, the e_r and e_ϕ components are coupled and the equation must be solved for both components simultaneously. However, the azimuthal ϕ dependence is handled analytically, and the eigenvalue problem (Equation 6.194) only requires a determination of the fields along the r-axis, which represents a huge advantage compared to the general 3D geometry.

The eigenproblem (Equation 6.194) can be solved either using a Fourier–Bessel expansion [14] similar to the plane-wave expansion in Cartesian coordinates discussed in Section 6.4.1 or using an SA approach [15] similar to procedure in Section 6.4.2.

EXERCISES

6.1 In this exercise, we analyse the properties of a 1D fibre Bragg grating (FBG) using the theory developed in Section 6.3. An FBG is a fibre where the refractive index of the fibre has been periodically increased by exposure to UV light. This exposure leads to a sinusoidal form of the index profile, which we will approximate to a step-index profile. Let us thus consider a 1D fibre with refractive index n_1 with an FBG section consisting of N alternating layer pairs of identical length L. The low- and high-index layers have refractive indices n_1 and n_2, respectively. The structure includes uniform fibre sections of index n_1 before and after the FBG.

　　1. Calculate and plot the reflected and transmitted intensities relative to the incoming intensity as function of wavelength. Is the incoming intensity equal to the sum of the reflected and transmitted intensity?

　　2. Compute and plot the field electric field profile for the total FBG at $\lambda = 1510$ nm and at $\lambda = 1550$ nm. What happens to the electric field amplitude as the wave propagates through the FBG at 1550 nm?

　　Use $n_1 = 1.44$, $n_2 = 1.45$, $L = 268$ nm, $N = 200$, and a wavelength interval from 1500 to 1600 nm for the reflection and transmission plots.

6.2 Let us now consider a 1D VCSEL made of GaAs and AlAs with refractive indices n_{GaAs} and n_{AlAs}, respectively. The VCSEL consists of a λ cavity made of GaAs of thickness L_{cav} surrounded by top and bottom DBRs consisting of p_{top} and p_{bot} pairs respectively of GaAs and AlAs layers of thicknesses L_{GaAs} and

L_{AlAs}. Above the structure, light is emitted into air, and the VCSEL rests on a GaAs substrate, which we can consider of infinite length.

1. Using the theory of Section 6.3.4, compute and plot the cavity field profile. You should evaluate the total transmission matrix $\overline{\overline{T}}_t$ for the entire structure including GaAs substrate, DBRs, cavity, and air as function of frequency, and look for a node of the matrix element $T_{t;22}(w)$ in the complex plane near $w_0 = 2\pi c/\lambda_0$. Use $p_{\text{top}} = p_{\text{bot}} = 20$. What is the Q factor of the cavity?

2. The out coupling ratio η of light propagating out of the VCSEL into air is given by $\eta = I_{\text{top}}/(I_{\text{top}} + I_{\text{bot}})$, where I_{top} and I_{bot} are the intensities of the forward and backward travelling fields in air and the substrate, respectively. What is η for $p_{\text{top}} = p_{\text{bot}} = 20$? We can improve the out coupling ratio by increasing the number of DBR layer pairs p_{bot} in the bottom DBR. How many layer pairs p_{bot} are needed to achieve $\eta > 0.99$?

Use $n_{\text{GaAs}} = 3.5$, $n_{\text{AlAs}} = 2.9$, $\lambda_0 = 980$ nm, $L_{\text{cav}} = \lambda_0/n_{\text{GaAs}}$, $L_{\text{GaAs}} = \lambda_0/(4n_{\text{GaAs}})$, and $L_{\text{AlAs}} = \lambda_0/(4n_{\text{AlAs}})$.

6.3 In this exercise, we will consider the 2D GaAs waveguide of diameter d surrounded by air as illustrated in Figure 6.13.

1. Using the plane-wave expansion technique described in Section 6.4.1, compute and plot the electric field fundamental mode profile for both TE and TM polarizations.

2. Now, use the SA approach described in Section 6.4.2 to compute and plot the same mode profiles.

3. Compare your mode profiles with those depicted in Figure 6.13. Which value of M is required in TM polarization in the plane-wave expansion technique for the mode to be approximately indistinguishable from that obtained using the SA approach?

Use $d = 250$ nm, $n_{\text{GaAs}} = 3.5$, $\lambda = 980$ nm, $L_x = 1.5$ µm, and periodic boundary conditions with $\mathbf{e}(L_x) = \mathbf{e}(0)$.

6.4 This and the subsequent two exercises are most easily solved using the plane-wave expansion technique.

We will now consider a 2D two-layer geometry for TE polarization consisting of a truncated GaAs waveguide of index n_{GaAs} surrounded by air as illustrated in Figure 6.16. The waveguide placed in the centre of the geometry has a diameter d. The interface is positioned at z_{12}, and the size of the computational domain is given by L_x and L_z. We will initially use periodic boundary conditions such that $e_y(L_x) = e_y(0)$.

1. Using the theory in Section 6.4, calculate and plot the real part of the field profile generated by illumination of the interface with the fundamental waveguide mode, that is $a_{1j} = \delta_{1j}$ and $b_{2j} = 0$. Use $d = 2$ µm. Do we observe any effects of the periodic boundary conditions on the field profile?

2. Perform the same calculation using $d = 500$ nm. What is now the effect of the periodic boundary conditions?

3. Using the theory in Section 6.4.5, implement PMLs in layers 1 and 2 to reduce the influence of the parasitic reflections for $d = 500$ nm. Use identical complex coordinate transformation (Equation 6.134) in the two layers characterized by the parameter s and the PML boundaries x_{12} and x_{45}, and calculate

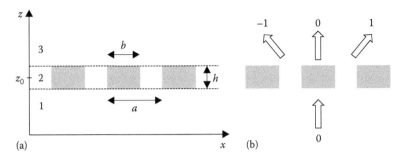

FIGURE 6.29 (a) The GaAs grating geometry surrounded by air. (b) Diffraction of incoming light into outgoing beams of orders -1, 0, and 1.

and plot the resulting field. How do the absorbing boundary conditions improve the field profile?
Use $n_{GaAs} = 3.5$, $\lambda = 980$ nm, $z_{12} = 1$ μm, $L_x = 10d$, $L_z = 5$ μm, $s = 0.4$, $x_{12} = 1.5d$, and $x_{45} = 8.5d$

6.5 We will now study the properties of the GaAs diffraction grating sketched in Figure 6.29a. The grating is uniform along the y-axis, periodic along the x-axis with period a, and positioned at $z = z_0$. The height is h and the duty cycle (DC) is given by $DC = b/a$. The grating is illuminated with TE polarized light. In the air layers 1 and 3, the eigenmodes are given by plane waves with lateral spatial dependence described by $e_y(x) = e^{imKx}$. Here, $K = 2\pi/a$ and m is the mode order. The power P_{qm} transmitted along the z axis for a mode m in the layer q is given by the integration of the z component of the Poynting vector as

$$P_{qm} = \frac{1}{2} \int \text{Re}(\mathbf{e}_{qm} \times \mathbf{h}_{qm}^*) \cdot \hat{\mathbf{z}} d\mathbf{r}_\perp, \tag{6.195}$$

and the relative power transmission T^P from the mode m in layer 1 to the mode n in layer 3 is then given by $T^P_{nm} = P_{3n}|t_{13,nm}|^2/P_{1m}$.
We now illuminate the grating from below with a plane wave of order $m = 0$ as illustrated in Figure 6.29b.

1. Compute and plot the transmissions T^P_{00} into the 0 order and $T^P_{\pm10}$ into the ±1 orders in the air layer 3 as well as their ratio $T^P_{\pm10}/T^P_{00}$ in the wavelength range from 500 to 600 nm. What happens at 558 nm?
2. Plot the electric field amplitude at the wavelength 558 nm in the intervals $z_0 - 3h \leq z \leq z_0 + 10h$. What field pattern is obtained above the grating? Such a structure is known as a first-order diffraction grating, and the field pattern can be used, for example in the UV writing of FBGs.

Use $n_{GaAs} = 3.5$, $a = 730$ nm, $DC = 0.6$, $h = 250$ nm, and periodic boundary conditions with $\mathbf{e}(a) = \mathbf{e}(0)$.

6.6 In this exercise, we again study the grating described in Exercise 6.5, but this time subject to illumination of TM polarized light.

1. Compute and plot the reflection $R^P_{00} = |r_{13,nm}|^2$ and transmission T^P_{00} and $T^P_{\pm10}$ of light into the 0 and ±1 orders in the wavelength range from 1300 to

1500 nm. How much light is transmitted into the ± 1 orders? Why is that? And what happens to the T_{00}^{P} transmission at 1400 nm?

2. Compare the reflection of TM and TE polarized lights at 1400 nm. Why do you think the two are so different?

This structure is known as a high-index-contrast grating and is used in lasers to obtain high tunable reflectivity.

6.7 We now consider the extension of the defining expressions (Equations 6.103 and 6.104) for the reflection and transmission matrices to the case where the orthogonality relation (Equation 2.90) does not hold, that is when $\langle e_{qm} | h_{qn} \rangle \neq \delta_{mn}$. This occurs when using periodic boundary conditions with $e^{i\alpha} \neq 1$ or when implementing absorbing boundary conditions with mode-dependent complex coordinate transformation [13]. Starting from Equations (6.90) and (6.91) and using a similar derivation, show that the reflection and transmission matrices in the non-orthogonal case are given by

$$\overline{\overline{t_{12}}} = 2 \left(\overline{\overline{\langle e_1 | h_1 \rangle}}^{T} \setminus \overline{\overline{\langle e_2 | h_1 \rangle}}^{T} + \overline{\overline{\langle e_1 | h_1 \rangle}} \setminus \overline{\overline{\langle e_1 | h_2 \rangle}} \right)^{-1} \qquad (6.196)$$

$$\overline{\overline{r_{12}}} = \frac{1}{2} \left(\overline{\overline{\langle e_1 | h_1 \rangle}}^{T} \setminus \overline{\overline{\langle e_2 | h_1 \rangle}}^{T} - \overline{\overline{\langle e_1 | h_1 \rangle}} \setminus \overline{\overline{\langle e_1 | h_2 \rangle}} \right) \overline{\overline{t_{12}}}. \qquad (6.197)$$

REFERENCES

1. M. G. Moharam, E. B. Grann, D. A. Pommet, and T. K. Gaylord, Formulation for stable and efficient implementation of the rigorous coupled-wave analysis of binary gratings, *J. Opt. Soc. Am. A*, 12, 1068–1076 (1995).

2. L. Li, New formulation of the Fourier modal method for crossed surface-relief gratings, *J. Opt. Soc. Am. A*, 14, 2758–2767 (1997).

3. P. Bienstman and R. Baets, Optical modelling of photonic crystals and VCSELs using eigenmode expansion and perfectly matched layers, *Opt. Quantum Electron.*, 33, 327–341 (2001).

4. J. P. Hugonin and P. Lalanne, Perfectly matched layers as nonlinear coordinate transforms: A generalized formalization, *J. Opt. Soc. Am. A*, 22, 1844–1849 (2005).

5. FIMMPROP, Photon Design, Oxford, U.K. http://www.photond.com/ (accessed May 2, 2014).

6. CAMFR, Ghent, Belgium, http://camfr.sourceforge.net/ (accessed May 2, 2014); Stanford Stratified Structure Solver, Stanford, CA, http://www.stanford.edu/group/fan/S4/ (accessed May 2, 2014).

7. H. Sagan, *Boundary and Eigenvalue Problems in Mathematical Physics*. New York: Dover Publications Inc., 1989.

8. M. Born and E. Wolf, *Principles of Optics, 7th ed.* Cambridge, U.K.: Cambridge University Press, 2003.

9. L. Li, Formulation and comparison of two recursive matrix algorithms for modeling layered diffraction gratings, *J. Opt. Soc. Am. A*, 13, 1024–1035 (1996).

10. L. Li, Use of Fourier series in the analysis of discontinuous periodic structures, *J. Opt. Soc. Am. A*, 13, 1870–1876 (1996).

11. J.-P. Berenger, A perfectly matched layer for the absorption of electromagnetic waves, *J. Comput. Phys.*, 114, 185–200 (1994).

12. A. J. Ward and J. B. Pendry, Refraction and geometry in Maxwell's equations, *J. Mod. Opt.*, 43, 773–793 (1996).

13. N. Gregersen and J. Mørk, An improved perfectly matched layer for the eigenmode expansion technique, *Opt. Quantum Electron.*, 40, 957–966 (2008).

14. N. Bonod, E. Popov, and M. Nevière, Differential theory of diffraction by finite cylindrical objects, *J. Opt. Soc. Am. A*, 22, 481–490 (2005).

15. A. Yariv and P. Yeh, *Photonics: Optical Electronics in Modern Communications.* New York: Oxford University Press, 2006.

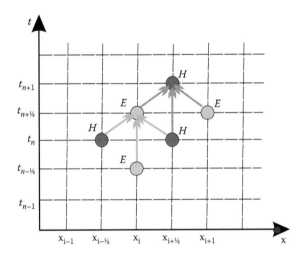

FIGURE 3.3 Leapfrog time updates of electric (green) and magnetic (red) fields are shown by arrows: blue for the electric fields update (3.19) and brown for the magnetic (3.23).

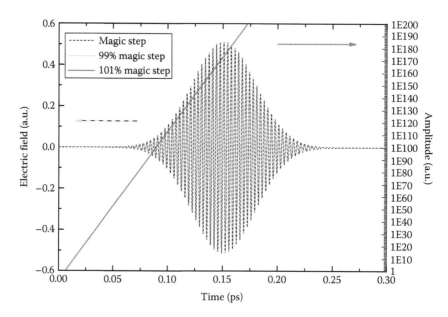

FIGURE 3.5 Fields in the FDTD scheme with different choice of the time step. Left scale: the electric fields in the pulse for $Q = 0.99$ and $Q = 1.00$ are unseparated. Right scale: the amplitude of the biggest electric field in the numerical domain for $Q = 1.01$.

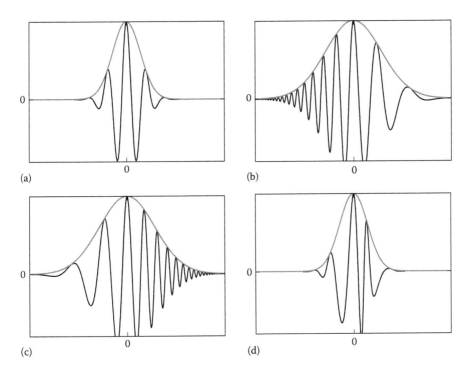

(a)   (b)

(c)   (d)

FIGURE 5.3 (a) Schematic of an unchirped few-cycle pulse with a Gaussian envelope. (b) The pulse is broadened after passage through a waveguide with negative β_2 –
short wavelengths travel faster. (c) The pulse is broadened in a waveguide with positive
β_2 – long wavelengths are faster. (d) The pulse is spectrally broadened by self-phase
modulation (SPM), but the temporal profile is unchanged.

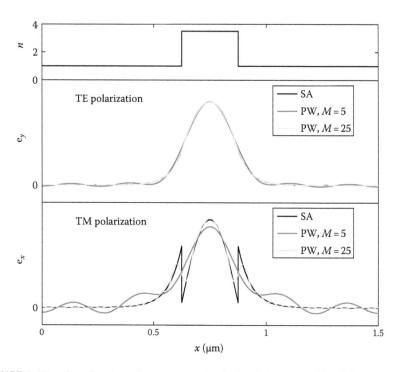

FIGURE 6.13 The refractive-index geometry (top) of a GaAs waveguide of diameter d surrounded by air. The corresponding fundamental mode computed using plane-wave expansion (PW) with M equaling 5 or 25 and using the SA approach in TE (centre) and TM polarizations (bottom). The geometrical parameters are $d = 250$ nm, $n_{\text{GaAs}} = 3.5$, $\lambda = 1$ µm, and $L_x = 1.5$ µm, and periodic boundary conditions with $\mathbf{e}(L_x) = \mathbf{e}(0)$ are employed.

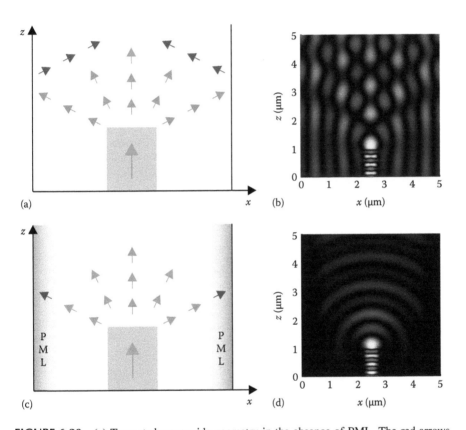

FIGURE 6.20 (a) Truncated waveguide geometry in the absence of PML. The red arrows indicate parasitic reflections. (b) Electric field profile without PML. (c) Same geometry with PML coatings and (d) corresponding electric field profile.

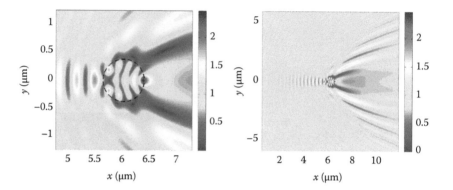

FIGURE 7.2 Magnitude of electric field for a TE-polarized plane wave of wavelength 633 nm being incident on a dielectric cylinder with dielectric constant $\varepsilon_2 = 4$ and radius $a = 375$ nm shown at two magnifications. The reference medium is vacuum (dielectric constant $\varepsilon_1 = 1$). A square region containing the cylinder was discretized in 30×30 area elements.

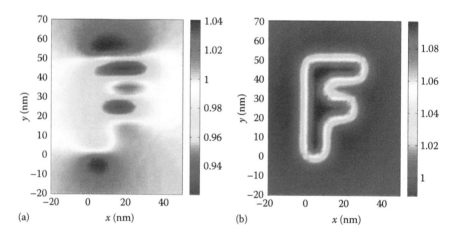

FIGURE 7.5 Magnitude of electric field in a plane located 5 nm below the letter F for the case of a plane wave of wavelength 633 nm propagating along the x-axis and being incident on the structure, where (a) the electric field is polarized along the y-axis and (b) the electric field is polarized along the z-axis.

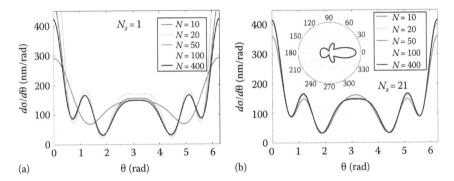

(a) θ (rad)   (b) θ (rad)

FIGURE 7.9 Differential scattering cross section calculated for a silver cylinder of radius 500 nm in air with a TM-polarized incident plane wave of wavelength 700 nm. A polar plot of the differential scattering cross section is shown as an insert.

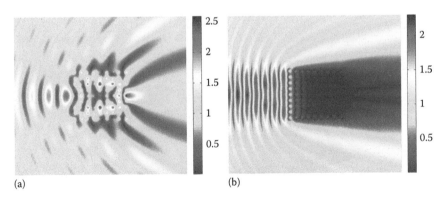

(a)   (b)

FIGURE 7.17 Magnitude of electric field for a plane TE-polarized wave incident on a photonic crystal with 9×9 square rods of size $a = \Lambda/3$, where Λ is the period and where the rod dielectric constant is $\varepsilon_2 = 12$ and the background dielectric constant is $\varepsilon_1 = 1$. The results correspond to using the frequency given by either (a) $\Lambda/\lambda = 0.2$ or (b) $\Lambda/\lambda = 0.35$.

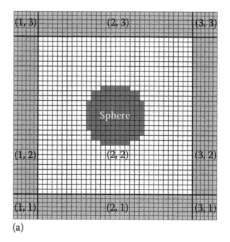

(a)

FIGURE 8.15 (a) Discretized computational domain containing a sphere. Discretization by rectangles. There are nine subdomains with possibly different discretizations $(1, 1), \ldots, (3, 3)$. The outer grey areas are the PML areas.

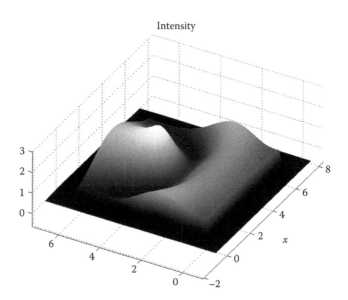

Intensity

FIGURE 8.21 Intensity distribution related to the mesh from Figure 8.15 for a lens with $n = 2$ embedded in air.

7 Green's Function Integral Equation Methods for Electromagnetic Scattering Problems

7.1 INTRODUCTION

Green's function integral equation methods (GFIEMs) that we will consider in this chapter relate the electric or magnetic field at any position in a scattering configuration to the fields inside or on the surface of a scattering object. The field at any position can be expressed as the sum of a field being incident on the scatterer and a scattered field, where the latter is expressed as an overlap integral between a Green's function [1,2] and the fields inside or on the surface of the scattering object. Green's function itself can be thought of as describing the field at a position **r** emitted by a point source at another position **r**'. The scattered field is thus described as a superposition of fields originating from a continuum of point sources either inside or on the surface of the scatterer whose strength is determined by the fields at that position.

From a numerical perspective, a main advantage of GFIEMs is that it is possible to restrict the numerical problem at first to either the inside or the surface of a scatterer being placed in a reference geometry and calculate the field self-consistently there by solving integral equations. Afterwards, the field can be calculated straightforwardly at all other positions via an overlap integral involving the obtained field inside or on the surface of the scatterer and Green's function. With other methods, such as the finite-difference time-domain (FDTD) method or the finite element method (FEM), it is also necessary to include in the numerical problem some region outside the scattering object, which will result in a larger numerical problem, and to handle, for example the radiating boundary condition with an absorbing and/or impedance-matching layer. This means that especially for a small scattering object, GFIEMs are very efficient. Furthermore, in the GFIEMs, the radiating boundary condition is automatically taken care of via the choice of Green's functions. A further advantage is that effects of a complicated reference geometry can also in principle be built into Green's function if Green's function is known for that geometry. Green's function is known analytically for a homogeneous medium, and it can be calculated numerically for stratified media via Sommerfeld integrals [3,4]. It is also possible to construct Green's functions via summations that can be used for modelling reflection and diffraction from periodic structures.

The chapter is organized as follows. In Section 7.2, we will briefly review the wave equations that we are going to solve with the GFIEMs and the underlying assumptions. In the next three sections we consider scattering from an object placed in a homogeneous dielectric reference medium with different GFIEMs, namely with Green's function area integral equation method (GFAIEM) in Section 7.3, Green's function volume integral equation method (GFVIEM) in Section 7.4, and Green's function surface integral equation method (GFSIEM) in Section 7.5. These cases are particularly simple since Green's function is known analytically. The theory of Green's functions for layered and periodic structures will be presented in Sections 7.6 and 7.7, respectively. We will then show how these Green's functions can be applied for calculating reflection from a periodic case and a layered geometry at the same time, namely for a periodically microstructured silicon surface, exemplified with the GFAIEM (Section 7.8). Finally, we will consider how it in some cases is possible to dramatically increase the calculation speed and the size of a problem that can be numerically managed by using an iterative solution scheme taking advantage of the fast Fourier transform (FFT) (Section 7.9).

Due to space limitations, this chapter can by no means be considered an exhaustive treatment of the various GFIEMs. However, the authors feel that the reader will be well equipped to adapt the material presented in this chapter to do calculations for many other cases than those considered here and to do research in this area. Suggestions for further reading are given in Section 7.10.

7.2   THEORETICAL FOUNDATION

We will assume throughout this chapter that our materials are non-magnetic, that is $\mathbf{M}(\mathbf{r}, t) = \mathbf{0}$ such that $\mathbf{B}(\mathbf{r}, t) = \mu_0 \mathbf{H}(\mathbf{r}, t)$, and that they are linear, local, and isotropic. It will be further assumed that the effect of any induced charge and current densities is incorporated into the dielectric constant of the materials, and we will assume the implicit time dependence of complex fields $\exp(i\omega t)$. In that case, the wave equations in the frequency domain become (see Chapter 2)

$$-\nabla \times \nabla \times \mathbf{E}(\mathbf{r}) + k_0^2 \varepsilon(\mathbf{r}) \mathbf{E}(\mathbf{r}) = i\omega \mu_0 \mathbf{J}_s(\mathbf{r}), \tag{7.1}$$

$$-\nabla \times \frac{1}{\varepsilon(\mathbf{r})} \nabla \times \mathbf{H}(\mathbf{r}) + k_0^2 \mathbf{H}(\mathbf{r}) = -\nabla \times \left(\frac{1}{\varepsilon(\mathbf{r})} \mathbf{J}_s(\mathbf{r}) \right), \tag{7.2}$$

where $\mathbf{J}_s(\mathbf{r})$ is a source current density. In the following sections, we will present a range of GFIEMs for solving the wave equations (7.1) and (7.2) in the case when light is scattered off an object or scattering geometry.

7.3   GREEN'S FUNCTION AREA INTEGRAL EQUATION METHOD

In this section, we will consider a 2D scattering problem where both the incident field and structure geometry can be considered invariant in the third dimension, along the z-axis, and we will further consider the case of transverse electric (TE) polarization where the electric field is polarized along the z-axis, that is $\mathbf{E}(\mathbf{r}) = \hat{\mathbf{z}} E(x, y)$ and $\varepsilon(\mathbf{r}) = \varepsilon(x, y)$. It is also common to refer to this polarization as s polarization.

Furthermore, we will assume that there is no source current density $(\mathbf{J}_s(\mathbf{r})=\mathbf{0})$. In that case, the wave equation for the electric field (Equation 7.1) reduces to the scalar-wave equation:

$$\left(\frac{\partial^2}{\partial x^2} + \frac{\partial^2}{\partial y^2} + k_0^2\varepsilon(x,y)\right)E(x,y) = 0. \tag{7.3}$$

Here, ε is the dielectric constant of the total structure including the scatterer. The field which is incident on the scatterer $[\mathbf{E}_0(\mathbf{r}) = \hat{z}E_0(x,y)]$ is a solution for a reference geometry, which we will here choose as a homogeneous dielectric, that is the incident field must satisfy the following wave equation:

$$\left(\frac{\partial^2}{\partial x^2} + \frac{\partial^2}{\partial y^2} + k_0^2\varepsilon_{\text{ref}}\right)E_0(x,y) = 0, \tag{7.4}$$

and we require as a boundary condition that the total field must be the sum of the given incident field and a scattered field component, where the latter propagates away from the scatterer. This boundary condition is known as the radiating boundary condition.

A Green's function for the operator on the left-hand side of Equation (7.4) can be defined as a solution to the equation

$$\left(\frac{\partial^2}{\partial x^2} + \frac{\partial^2}{\partial y^2} + k_0^2\varepsilon_{\text{ref}}\right)g(\mathbf{r},\mathbf{r}') = -\delta(\mathbf{r}-\mathbf{r}'), \tag{7.5}$$

where $\mathbf{r} = \hat{x}x + \hat{y}y$ and $\mathbf{r}' = \hat{x}x' + \hat{y}y'$ are positions in the xy-plane with $\hat{x}$ and $\hat{y}$ being unit vectors along the x- and y-axis, respectively. Any solution of Equation (7.5) can be considered a Green's function for the operator on the left-hand side in Equation (7.5), and both Equation (7.5) and boundary conditions are required to uniquely define a Green's function. With the chosen sign convention, the following Green's function

$$g(\mathbf{r},\mathbf{r}') = \frac{1}{4i}H_0^{(2)}(k_0 n_{\text{ref}}|\mathbf{r}-\mathbf{r}'|) \tag{7.6}$$

is an outgoing cylindrical wave. Here, $H_0^{(2)}$ is the Hankel function of order zero and second kind, which is defined in many mathematical textbooks and handbooks of formulas under the subject Bessel functions (see e.g. [5]), and $n_{\text{ref}}^2 = \varepsilon_{\text{ref}}$. This Green's function is equivalent to the field generated by a point source located at position $\mathbf{r}'$ in a homogeneous reference medium with refractive index n_{ref}. Note that this field propagates away from the point source and thus satisfies the radiating boundary condition.

It is straightforward to show that Green's function (7.6) satisfies Equation (7.5) by considering the case of $\mathbf{r}' = \mathbf{0}$ and $\mathbf{r} \neq \mathbf{0}$ and then look for solutions in the form $g(\mathbf{r},\mathbf{r}') = g(r)$, where $r = |\mathbf{r}|$. Equation (7.5) then reduces to the Bessel equation of zero order, that is

$$(k_0 n_{\text{ref}}r)^2\frac{\partial^2 g}{\partial(k_0 n_{\text{ref}}r)^2} + (k_0 n_{\text{ref}}r)\frac{\partial g}{\partial(k_0 n_{\text{ref}}r)} + (k_0 n_{\text{ref}}r)^2 g = 0, \tag{7.7}$$

which is satisfied by zero-order Bessel functions such as (7.6). It now remains to show that the integral of the left-hand side in Equation (7.5) over the singularity gives -1 for the particular function (7.6). This follows from a calculation of the area integral over a small cylindrical disk centred at the origin and with radius a in the limit of $a \to 0$, that is

$$\lim_{a \to 0} \int \left(\nabla^2 + k_0^2 \varepsilon_{\text{ref}} \right) \frac{1}{4i} H_0^{(2)}(k_0 n_{\text{ref}} r) dA = \lim_{a \to 0} \oint \frac{1}{4i} \hat{\mathbf{n}} \cdot \nabla H_0^{(2)}(k_0 n_{\text{ref}} r) dl$$

$$= \lim_{a \to 0} \int_{\phi=0}^{2\pi} \frac{1}{4i} \hat{\mathbf{n}} \cdot (-H_1^{(2)}(k_0 n_{\text{ref}} a)) k_0 n_{\text{ref}} a \hat{\mathbf{r}} d\phi$$

$$= -1. \tag{7.8}$$

Here, we used the Gauss law to convert the area integral into a surface integral over a cylinder, with surface normal vector $\hat{\mathbf{n}} = \hat{\mathbf{r}}$, and the approximation $H_1^{(2)}(x) \approx -i(\frac{-1}{\pi})\frac{2}{x}$ for small x for the Hankel function of order 1 and type 2. We then find that the integral over the singularity gives -1 in accordance with Equation (7.5). In the case of another choice of $\mathbf{r}'$, we can first apply a coordinate transformation and then apply the same approach to show that (7.6) satisfies Equation (7.5).

If we first consider the case of a given source current density distribution in the reference geometry $(\mathbf{J}_s(\mathbf{r}) = \hat{\mathbf{z}} J_s(x, y))$, the resulting wave equation governing the generated electric field is given by

$$\left(\frac{\partial^2}{\partial x^2} + \frac{\partial^2}{\partial y^2} + k_0^2 \varepsilon_{\text{ref}} \right) E(\mathbf{r}) = i\omega \mu_0 J_s(\mathbf{r}). \tag{7.9}$$

A particular solution to this equation can be obtained straightforwardly from Green's function and the current density by an overlap integral, that is

$$E(\mathbf{r}) = -i\omega \mu_0 \int g(\mathbf{r}, \mathbf{r}') J_s(\mathbf{r}') dA'. \tag{7.10}$$

It follows that (7.10) is a solution to Equation (7.9) by using that Green's function satisfies Equation (7.5). With our choice of Green's function, this solution will also satisfy the radiating boundary condition such that the field generated by the sources propagates away from the source region.

In a similar way, we can also see that if we combine Equations (7.3) and (7.4) into

$$\left(\frac{\partial^2}{\partial x^2} + \frac{\partial^2}{\partial y^2} + k_0^2 \varepsilon_{\text{ref}} \right) (E(x, y) - E_0(x, y)) = -k_0^2 \left(\varepsilon(x, y) - \varepsilon_{\text{ref}} \right) E(x, y) \tag{7.11}$$

we can likewise obtain a solution for $(E(\mathbf{r}) - E_0(\mathbf{r}))$ satisfying the radiating boundary condition by solving

$$E(\mathbf{r}) = E_0(\mathbf{r}) + \int g(\mathbf{r}, \mathbf{r}') k_0^2 \left(\varepsilon(\mathbf{r}') - \varepsilon_{\text{ref}} \right) E(\mathbf{r}') dA'. \tag{7.12}$$

This integral equation has the property that when solving it, the resulting field will exactly have the form of a sum of the given incident and a scattered field component, where the latter will propagate away from the scatterer due to the properties of the chosen Green's function.

Note that it is clearly seen from Green's function integral equation (Equation 7.12) that it is sufficient to initially restrict the numerical problem to the region where $\varepsilon(\mathbf{r}) - \varepsilon_{\text{ref}} \neq 0$, that is to the region of the scattering object. Once the field in this region has been determined, the field can be obtained straightforwardly at all other positions via the overlap integral between Green's function and the field inside the scatterer.

The integral equation (7.12) can be discretized and solved numerically. The simplest approach is to divide the scatterer into N small elements where element i has the area A_i and centre at position $\mathbf{r}_i$, and if the elements are small enough, we can approximate the electric field in each element with a constant value E_i. In that case, we obtain the following system of linear equations:

$$E_i = E_{0,i} + \sum_j g_{ij} k_0^2 \left(\varepsilon_j - \varepsilon_{\text{ref}} \right) E_j A_j, \tag{7.13}$$

where we can use the approximation $g_{ij} = g(\mathbf{r}_i, \mathbf{r}_j)$ when $i \neq j$, and $\varepsilon_i = \varepsilon(\mathbf{r}_i)$. The interaction of an element with itself governed by g_{ii} requires that we find a reasonable way to handle the singular nature of Green's function. For the particular case, we consider a Green's function with a logarithmic and thus rather weak singularity. For the self-interaction Green's function term g_{ii}, we can use

$$g_{ii} = \frac{1}{A_i} \int_{A_i} g(\mathbf{r}_i, \mathbf{r}') dA'. \tag{7.14}$$

Typically, one will consider using an array of square-shaped elements with the same size ΔA. Rather than calculating the integral (7.14) numerically, it can be acceptable to approximate the square-shaped element with a circular element of radius a with the same area and same centre position, in which case, the integral (7.14) can be evaluated analytically resulting in [6]

$$g_{ii} \approx \frac{1}{2i(ka)^2} \left(ka H_1^{(2)}(ka) - i\frac{2}{\pi} \right), \tag{7.15}$$

where $k = k_0 n_{\text{ref}}$ and $\pi a^2 = A_i$.

While the integral equation (7.12) can be used straightforwardly to calculate the field at all positions using the exact Green's function, it is preferable to use another simpler expression for Green's function when calculating the scattered far field. Consider an object located near the origin such that when $\mathbf{r}$ is a position far from the object, the involved source positions $\mathbf{r}'$ in Equation (7.14) will be at relatively small distances from the origin in comparison. For positions in the far field, $k|\mathbf{r} - \mathbf{r}'| \gg 1$, and Green's function can be approximated with

$$g^{ff}(\mathbf{r}, \mathbf{r}') \approx \frac{1}{4i} \sqrt{\frac{2}{\pi k |\mathbf{r} - \mathbf{r}'|}} e^{i\pi/4} e^{-ik|\mathbf{r} - \mathbf{r}'|}, \tag{7.16}$$

which, since $|\mathbf{r}'|/|\mathbf{r}| \ll 1$, can be further approximated with

$$g^{ff}(\mathbf{r}, \mathbf{r}') = \frac{1}{4}\sqrt{\frac{2}{\pi k r}}e^{-i\pi/4}e^{-ikr}e^{ik\frac{\mathbf{r}}{r}\cdot\mathbf{r}'}. \tag{7.17}$$

The scattered far field can then be obtained by integration as

$$E^{ff}_{sc}(\mathbf{r}) = \frac{1}{4}\sqrt{\frac{2}{\pi k r}}e^{-i\pi/4}e^{-ikr}\int k_0^2\left(\varepsilon(\mathbf{r}') - \varepsilon_{ref}\right)E(\mathbf{r}')e^{ik\frac{\mathbf{r}}{r}\cdot\mathbf{r}'}dA' = \frac{e^{-ikr}}{\sqrt{r}}f(\theta), \tag{7.18}$$

where θ represents the direction of the vector $\mathbf{r}/r$, and we note that the final far-field expression is on the form of a cylindrical-wave term $e^{-ikr}/\sqrt{r}$ times a function f that depends only on θ.

If we note that the far-field time-averaged Poynting vector is given by

$$\mathbf{S}^{ff} = \frac{1}{2}\mathrm{Re}\left(\mathbf{E}^{ff}_{sc} \times \mathbf{H}^{ff*}_{sc}\right) = \frac{\mathbf{r}}{r}\frac{1}{2}\mathrm{Re}(n_{ref})\sqrt{\frac{\varepsilon_0}{\mu_0}}|E|^2, \tag{7.19}$$

and if we assume that the incident field is a plane wave with unity amplitude of the electric field, that is $|\mathbf{E}_0(\mathbf{r})| = 1$, then the *scattering cross section* can be calculated as

$$\sigma_{scat} = \frac{\oint \mathbf{S}^{ff}(\mathbf{r}) \cdot \hat{\mathbf{n}}dl}{|\mathbf{S}_0|} = \int |f(\theta)|^2 d\theta, \tag{7.20}$$

where $\mathbf{S}_0 = \frac{1}{2}\mathrm{Re}\left(\mathbf{E}_0 \times \mathbf{H}_0^*\right)$ is the time-averaged Poynting vector of the incident field or the power per unit area of the incident field. Note that the scattering cross section has units of *length* because this is a 2D problem. The *differential scattering cross section* for radiation in the direction θ per unit angle is correspondingly given by $|f(\theta)|^2$.

As an example, we will consider the scattering situation illustrated in Figure 7.1 where a dielectric cylinder placed in vacuum is illuminated with a plane TE-polarized wave propagating along the x-axis ($E_0(x, y) = E_0 e^{-ik_0 x}$). We will choose

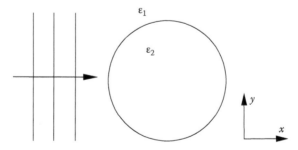

FIGURE 7.1 Schematic of scatterer with dielectric constant ε_2 embedded in a homogeneous dielectric with dielectric constant ε_1 being illuminated with a TE-polarized plane wave propagating along the x-axis.

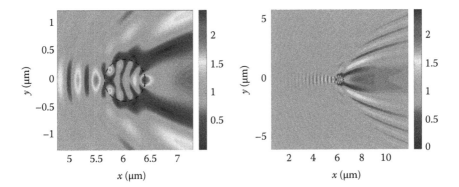

FIGURE 7.2 (**See colour insert.**) Magnitude of electric field for a TE-polarized plane wave of wavelength 633 nm being incident on a dielectric cylinder with dielectric constant $\varepsilon_2 = 4$ and radius $a = 375$ nm shown at two magnifications. The reference medium is vacuum (dielectric constant $\varepsilon_1 = 1$). A square region containing the cylinder was discretized in 30×30 area elements.

the wavelength 633 nm, cylinder radius 375 nm, and cylinder dielectric constant $\varepsilon_2 = 4$ ($\varepsilon_1 = 1$).

The resulting distribution of the electric field magnitude inside and outside the scattering object calculated using Equation (7.13) and 30×30 discretization elements for a square region containing the cylinder is shown in Figure 7.2. Depending on whether the centre of each area element is inside or outside the cylinder, we assigned either the dielectric constant ε_2 or ε_1. This results in an approximation to the actual geometry with a staircased surface. Note that the field inside the cylinder was obtained in the first step, and then the field for a large region outside the cylinder was calculated afterwards via simple integration over the cylinder using Equation (7.12). The differential scattering cross section for the same situation calculated using Equations (7.18) and (7.20) is shown in Figure 7.3. For this example, it is clear that scattering in the forward direction is dominant. By integrating the differential scattering cross section according to Equation (7.20), we find the total scattering cross section $\sigma = 1.206$ µm. If we use the GFSIEM being presented later on for the same situation, but with 200 points on the surface, we find practically the same results but with some very small differences (a few percent) due to a better representation (no staircasing) of the actual structure. The total scattering cross section with that method was found to be $\sigma = 1.233$ µm. If we double the resolution with the GFAIEM and use 60×60 points, we find the scattering cross section $\sigma = 1.223$ µm. Another possibility is to have a better representation of the actual geometry by assigning a dielectric constant to each area element given by the spatial average over that element, that is

$$\varepsilon_{\text{eff},i} = \frac{1}{A_i} \int_{A_i} \varepsilon(\mathbf{r}) dA, \tag{7.21}$$

or to use elements that are shaped in a way such that the staircasing at the surface is avoided (see Figure 4.5). If we use 30×30 elements and apply Equation (7.21)

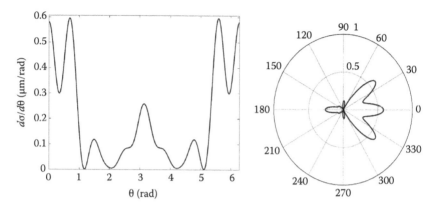

FIGURE 7.3 Differential scattering cross section for a cylinder with dielectric constant $\varepsilon_2 = 4$ and radius 375 nm surrounded by vacuum $\varepsilon_1 = 1$ and being illuminated with a TE-polarized plane wave propagating in the direction $\theta = 0$. The reference medium is vacuum (dielectric constant $\varepsilon_1 = 1$). A square region containing the cylinder was discretized in 30×30 area elements.

to obtain a better representation of the surface, we find the scattering cross section $\sigma = 1.251$ μm, and with 60×60 elements and applying Equation (7.21), we find $\sigma = 1.237$ μm.

Note that while we considered a simple geometry here, it is straightforward to use the same method for doing calculations for another scattering object having a different shape and size and possibly having a varying dielectric constant inside the scatterer. It is also possible to use elements that have a different shape such as triangles. Here, we note that rectangular elements on a quadratic grid have the advantage that the summation in Equation (7.13) becomes a discrete 2D convolution, which can be calculated very fast using the FFT algorithm, and thus, this equation can be solved with very efficient iterative methods as will be shown in Section 7.9 allowing easily the usage of hundreds of thousands of discretization elements.

7.4 GREEN'S FUNCTION VOLUME INTEGRAL EQUATION METHOD

The GFVIEM originates from astrophysics where a related method known as the discrete dipole approximation (DDA) was developed for calculating scattering and absorption of light from interstellar dust grains [7–10]. In the DDA, a 3D scattering object of arbitrary shape is divided into a finite number of discrete dipoles with polarizabilities determined by the volume that each dipole replaces. Each dipole is driven by the external field at the dipole position, which may, for example be a superposition of a plane wave incident on the whole dust grain and the field at that position generated by all the other dipoles. Thus, the radiation emitted from each dipole is determined by the radiation emitted by all the other dipoles, such that in the numerical problem, all dipoles are coupled via their radiation fields. In the limit where the number of dipoles increases towards infinity, the DDA and the GFVIEM are equivalent, but in the latter, total fields are calculated instead of external fields. A real physical structure can also be thought of as consisting of a large number of dipoles, one dipole for each atom, and so the DDA is closely following the physics of a real scattering problem, except that for numerical reasons, the geometry must be divided into a lot fewer elements than there are atoms. A main concern with the DDA is that the representation of a geometry in terms of discrete dipoles of, for example the same polarizability, and being placed on a cubic lattice, results in a staircased representation of the geometry surface, which in the case of large refractive-index contrasts may result in severe convergence problems [7,11]. The convergence problem can be effectively solved by avoiding staircasing either by using specially shaped elements near the surface [11] or by using triangular-shaped elements that also improves the description of the surface [12]. A description of such elements is available in Section 8.5. It is also necessary in some cases to avoid sharp corners due to numerical issues related to the presence of singular electromagnetic fields there [13]. Instead of rounding sharp corners, another possibility is to use a gradually varying dielectric constant near surfaces rather than having a sharp transition [11].

In this chapter, we will consider scattering of light by a general 3D object placed in a homogeneous dielectric. Specifically, we will consider an example from Ref. [14] with an object of dielectric constant 2.25 in vacuum. In this case, the difference in dielectric constant is not very large, and reasonable convergence is possible without using advanced discretization schemes.

The wave equation of interest for the total electric field is given by

$$\left(-\nabla \times \nabla \times +k_0^2 \varepsilon(\mathbf{r})\right) \mathbf{E}(\mathbf{r}) = \mathbf{0}, \tag{7.22}$$

where $\varepsilon(\mathbf{r})$ is the dielectric constant of the total structure and $\mathbf{r} = \hat{\mathbf{x}} x + \hat{\mathbf{y}} y + \hat{\mathbf{z}} z$. We will assume that the scattering object is illuminated with light that can be described with a solution $\mathbf{E}_0(\mathbf{r})$ of the wave equation in the case where the object is absent, that is

$$\left(-\nabla \times \nabla \times +k_0^2 \varepsilon_{\text{ref}}\right) \mathbf{E}_0(\mathbf{r}) = \mathbf{0}. \tag{7.23}$$

A dyadic Green's tensor for the operator on the left-hand side of Equation (7.23) that is also satisfying the radiating boundary condition is given by

$$\mathbf{G}(\mathbf{r}, \mathbf{r}') = \left(\mathbf{I} + \frac{1}{k_0^2 \varepsilon_{\text{ref}}} \nabla\nabla\right) g(\mathbf{r}, \mathbf{r}')$$

$$= \left[\mathbf{I}\left(1 - \frac{i}{kR} - \frac{1}{(kR)^2}\right) - \frac{\mathbf{R}\mathbf{R}}{R^2}\left(1 - \frac{3i}{kR} - \frac{3}{(kR)^2}\right)\right] g(\mathbf{r}, \mathbf{r}'), \tag{7.24}$$

where

$$g(\mathbf{r}, \mathbf{r}') = \frac{e^{-ikR}}{4\pi R}, \tag{7.25}$$

and $k = k_0 n_{\text{ref}}$, $\mathbf{R} = \mathbf{r} - \mathbf{r}'$, and $R = |\mathbf{R}|$, and $\mathbf{I} = \hat{\mathbf{x}}\hat{\mathbf{x}} + \hat{\mathbf{y}}\hat{\mathbf{y}} + \hat{\mathbf{z}}\hat{\mathbf{z}}$ is the unit dyad. Often when the dyads are applied to a vector, this will be with a dot product. If we apply the two dyads involved in the expression for the dyadic Green's tensor to a current density vector, this means that $\mathbf{I} \cdot \mathbf{J}_s = \mathbf{J}_s$, and $\mathbf{RR} \cdot \mathbf{J}_s = \mathbf{R}(\mathbf{R} \cdot \mathbf{J}_s)$. In general, $(\mathbf{AB}) \cdot \mathbf{C} = \mathbf{A}(\mathbf{B} \cdot \mathbf{C})$. It is also possible to apply a vector from the other side, in which case $\mathbf{C} \cdot (\mathbf{AB}) = (\mathbf{C} \cdot \mathbf{A})\mathbf{B}$. If we, for example take the divergence of $\mathbf{G}$, this means $\nabla \cdot \mathbf{G} = \frac{\partial}{\partial x}(\hat{\mathbf{x}} \cdot \mathbf{G}) + \frac{\partial}{\partial y}(\hat{\mathbf{y}} \cdot \mathbf{G}) + \frac{\partial}{\partial z}(\hat{\mathbf{z}} \cdot \mathbf{G})$, etc.

The dyadic Green's tensor (7.24) is a solution to the equation

$$\left(-\nabla \times \nabla \times + k_0^2 \varepsilon_{\text{ref}}\right) \mathbf{G}(\mathbf{r}, \mathbf{r}') = -\mathbf{I}\delta(\mathbf{r} - \mathbf{r}'), \tag{7.26}$$

whereas the scalar Green's function $g(\mathbf{r}, \mathbf{r}')$ is a solution to the equation

$$\left(\nabla^2 + k_0^2 \varepsilon_{\text{ref}}\right) g(\mathbf{r}, \mathbf{r}') = -\delta(\mathbf{r} - \mathbf{r}'). \tag{7.27}$$

We can straightforwardly show that (7.25) is a solution to Equation (7.27). If we consider the case of $\mathbf{r}' = 0$ and $\mathbf{r} \neq 0$ and look for solutions in the form $g(\mathbf{r}, \mathbf{r}') = g(r)$, where $r = |\mathbf{r}|$, Equation (7.27) reduces to

$$\frac{\partial^2 g}{\partial r^2} + \frac{2}{r}\frac{\partial g}{\partial r} + (k_0^2 n_{\text{ref}}^2 g) = 0, \tag{7.28}$$

which can be further rewritten into

$$\left(\frac{\partial^2}{\partial r^2} + k_0^2 n_{\text{ref}}^2\right)(rg) = 0, \tag{7.29}$$

which is similar to the 1D wave equation in rg, and thus, the spherical wave (7.25) is a solution. It now remains to show that the integral of the left-hand side in Equation (7.27) gives -1 when inserting (7.25). This follows from evaluating the volume integral over a small sphere centred at the origin and with radius a in the limit of $a \to 0$, that is

$$\lim_{a \to 0} \int \left(\nabla^2 + k_0^2 \varepsilon_{\text{ref}}\right) \frac{e^{-ik_0 n_{\text{ref}} r}}{4\pi r} dV$$

$$= \lim_{a \to 0} \oint \hat{\mathbf{n}} \cdot \nabla \frac{e^{-ik_0 n_{\text{ref}} r}}{4\pi r} dA$$

$$= \lim_{a \to 0} \oint -\hat{\mathbf{n}} \cdot \frac{e^{-ik_0 n_{\text{ref}} a}}{4\pi a^2} \hat{\mathbf{r}} dA = -1, \tag{7.30}$$

where we used Gauss law to convert the volume integral into a surface integral over the small sphere, in which case the outward normal unit vector $\hat{\mathbf{n}} = \hat{\mathbf{r}}$, and we used that the total surface area of the small sphere is $4\pi a^2$. In the case of another choice of $\mathbf{r}'$, we can apply a coordinate transformation and apply the same approach. It now follows after some algebra that (7.24) is a solution to Equation (7.26) by simply inserting (7.24) into the left-hand side of Equation (7.26) and using the property of the scalar Green's function (Equation 7.27). It is convenient to use $-\nabla \times \nabla \times = -\nabla\nabla \cdot + \nabla^2$, and that $(\nabla\nabla \cdot)(\nabla\nabla g) = \nabla^2 \nabla\nabla g$ and $(\nabla\nabla \cdot)\mathbf{I}g = \nabla\nabla g$.

A particular solution of the wave equation

$$\left(-\nabla \times \nabla \times +k_0^2 \varepsilon_{\text{ref}}\right) \mathbf{E}(\mathbf{r}) = i\omega\mu_0 \mathbf{J}_s(\mathbf{r}), \tag{7.31}$$

where $\mathbf{J}_s(\mathbf{r})$ is a given source current density, is straightforwardly given by the integral

$$\mathbf{E}(\mathbf{r}) = -i\omega\mu_0 \int \mathbf{G}(\mathbf{r}, \mathbf{r}') \cdot \mathbf{J}_s(\mathbf{r}') d^3 r'. \tag{7.32}$$

This can be seen by using Equation (7.26). Furthermore, since $\mathbf{G}(\mathbf{r}, \mathbf{r}')$ satisfies the radiating boundary condition, so will the solution obtained in Equation (7.32), meaning that the solution will describe light propagating away from the region of the source current density.

If we now combine Equations (7.22) and (7.23) into

$$\left(-\nabla \times \nabla \times +k_0^2 \varepsilon_{\text{ref}}\right) [\mathbf{E}(\mathbf{r}) - \mathbf{E}_0(\mathbf{r})] = -k_0^2 \left(\varepsilon(\mathbf{r}) - \varepsilon_{\text{ref}}\right) \mathbf{E}(\mathbf{r}), \tag{7.33}$$

then similar to the solution for the given current density, we find the integral equation

$$\mathbf{E}(\mathbf{r}) = \mathbf{E}_0(\mathbf{r}) + \int \mathbf{G}(\mathbf{r}, \mathbf{r}') k_0^2 \left(\varepsilon(\mathbf{r}') - \varepsilon_{\text{ref}}\right) \cdot \mathbf{E}(\mathbf{r}') d^3 r'. \tag{7.34}$$

Clearly, the integrand on the right-hand side is only non-zero for positions $\mathbf{r}'$ inside the scatterer due to the term $(\varepsilon(\mathbf{r}') - \varepsilon_{\text{ref}})$. Thus, if the electric field is known inside the scatterer, the integral expression (7.34) can be immediately applied to obtain the electric field at all other positions by simple integration similar to Equation (7.32) for a given source current density. First, however, it is necessary to obtain the field inside the scatterer, and due to the properties of the integrand, this can be done by initially restricting the numerical problem to those positions $\mathbf{r}$ that are inside the scatterer. Green's function volume integral equation can be solved by, for example subdividing the scatterer into a number of small volume elements in which the electric field vector is approximated with a constant value. This leads to the following discretized version of Equation (7.34), that is

$$\mathbf{E}_i = \mathbf{E}_{0,i} + \sum_j \mathbf{G}_{ij} \left(\varepsilon_j - \varepsilon_{\text{ref}}\right) \cdot \mathbf{E}_j, \tag{7.35}$$

where we have introduced the field and dielectric constants in volume element i given by $\mathbf{E}_i$ and ε_i, respectively, and where

$$\mathbf{G}_{ij} = \int_{V_j} \mathbf{G}(\mathbf{r}_i, \mathbf{r}') k_0^2 d^3 r', \tag{7.36}$$

where the position $\mathbf{r}_i$ can represent the centre of volume element i and V_j denotes that the integral is over volume element j. In the case when $i \neq j$, it can be acceptable to use the approximation

$$\mathbf{G}_{ij} = \mathbf{G}(\mathbf{r}_i, \mathbf{r}_j) k_0^2 V_j. \tag{7.37}$$

In the case when $i = j$, it is necessary to use a more precise method. In order to avoid having to integrate directly over the very strong singularity of $\mathbf{G}$, we will suggest similar to Refs. [15,16] that the volume integral in (7.36) should be transformed to a surface integral away from the singularity. This can be done by exploiting the expressions (7.24) and (7.27) and Gauss law, resulting in the relation

$$\mathbf{G}_{ii} = -\frac{1}{\varepsilon_{\text{ref}}}\mathbf{I} + \frac{1}{\varepsilon_{\text{ref}}}\oint_{\partial V_i}\left(\hat{\mathbf{n}}'\nabla' - \mathbf{I}\hat{\mathbf{n}}'\cdot\nabla'\right)g(\mathbf{r}_i,\mathbf{r}')dA', \qquad (7.38)$$

where

∂V_i represents the surface of volume element i

$\hat{\mathbf{n}}'$ is the outward surface normal vector of volume element i at the position on the surface $\mathbf{r}'$

dA' is the infinitesimal surface area

It is relevant to note that even in the limit where the volume of the elements becomes infinitesimal, $\mathbf{G}_{ii}$ will not vanish. This is due to the strong singularity in $\mathbf{G}(\mathbf{r},\mathbf{r}')$. In this limit, $\mathbf{G}_{ii}$ depends only on the shape of element i (and possibly the position chosen within this element $\mathbf{r}_i$). In this limit, $\mathbf{G}_{ii}$ has been tabulated for a wide range of different shapes in Ref. [15]. In the simplest case of a cubic volume element with $\mathbf{r}_i$ in the centre, we find in the limit of very small volume elements that $\mathbf{G}_{ii} \approx -\frac{1}{3\varepsilon_{\text{ref}}}\mathbf{I}$. This is not surprising since in this limit, $\mathbf{G}_{ii} = \int_{V_i}\mathbf{G}(\mathbf{r},\mathbf{r}')k_0^2 d^3r' \approx \frac{1}{\varepsilon_{\text{ref}}}\int_{V_i}\nabla'\nabla'g(\mathbf{r},\mathbf{r}')d^3r'$ and the trace (Tr) given by the sum of the diagonal elements $\text{Tr}(\mathbf{G}_{ii}) \approx \int_{V_i}\nabla'^2 g(\mathbf{r},\mathbf{r}')d^3r' \approx -1$, where we have used Equation (7.27). For other more complex shapes (and possibly slightly larger sizes than those of the infinitesimal limit), the expression (7.38) can be straightforwardly applied. Naturally, since this self-interaction element is very sensitive to the shape of the element, there can be some tricky issues when making the numerical integration in Equation (7.34).

We will now consider an example of scattering by a geometry shaped like the letter F shown in Figure 7.4 with refractive index 1.5, a thickness of 7.5 nm, and that fits in an area of 30 nm × 50 nm. This instructive example is borrowed from Ref. [14], and it has the advantage that the geometry is sufficiently small in size

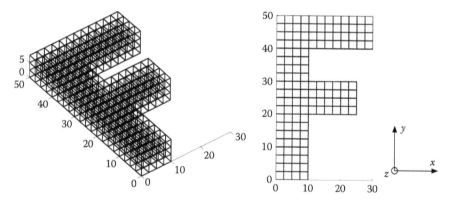

FIGURE 7.4 Schematic of a nano-size letter F with thickness 7.5 nm and fitting within 30 nm × 50 nm in the xy-plane and being discretized in volume elements of size $\Delta x = \Delta y = \Delta z = 2.5$ nm.

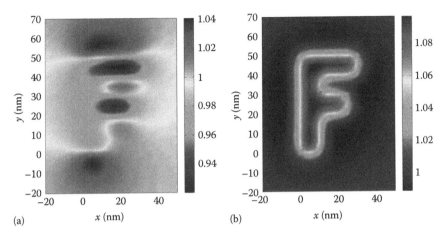

FIGURE 7.5 (See colour insert.) Magnitude of electric field in a plane located 5 nm below the letter F for the case of a plane wave of wavelength 633 nm propagating along the x-axis and being incident on the structure, where (a) the electric field is polarized along the y-axis and (b) the electric field is polarized along the z-axis.

and that the dielectric constant is sufficiently small that the numerical calculation is fairly straightforward. Calculations of the magnitude of the electric field in a plane located 5 nm below the letter F are shown in Figure 7.5 for illumination with a plane wave propagating along the x-axis and having the electric field polarized either along the y- or z-axis. In the calculation, we used discretization elements of the size $\Delta x = \Delta y = \Delta z = 2.5$ nm.

The example shows that the polarization of the light can make a very large difference regarding what will be the result of a near-field measurement and that there can be a significant difference regarding how closely the near field reveals the actual shape of the object depending on the polarization of the incident light.

7.5   GREEN'S FUNCTION SURFACE INTEGRAL EQUATION METHOD (2D)

The GFSIEM [11,17–20] is based on identities where the field at any position inside a closed domain can be obtained from an overlap integral between a Green's function and the fields and their normal derivatives at the domain boundaries. This means that it is sufficient to know the fields and normal derivatives at the domain boundaries since then the fields are known at all other positions from the identities. The first step in this method is thus to set up equations for calculating self-consistently the field and normal derivatives at the domain boundaries, and once this has been achieved, the field can be obtained everywhere else via simple integration.

7.5.1   SURFACE INTEGRAL EQUATIONS

Consider a geometry such as Figure 7.6, where a scatterer with dielectric constant ε_2 is embedded in a homogeneous dielectric with dielectric constant ε_1. Consider

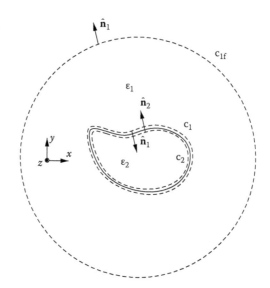

FIGURE 7.6 Schematic of scatterer with dielectric constant ε_2 embedded in a homogeneous dielectric with dielectric constant ε_1. Just inside the boundary of the scatterer is placed an imaginary surface c_2 with outward normal vector $\hat{\mathbf{n}}_2$, and outside the scatterer, a closed domain in the region with dielectric constant ε_1 is specified by the boundaries c_1 and c_{1f}, where c_1 is located just outside the boundary of the scatterer. The outward normal vector of this closed domain is denoted $\hat{\mathbf{n}}_1$.

also an imaginary boundary just inside the surface of the scatterer c_2 with outward normal vector $\hat{\mathbf{n}}_2$ and a position $\mathbf{r}$ inside the scatterer. We consider a 2D situation and transverse magnetic (TM) polarization where the magnetic field is oriented along the z-axis and where the fields and the geometry are invariant along the z-axis, that is $\mathbf{H}(\mathbf{r}) = \hat{\mathbf{z}}H(x, y)$ and $\varepsilon(\mathbf{r}) = \varepsilon(x, y)$. This polarization is also commonly referred to as p polarization. The domain of the scatterer is denoted Ω_2, and the domain outside the scatterer is denoted Ω_1. For the situation here, we thus have that

$$\varepsilon(\mathbf{r}) = \begin{cases} \varepsilon_1 & \text{if } \mathbf{r} \in \Omega_1 \\ \varepsilon_2 & \text{if } \mathbf{r} \in \Omega_2 \end{cases}$$

We also assume that we know two Green's functions ($i = 1, 2$) satisfying both

$$\left(\nabla^2 + k_0^2 \varepsilon_i\right) g_i(\mathbf{r}; \mathbf{r}') = -\delta(\mathbf{r} - \mathbf{r}'), \tag{7.39}$$

and

$$\left(\nabla'^2 + k_0^2 \varepsilon_i\right) g_i(\mathbf{r}; \mathbf{r}') = -\delta(\mathbf{r} - \mathbf{r}'), \tag{7.40}$$

where $\nabla^2 = \frac{\partial^2}{\partial x^2} + \frac{\partial^2}{\partial y^2}$ and $\nabla'^2 = \frac{\partial^2}{\partial x'^2} + \frac{\partial^2}{\partial y'^2}$. For reasons that will become clear later, it is desirable that at least Green's function g_1 satisfies the radiating boundary condition.

Thus, for the two Green's functions, we may choose the same Green's function that we used in Section 7.3:

$$g_i(\mathbf{r}, \mathbf{r}') = \frac{1}{4i} H_0^{(2)}(k_0 n_i |\mathbf{r} - \mathbf{r}'|), \tag{7.41}$$

where $n_i = \sqrt{\varepsilon_i}$ and $H_0^{(2)}$ is a Hankel function. The magnetic field in each of the two regions must satisfy

$$\left(\nabla^2 + k_0^2 \varepsilon_i\right) H(\mathbf{r}) = 0. \tag{7.42}$$

Consider a position $\mathbf{r} \in \Omega_2$. In that case, we can obtain the following integral relation between the field at position $\mathbf{r}$ and the field and its normal derivative at the imaginary surface c_2:

$$H(\mathbf{r}) = \oint_{c_2} \left[g_2(\mathbf{r}, \mathbf{r}')\hat{\mathbf{n}}' \cdot \nabla' H(\mathbf{r}') - H(\mathbf{r}')\hat{\mathbf{n}}' \cdot \nabla' g_2(\mathbf{r}, \mathbf{r}') \right] dl', \tag{7.43}$$

where the normal vector $\hat{\mathbf{n}} = \hat{\mathbf{n}}_2$ (see Figure 7.6).

This expression can be derived by first converting the right-hand side to an area integral over the domain Ω_2 using the Gauss law and then using the relations (7.39) and (7.40) and the properties of the δ-function, that is

$$\oint_{c_2} \left[g_2(\mathbf{r}, \mathbf{r}')\hat{\mathbf{n}}' \cdot \nabla' H(\mathbf{r}') - H(\mathbf{r}')\hat{\mathbf{n}}' \cdot \nabla' g_2(\mathbf{r}, \mathbf{r}') \right] dl'$$

$$= \int_{\Omega_2} \nabla' \cdot \left[g_2(\mathbf{r}, \mathbf{r}')\nabla' H(\mathbf{r}') - H(\mathbf{r}')\nabla' g_2(\mathbf{r}, \mathbf{r}') \right] dA'$$

$$= \int_{\Omega_2} \left[g_2(\mathbf{r}, \mathbf{r}')\nabla'^2 H(\mathbf{r}') - H(\mathbf{r}')\nabla'^2 g_2(\mathbf{r}, \mathbf{r}') \right] dA'$$

$$= \int_{\Omega_2} \left[g_2(\mathbf{r}, \mathbf{r}') \left(-k_0^2 \varepsilon_2 H(\mathbf{r}') \right) - H(\mathbf{r}') \left(-k_0^2 \varepsilon_2 g_2(\mathbf{r}, \mathbf{r}') - \delta(\mathbf{r} - \mathbf{r}') \right) \right] dA' = H(\mathbf{r}).$$

Consider now a position $\mathbf{r} \in \Omega_1$. In that case, we can obtain the following integral relation between the total field at position $\mathbf{r}$, the incident field $H_0(\mathbf{r})$, and the field and its normal derivative at the imaginary surface c_1:

$$H(\mathbf{r}) = H_0(\mathbf{r}) - \oint_{c_1} \left[g_1(\mathbf{r}, \mathbf{r}')\hat{\mathbf{n}}' \cdot \nabla' H(\mathbf{r}') - H(\mathbf{r}')\hat{\mathbf{n}}' \cdot \nabla' g_1(\mathbf{r}, \mathbf{r}') \right] dl', \tag{7.44}$$

where here the normal vector is also $\hat{\mathbf{n}} = \hat{\mathbf{n}}_2$, that is is also pointing in the direction out of the scatterer. This expression can be obtained by first following the same approach as before but for the domain Ω_1 and the boundaries of that domain c_1 and c_{1f}, which leads to

$$H(\mathbf{r}) = \oint_{c_1 + c_{1f}} \left[g_1(\mathbf{r}, \mathbf{r}')\hat{\mathbf{n}}_1' \cdot \nabla' H(\mathbf{r}') - H(\mathbf{r}')\hat{\mathbf{n}}_1' \cdot \nabla' g_1(\mathbf{r}, \mathbf{r}') \right] dl'. \tag{7.45}$$

Since this is valid for any arbitrary choice of curve c_{1f} as long as the position $\mathbf{r}$ is between the curves c_1 and c_{1f}, we will consider a circular curve for c_{1f} placed at a very large distance from the scatterer. The total field must satisfy the boundary condition that it is the sum of the given incident field H_0 and a scattered field H_{sc}, where the latter must satisfy the radiating boundary condition, that is in the far field, and thus at c_{1f}, the scattered field component must be on the form

$$H_{sc}^{ff}(\mathbf{r}') \approx \frac{e^{-ikr'}}{\sqrt{r'}} f_H(\theta'), \tag{7.46}$$

and now it is convenient that the chosen Green's function (7.41) also satisfies the radiating boundary condition such that at c_{1f}, it is on the form

$$g_1^{ff}(\mathbf{r}, \mathbf{r}') \approx \frac{e^{-ikr'}}{\sqrt{r'}} f_g(\mathbf{r}, \theta'). \tag{7.47}$$

Thus, by inserting these far-field expressions for the scattered field and Green's function, we find that

$$\oint_{c_{1f}} \left[g_1(\mathbf{r}, \mathbf{r}') \hat{\mathbf{n}}_1' \cdot \nabla' H_{sc}(\mathbf{r}') - H_{sc}(\mathbf{r}') \hat{\mathbf{n}}_1' \cdot \nabla' g_1(\mathbf{r}, \mathbf{r}') \right] r' d\theta' = 0, \tag{7.48}$$

since the integrand vanishes.

On the other hand, similar to this equation,

$$\oint_{c_{1f}} \left[g_1(\mathbf{r}, \mathbf{r}') \hat{\mathbf{n}}_1' \cdot \nabla' H_0(\mathbf{r}') - H_0(\mathbf{r}') \hat{\mathbf{n}}_1' \cdot \nabla' g_1(\mathbf{r}, \mathbf{r}') \right] dl' = H_0(\mathbf{r}). \tag{7.49}$$

Finally, if we notice that $\hat{\mathbf{n}}_1 = -\hat{\mathbf{n}}_2 = -\hat{\mathbf{n}}$ and combine the results, we end up with Green's function surface integral equation in Equation (7.44). In summary, we have obtained Green's function surface integral equations

$$H(\mathbf{r}) = \begin{cases} H_0(\mathbf{r}) - \oint_{c_1} \left[g_1(\mathbf{r}, \mathbf{r}') \hat{\mathbf{n}}' \cdot \nabla' H(\mathbf{r}') - H(\mathbf{r}') \hat{\mathbf{n}}' \cdot \nabla' g_1(\mathbf{r}, \mathbf{r}') \right] dl', & \text{if } \mathbf{r} \in \Omega_1 \\ \oint_{c_2} \left[g_2(\mathbf{r}, \mathbf{r}') \hat{\mathbf{n}}' \cdot \nabla' H(\mathbf{r}') - H(\mathbf{r}') \hat{\mathbf{n}}' \cdot \nabla' g_2(\mathbf{r}, \mathbf{r}') \right] dl', & \text{if } \mathbf{r} \in \Omega_2 \end{cases} \tag{7.50}$$

It is clear that if we know the magnetic field and its normal derivative at the boundaries c_1 and c_2, we can straightforwardly obtain the field at all other positions via simple integration. However, in order to do that, we must then first calculate the field and normal derivative at the boundary. We will consider how that can be done in the next section.

7.5.2 Calculating the Field and Normal Derivative at the Boundary

In order to obtain the field and its normal derivative at the boundaries, the approach is to consider the integral equations (7.50) in the limit when the position $\mathbf{r}$ approaches the surface of the scatterer from one and the other side. Thereby we end up with two integral equations that are only concerned with fields at positions on the inner and outer surface of the scatterer, that is at c_1 and c_2. While there are two equations, there are four unknowns at the two boundaries of the scatterer, namely the magnetic field and its normal derivative, on each side of the boundary. However, here, we can exploit that the electromagnetics boundary conditions across an interface give that the tangential electric and magnetic fields are conserved across an interface resulting in $H_1 = H_2$ and $(\hat{n} \cdot \nabla H_1)/\varepsilon_1 = (\hat{n} \cdot \nabla H_2)/\varepsilon_2$, where H_1 and H_2 represent the magnetic field on each side of the interface. Thus, when applying these boundary conditions, the number of equations and unknowns is the same. Another consideration is that it is necessary to find a way to deal with the singularity of the normal derivative of Green's function when taking the limit of $\mathbf{r}$ approaching the surface.

In the case of a smooth surface, where all corners are rounded, the surface will always appear flat in the vicinity of the position $\mathbf{r}$ when $\mathbf{r}$ is very close to the surface. A position $\mathbf{r}$ being very close to the surface and approaching it from one side can be expressed by $\mathbf{r} = \mathbf{s}_0 + \delta\hat{n}$, where $\mathbf{s}_0$ is a position on the surface, δ is a very small distance, and $\hat{n}$ is the surface normal vector. This is illustrated in Figure 7.7.

We will consider a line segment of the boundary c which is small enough that we can assume that the magnetic field $H(\mathbf{r}')$ at the boundary is constant and that the boundary can be approximated with a straight line such that the normal vector is also constant. Furthermore, Green's function can be replaced with its approximation for short distances given by [5]

$$g(\mathbf{r}, \mathbf{r}') = \frac{1}{4i}H_0^{(2)}(k|\mathbf{r} - \mathbf{r}'|) \approx \frac{-1}{2\pi}\ln(k|\mathbf{r} - \mathbf{r}'|/2). \qquad (7.51)$$

Since this is a logarithmic singularity, we can replace the integral

$$\lim_{\delta \to 0^+} \int \hat{n}' \cdot \nabla'H(\mathbf{r}')g(\mathbf{r}, \mathbf{r}')dl' \approx \lim_{\delta \to 0^+} \hat{n} \cdot \nabla H(\mathbf{r})|_{\mathbf{r}=\mathbf{s}_0} \int_{\mathbf{s}_0 - \Delta\hat{t}}^{\mathbf{s}_0 + \Delta\hat{t}} g(\mathbf{s}_0 + \delta\hat{n}, \mathbf{r}')dl', \qquad (7.52)$$

where $\hat{t}$ is a unit vector parallel with the surface, with the following principal value integral excluding the singularity point from the integration

$$\lim_{\delta \to 0^+} \int \hat{n}' \cdot \nabla'H(\mathbf{r}')g(\mathbf{r}, \mathbf{r}')dl' \approx P\int \hat{n}' \cdot \nabla'H(\mathbf{r}')g(\mathbf{s}_0, \mathbf{r}')dl'. \qquad (7.53)$$

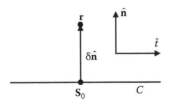

FIGURE 7.7 Schematic of situation with the position $\mathbf{r}$ being close to the surface c of the scattering object. The position $\mathbf{r} = \mathbf{s}_0 + \delta\hat{n}$, where δ is a very small distance and $\hat{n}$ is the surface normal vector.

Consider now the integral

$$A = \int H(\mathbf{r}')\hat{\mathbf{n}}' \cdot \nabla' g(\mathbf{r}, \mathbf{r}')dl' \approx H(\mathbf{s}_0) \int_{\mathbf{s}_0 - \Delta\hat{\mathbf{t}}}^{\mathbf{s}_0 + \Delta\hat{\mathbf{t}}} \hat{\mathbf{n}} \cdot \nabla' g(\mathbf{s}_0 + \delta\hat{\mathbf{n}}, \mathbf{r}')dl'. \tag{7.54}$$

If we make a change of variables and use the short-distance approximation of $g(\mathbf{r}, \mathbf{r}')$, this can be rewritten as

$$A \approx H(\mathbf{s}_0) \int_{-\Delta}^{\Delta} \frac{-1}{2\pi} \frac{-\delta}{\delta^2 + x^2} dx \approx H(\mathbf{s}_0) \frac{-1}{2\pi} (-\pi) = \frac{1}{2} H(\mathbf{s}_0). \tag{7.55}$$

Thus, in the limit where $\mathbf{r} = \mathbf{s}_0$, we can rewrite the integral in the form of a principal value integral, that is

$$\lim_{\delta \to 0^+} \int H(\mathbf{r}')\hat{\mathbf{n}}' \cdot \nabla' g(\mathbf{s}_0 + \delta\hat{\mathbf{n}}, \mathbf{r}')dl' = \frac{1}{2} H(\mathbf{s}_0) + P \int H(\mathbf{r}')\hat{\mathbf{n}}' \cdot \nabla' g(\mathbf{r}, \mathbf{r}')dl'. \tag{7.56}$$

If we carry out the same procedure with the position $\mathbf{r}$ approaching the surface from the other side of the boundary instead, we find a similar result except for a sign, that is

$$\lim_{\delta \to 0^-} \int H(\mathbf{r}')\hat{\mathbf{n}}' \cdot \nabla' g(\mathbf{s}_0 + \delta\hat{\mathbf{n}}, \mathbf{r}')dl' = -\frac{1}{2} H(\mathbf{s}_0) + P \int H(\mathbf{r}')\hat{\mathbf{n}}' \cdot \nabla' g(\mathbf{r}, \mathbf{r}')dl'. \tag{7.57}$$

Note: In the case that $\mathbf{s}_0$ is not placed near a flat part of the surface but instead right at a corner, the only difference will be that the factor $1/2$ will be replaced by $\theta/2\pi$, where θ is the angle between the two lines meeting in the corner on the inside of the corner (region Ω_2), and $-1/2$ will be replaced by $-(2\pi - \theta)/2\pi$ [20]. Here, we will assume that our geometry does not have sharp corners.

It is now clear in the limit of $\mathbf{r} = \mathbf{s}$ being on just one or the other side of the surface of the scatterer, the integral equations (7.50) can be written as

$$\frac{1}{2} H_{\Omega_1}(\mathbf{s}) = H_0(\mathbf{s}) - P \oint_c \left[g_1(\mathbf{s}, \mathbf{s}')\hat{\mathbf{n}}' \cdot \nabla' H_{\Omega_1}(\mathbf{s}') - H_{\Omega_1}(\mathbf{s}')\hat{\mathbf{n}}' \cdot \nabla' g_1(\mathbf{s}, \mathbf{s}') \right] dl'$$

$$\frac{1}{2} H_{\Omega_2}(\mathbf{s}) = P \oint_c \left[g_2(\mathbf{s}, \mathbf{s}')\hat{\mathbf{n}}' \cdot \nabla' H_{\Omega_2}(\mathbf{s}') - H_{\Omega_2}(\mathbf{s}')\hat{\mathbf{n}}' \cdot \nabla' g_2(\mathbf{s}, \mathbf{s}') \right] dl', \tag{7.58}$$

where Ω_1 and Ω_2 refer to that the field and normal derivative must be taken just outside the boundary (or the scatterer) in medium Ω_1 or just inside in medium Ω_2. If we use the boundary conditions across the interface, we can eliminate H_{Ω_2} and $\hat{n} \cdot \nabla H_{\Omega_2}(\mathbf{r})$, resulting finally in the equations

$$\frac{1}{2} H_{\Omega_1}(\mathbf{s}) = \begin{cases} H_0(\mathbf{s}) - P \oint_c \left[g_1(\mathbf{s}, \mathbf{s}')\hat{\mathbf{n}}' \cdot \nabla' H_{\Omega_1}(\mathbf{s}') - H_{\Omega_1}(\mathbf{s}')\hat{\mathbf{n}}' \cdot \nabla' g_1(\mathbf{s}, \mathbf{s}') \right] dl' \\ P \oint_c \left[\frac{\varepsilon_2}{\varepsilon_1} g_2(\mathbf{s}, \mathbf{s}')\hat{\mathbf{n}}' \cdot \nabla' H_{\Omega_1}(\mathbf{s}') - H_{\Omega_1}(\mathbf{s}')\hat{\mathbf{n}}' \cdot \nabla' g_2(\mathbf{s}, \mathbf{s}') \right] dl' \end{cases}$$

$$\tag{7.59}$$

These equations can be discretized by, for example dividing the boundary into a number of small curve segments and by assuming that the field and its normal derivative are constant in each segment. The discretized equations can then be put on matrix form, where the unknowns are the field and normal derivative in each segment and the knowns are the incident field H_0 in each segment. Once these equations have been solved, the field can be calculated at all other positions using the field and normal derivative at the boundary and Equation (7.50).

In order to simplify the notation, we will let H_j and ϕ_j represent the value of the magnetic field and its normal derivative just outside the scatterer $\hat{\mathbf{n}} \cdot \nabla H_{\Omega_1}$, respectively, at curve segment j. Then the discretized version of Equation (7.59) can be written as

$$\frac{1}{2}H_i + \sum_j \left(A_{i,j}^{(1)}\phi_j - B_{i,j}^{(1)}H_j\right) = H_{0,i}, \tag{7.60}$$

$$\frac{1}{2}H_i - \sum_j \left(A_{i,j}^{(2)}\phi_j - B_{i,j}^{(2)}H_j\right) = 0, \tag{7.61}$$

where

$$A_{i,j}^{(1)} = P \int_j g_1(\mathbf{s}_i, \mathbf{s}')dl', \tag{7.62}$$

$$B_{i,j}^{(1)} = P \int_j \hat{\mathbf{n}}' \cdot \nabla' g_1(\mathbf{s}_i, \mathbf{s}')dl', \tag{7.63}$$

$$A_{i,j}^{(2)} = \frac{\varepsilon_2}{\varepsilon_1}P \int_j g_2(\mathbf{s}_i, \mathbf{s}')dl', \tag{7.64}$$

$$B_{i,j}^{(2)} = P \int_j \hat{\mathbf{n}}' \cdot \nabla' g_2(\mathbf{s}_i, \mathbf{s}')dl', \tag{7.65}$$

Note that the integrals over line segment j involved in obtaining these coefficients are not restricted to linear line segments but allow discretization of a surface into a number of curved line segments resulting in a description of the actual surface that does not have staircasing. Thus, while we consider constant fields in each segment as an approximation, the actual structure surface can be described exactly. This is especially important for large contrasts in refractive indices between media Ω_1 and Ω_2.

As an example, we will apply the GFSIEM to calculate scattering of light by a silver cylinder of diameter $D = 500$ nm located in air ($\varepsilon_1 = 1$) when a plane wave of wavelength $\lambda = 700$ nm propagating along the x-axis ($H_0(\mathbf{r}) = e^{-ik_0 x}$) is incident on the cylinder. We will use the silver dielectric constant for this particular wavelength from Ref. [21] given by $\varepsilon_2 = -22.9915 - i0.395194$. The first step in the calculation is to find the field and the normal derivative of the field on the outer boundary of the cylinder. In a first implementation, when $i \neq j$, we can consider using the approximations

$$A_{i,j}^{(1)} \approx g_1(\mathbf{s}_i, \mathbf{s}_j)\Delta l_{j}, \tag{7.66}$$

$$B_{i,j}^{(1)} \approx \left[\hat{\mathbf{n}}' \cdot \nabla' g_1(\mathbf{s}_i, \mathbf{s}')\right]_{\mathbf{s}'=\mathbf{s}_j} \Delta l_{j}, \tag{7.67}$$

$$A_{i,j}^{(2)} \approx \frac{\varepsilon_2}{\varepsilon_1} g_2(\mathbf{s}_i, \mathbf{s}_j)\Delta l_{j}, \tag{7.68}$$

$$B_{i,j}^{(2)} \approx \left[\hat{\mathbf{n}}' \cdot \nabla' g_2(\mathbf{s}_i, \mathbf{s}')\right]_{\mathbf{s}'=\mathbf{s}_j} \Delta l_{j}, \tag{7.69}$$

and when $i = j$, we can use that for small distances $H_0^{(2)}(x) \approx 1 - i\frac{2}{\pi}\left(\ln\frac{x}{2} + \gamma\right)$, where $\gamma = 0.5772156$ is Euler's constant, and further approximate the line segment with a straight segment of the same length as the curved element, in which case, the resulting principal value integrals are given by

$$A_{i,i}^{(1)} \approx \frac{1}{4i}\left(1 - i\frac{2}{\pi}(\gamma - \ln 2)\right)\Delta l_i - \frac{1}{2\pi}\Delta l_i\left(\ln\frac{k_1\Delta l_i}{2} - 1\right), \tag{7.70}$$

$$A_{i,i}^{(2)} \approx \left[\frac{1}{4i}\left(1 - i\frac{2}{\pi}(\gamma - \ln 2)\right)\Delta l_i - \frac{1}{2\pi}\Delta l_i\left(\ln\frac{k_2\Delta l_i}{2} - 1\right)\right]\frac{\varepsilon_2}{\varepsilon_1}, \tag{7.71}$$

$$B_{i,i}^{(1)} = B_{i,i}^{(2)} = 0. \tag{7.72}$$

Here $k_i = k_0 n_i$. These are of course rather crude approximations, and the price to be paid is that it is necessary to compensate by increasing the number of elements in which the scatterer surface is discretized. The number of elements required to obtain the same precision can be reduced by using the relations (7.62) through (7.65) with an accurate description of the boundary in each element.

A discretization scheme with a better approximation to the integrals in Equations (7.62) through (7.65) can be obtained by subdividing each of the N curved surface elements into N_s subelements determining how many sampling points that should be used in the numerical integration. In that case, the coefficients will instead be approximated with

$$A_{i,j}^{(1)} \approx \sum_{u=1}^{N_s} g_1(\mathbf{s}_i, \mathbf{s}_{j,u})\Delta l_{j,u}, \quad i \neq j \tag{7.73}$$

$$A_{i,i}^{(1)} \approx \sum_{u\neq u_c, u=1}^{N_s} g_1(\mathbf{s}_i, \mathbf{s}_{i,u})\Delta l_{i,u}$$
$$+ \left[\frac{1}{4i}\left(1 - i\frac{2}{\pi}(\gamma - \ln 2)\right)\Delta l_{i,u_c} - \frac{1}{2\pi}\Delta l_{i,u_c}\left(\ln\frac{k_1\Delta l_{i,u_c}}{2} - 1\right)\right], \tag{7.74}$$

$$A_{i,j}^{(2)} \approx \sum_{u=1}^{N_s} \frac{\varepsilon_2}{\varepsilon_1} g_2(\mathbf{s}_i, \mathbf{s}_{j,u})\Delta l_{j,u}, \quad i \neq j \tag{7.75}$$

$$A_{i,i}^{(2)} \approx \sum_{u\neq u_c, u=1}^{N_s} g_2(\mathbf{s}_i, \mathbf{s}_{i,u})\Delta l_{i,u}\frac{\varepsilon_2}{\varepsilon_1}$$
$$+ \left[\frac{1}{4i}\left(1 - i\frac{2}{\pi}(\gamma - \ln 2)\right)\Delta l_{i,u_c} - \frac{1}{2\pi}\Delta l_{i,u_c}\left(\ln\frac{k_2\Delta l_{i,u_c}}{2} - 1\right)\right]\frac{\varepsilon_2}{\varepsilon_1}, \tag{7.76}$$

$$B_{i,j}^{(1)} \approx \sum_{u=1}^{N_s} \left[\hat{\mathbf{n}}'_{j,u} \cdot \nabla' g_1(\mathbf{s}_i, \mathbf{s}_{j,u}) \right]_{\mathbf{s}'=\mathbf{s}_{j,u}} \Delta l_{j,u}, \quad i \neq j \qquad (7.77)$$

$$B_{i,i}^{(1)} \approx \sum_{u \neq u_c, u=1}^{N_s} \left[\hat{\mathbf{n}}'_{i,u} \cdot \nabla' g_1(\mathbf{s}_i, \mathbf{s}_{i,u}) \right]_{\mathbf{s}'=\mathbf{s}_{i,u}} \Delta l_{i,u}, \qquad (7.78)$$

$$B_{i,j}^{(2)} \approx \sum_{u=1}^{N_s} \left[\hat{\mathbf{n}}'_{j,u} \cdot \nabla' g_2(\mathbf{s}_i, \mathbf{s}_{j,u}) \right]_{\mathbf{s}'=\mathbf{s}_{j,u}} \Delta l_{j,u}, \quad i \neq j \qquad (7.79)$$

$$B_{i,i}^{(2)} \approx \sum_{u \neq u_c, u=1}^{N_s} \left[\hat{\mathbf{n}}'_{i,u} \cdot \nabla' g_2(\mathbf{s}_i, \mathbf{s}_{i,u}) \right]_{\mathbf{s}'=\mathbf{s}_{i,u}} \Delta l_{i,u}, \qquad (7.80)$$

where $\mathbf{s}_{i,u}$ is the centre position of subelement u within the element i. The sampling position for each element is related to the centre of one of the subelements denoted u_c, that is $\mathbf{s}_i = \mathbf{s}_{i,u_c}$. The same approximation used earlier when not including subsectioning for the elements $A_{i,i}^{(1)}$ and $A_{i,i}^{(2)}$ is still used but now only to the subelement (i, u_c) within the element i.

A calculation of the magnitude of the total magnetic field on the cylinder boundary using different numbers of discretization elements and either using subsectioning ($N_s = 21$) or not using subsectioning ($N_s = 1$) is shown in Figure 7.8. It appears that without subsectioning, $N = 50$ surface elements are fairly good, while $N = 10$ are too few elements. In the case when subsectioning is applied, the very small number of elements $N = 10$ is already fairly good. Convergence is clearly significantly improved by the subsectioning scheme which can be ascribed to a better representation of the surface of the scatterer.

The calculated field and normal derivative on the cylinder boundary can now be used to calculate the differential scattering cross section. Here, we will use that the

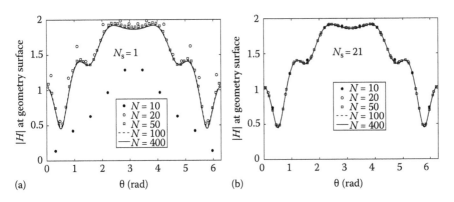

FIGURE 7.8 Total magnetic field calculated with the GFSIEM on the surface of a silver cylinder of diameter 500 nm in air, when a TM-polarized plane wave of wavelength 700 nm is incident along the x-axis. Results are presented for different numbers of discretization elements N (a) without ($N_s = 1$) and (b) with ($N_s = 21$) subsectioning of each element.

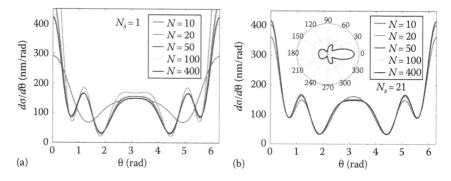

FIGURE 7.9 (See colour insert.) Differential scattering cross section calculated for a silver cylinder of radius 500 nm in air with a TM-polarized incident plane wave of wavelength 700 nm. A polar plot of the differential scattering cross section is shown as an insert.

far-field approximation of Green's function Equation (7.17) leads to the following expression for the scattered far field:

$$H_{\text{sc}}^{\text{ff}} = -\frac{1}{4}\sqrt{\frac{2}{\pi k_1 r}} e^{-i\pi/4} e^{-ik_1 r} \oint_{c_1} e^{ik_1 \frac{\mathbf{r}}{r} \cdot \mathbf{r}'} \left(\hat{\mathbf{n}}' \cdot \nabla' H(\mathbf{r}') - H(\mathbf{r}')ik_1 \hat{\mathbf{n}}' \cdot \frac{\mathbf{r}}{r} \right) dl'. \quad (7.81)$$

This can then be used to obtain both the differential scattering cross section and the total scattering cross section, that is

$$\sigma_{\text{diff}}(\theta) = \frac{r|H_{\text{sc}}^{\text{ff}}(r,\theta)|^2}{|H_0|^2}, \quad (7.82)$$

$$\sigma_{\text{scat}} = \int_0^{2\pi} \sigma_{\text{diff}}(\theta) d\theta, \quad (7.83)$$

where θ is related to the observation point in the far field $\mathbf{r} = r\left(\hat{\mathbf{x}}\cos\theta + \hat{\mathbf{y}}\sin\theta\right)$. The calculation of the differential scattering cross section for different N and N_s is shown in Figure 7.9 with a similar conclusion regarding convergence. Finally, the scattering cross section obtained by integrating the curves in Figure 7.9 is shown in Figure 7.10. In this case, with subsectioning, a noticeable error is seen with $N = 20$ elements, whereas a close-to-exact value is found with $N = 50$ elements.

7.6 CONSTRUCTION OF 2D GREEN'S FUNCTIONS FOR LAYERED STRUCTURES

The approach we will use to obtain Green's function for layered structures is that firstly, we will expand the free-space Green's function into plane waves. Each plane wave incident on a layered geometry will then result in a reflected wave, which must be added, and a transmitted wave on the other side of the layered geometry, where the

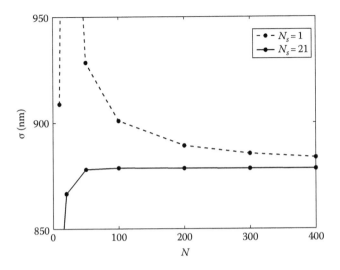

FIGURE 7.10 Scattering cross section calculated for a silver cylinder of radius 500 nm in air with a TM-polarized incident plane wave of wavelength 700 nm.

reflected and transmitted parts are obtained by the standard Fresnel reflection from a layered structure. Because the reflection and transmission coefficients vary with the wave number along the interface of the incident plane waves, the resulting Green's function will not be a simple analytical expression but will have to be expressed as an integral.

7.6.1 PLANE-WAVE EXPANSION OF THE FREE-SPACE GREEN'S FUNCTION

The 2D Green's function for layered structures can be constructed by starting with the 2D free-space Green's function considered previously, that is

$$g(\mathbf{r}, \mathbf{r}') = \frac{1}{4i} H_0^{(2)}(k_0 n |\mathbf{r} - \mathbf{r}'|). \tag{7.84}$$

Consider a source position $\mathbf{r}'$ in the semi-infinite homogeneous region placed above a layered structure. Then for positions $\mathbf{r}$ that are also in that region, the function (7.84) will still satisfy the equation

$$\left(\frac{\partial^2}{\partial x^2} + \frac{\partial^2}{\partial y^2} + k_0^2 \varepsilon_{\text{ref}}(\mathbf{r}) \right) g(\mathbf{r}; \mathbf{r}') = -\delta(\mathbf{r} - \mathbf{r}'). \tag{7.85}$$

However, the boundary conditions are now different from the free-space case due to the presence of the layered structure. Green's function satisfying the appropriate boundary conditions must of course also be a solution to Equation (7.85) and can thus be constructed by starting out with the particular solution (7.84) and adding homogeneous solutions to Equation (7.85) such that appropriate boundary conditions will be satisfied. For this purpose, it is more convenient to have the free-space Green's function in the form of an expansion over plane waves, since the reflection of a plane wave from a layered geometry is well known.

The plane-wave expansion of the free-space Green's function can be obtained with the mode-expansion method, where we consider a complete set of eigenmodes to the left-hand-side operator in Equation (7.85), that is

$$\left(\nabla^2 + k_0^2 \varepsilon_{\text{ref}}\right) E_\lambda(\mathbf{r}) = \lambda E_\lambda(\mathbf{r}). \tag{7.86}$$

This leads to the expression

$$\left(\nabla^2 + k^2\right) E_\mathbf{k}(\mathbf{r}) = 0, \tag{7.87}$$

where $k^2 = k_0^2 \varepsilon_{\text{ref}} - \lambda_\mathbf{k}$. This expression has a complete set of solutions of the form

$$E_\mathbf{k}(\mathbf{r}) = e^{-i\mathbf{k}\cdot\mathbf{r}}. \tag{7.88}$$

These modes satisfy the following normalization and orthogonality condition:

$$\int E_\mathbf{k}(\mathbf{r}) \left(E_{\mathbf{k}'}(\mathbf{r})\right)^* dA = \int e^{i(\mathbf{k}'-\mathbf{k})\cdot\mathbf{r}} dA = (2\pi)^2 \delta(\mathbf{k}' - \mathbf{k}) = N_\mathbf{k} \delta(\mathbf{k}' - \mathbf{k}), \tag{7.89}$$

where $N_k = (2\pi)^2$. A Green's function satisfying Equation (7.85) with $\varepsilon_{\text{ref}}(\mathbf{r}) = \varepsilon_{\text{ref}}$ can now be constructed with the mode-expansion method, that is

$$g(\mathbf{r}, \mathbf{r}') = -\int_\mathbf{k} \frac{E_\mathbf{k}(\mathbf{r}) \left(E_\mathbf{k}(\mathbf{r}')\right)^*}{N_\mathbf{k} \lambda_\mathbf{k}} d^2k = -\int \frac{e^{-i\mathbf{k}\cdot(\mathbf{r}-\mathbf{r}')}}{(2\pi)^2 (k_0^2 n_{\text{ref}}^2 - k^2)} d^2k. \tag{7.90}$$

Note that if we apply the operator $(\nabla^2 + k_0^2 \varepsilon_{\text{ref}})$ to this expression, then the resulting expression will be similar but without the eigenvalue $\lambda_\mathbf{k}$ in the denominator. From the normalization and orthogonality condition, it now follows that the resulting expression is minus the identity operator. This can be seen by making an overlap integral between this expression and any function $f(\mathbf{r}')$ being expanded in the complete set of eigenfunctions. By using the normalization and orthogonality condition, the result is $-f(\mathbf{r})$. This shows that we have constructed a $g(\mathbf{r}, \mathbf{r}')$ that satisfies Equation (7.85).

As will become clear later, we can obtain the particular Green's function that satisfies the radiating wave boundary condition by adding a small imaginary part in the denominator in the following way:

$$g(\mathbf{r}, \mathbf{r}') = -\int \frac{e^{-i\mathbf{k}\cdot(\mathbf{r}-\mathbf{r}')}}{(2\pi)^2 (k_0^2 n_{\text{ref}}^2 - k^2 - i\varepsilon')} d^2k. \tag{7.91}$$

The only effect of this change is to add a homogeneous solution to Equation (7.85), and thus, Equation (7.85) is still satisfied. In the case of a layered structure instead of a homogeneous medium, Green's function can be obtained straightforwardly if the free-space Green's function is decomposed into its in-plane spectral components. Thus, we rewrite Equation (7.91) into

$$g(\mathbf{r}, \mathbf{r}') = \int \frac{e^{-ik_x(x-x')}e^{-ik_y(y-y')}}{(2\pi)^2 (k_y - \sqrt{k_0^2 n_{ref}^2 - k_x^2} + i\varepsilon)(k_y + \sqrt{k_0^2 n_{ref}^2 - k_x^2} - i\varepsilon)} dk_x dk_y. \quad (7.92)$$

We can carry out the integration over k_y by using the following residue theorem [22]:

Residue theorem: Let $f(z)$ be analytic inside a simple closed path C and on C, except for finitely many singular points $z_1, z_2, \ldots, z_n$ inside C. Then the integral of $f(z)$ taken counterclockwise around C equals $2\pi i$ times the sum of residues of $f(z)$ at $z_1, z_2, \ldots, z_n$:

$$\oint_C f(z)dz = 2\pi i \sum_{j=1}^{n} [\text{Res} f(z)]_{z=z_j}.$$

Residue at simple pole:

$$[\text{Res} f(z)]_{z=z_0} = \lim_{z \to z_0} (z - z_0) f(z).$$

For a simple pole, the above limit will converge to a finite non-zero number as $z \to z_0$. It is also possible to find the residue at poles of higher order but that will not be needed in this text.

Analyticity: A complex function $f(z)$ is said to be *analytic* in a domain D if $f(z)$ is defined and differentiable at all points of D. The function $f(z)$ is said to be *analytic at a point $z = z_0$ in D* if $f(z)$ is analytic in a neighbourhood of z_0.

In order to use the residue theorem to carry out the integral over k_y in (7.92) along the real axis, we can use that it is possible to replace this integral with an integral along a closed curve in the complex plane giving the same result, and then the residue theorem applies. The integration strategy is illustrated in Figure 7.11. In the figure, we start with an integral along the real axis from $-K$ to $+K$. If we add to the integral an extra integral over a semi-circle going into the complex plane starting and ending at $k_y = \pm K$ then the resulting integration path will be closed, and the total integral will be given by the residues of poles enclosed inside the integration curve. In our case we should choose the semi-circle going to the side in the complex plane where the magnitude of $\exp(-ik_y(y - y'))$ will exponentially decrease with distance to the real axis, which depends on the sign of $(y - y')$, since then this extra integral will be vanishing in the limit of $K \to \infty$. The resulting integral along the whole of the real axis will thus be given from the residues of poles in either the upper or the lower complex half-plane depending on whether $y > y'$ or $y < y'$. The two poles in (7.92) are illustrated in Figure 7.11 with two dots, and depending on whether the semi-circle is taken to one or the other side in the complex plane, it will be a different pole that

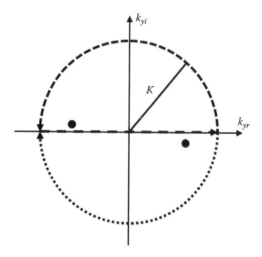

FIGURE 7.11 Illustration of integration strategy for applying the residue theorem. In the limit where the radius $K \to \infty$, the integral along the real axis can be replaced with an integral over a closed curve in the complex plane by adding either an integral over a semi-circle going into the upper or lower half-plane if that added integral will vanish when $K \to \infty$. Then the integral will be given by the residues of poles of the integrand enclosed by the integration path.

is enclosed within the curve of integration. The result for the two cases of $y > y'$ and $y < y'$ can after some algebra be combined into the following single expression:

$$g(\mathbf{r}, \mathbf{r}') = -\frac{i}{2\pi} \int\limits_{0}^{+\infty} \frac{\cos(k_x[x - x'])e^{-ik_y|y-y'|}}{k_y} dk_x, \tag{7.93}$$

where here $k_y = \sqrt{k_0^2 n_{\text{ref}}^2 - k_x^2}$ with $\text{Im}(k_y) \leq 0$. It is clear from this expression that each wave component will propagate into the upper half-plane if $y > y'$ and into the lower half-plane if $y < y'$, as we will require from our Green's function.

Note: If we had added an imaginary part in the denominator of Equation (7.91) with opposite sign, we would find a similar expression but with $-ik_y|y - y'|$ replaced with $+ik_y|y - y'|$, which would also be a Green's function but one describing light propagating towards the point source at $\mathbf{r}'$ instead of away from the point source, and thus, such a Green's function would not satisfy the radiating boundary condition.

7.6.2 2D TE-Polarized Scalar Green's Function for a Layered Structure

In this section, we will construct Green's function for a case with TE-polarized light, having the electric field oriented along the z-axis and propagating in the xy-plane. Both structure and fields can be assumed invariant along the z-axis, that is $\mathbf{E}(\mathbf{r}) = \hat{z}E(x, y)$ and $\varepsilon(\mathbf{r}) = \varepsilon(x, y)$. A layered structure exists along the y-axis as illustrated in Figure 7.12. It is assumed that the source position $\mathbf{r}'$ and the observation point $\mathbf{r}$ are placed outside the layered structure ($y' > 0$).

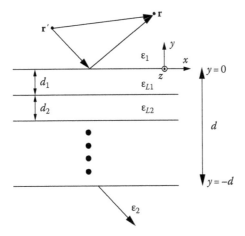

FIGURE 7.12 Schematic of a radiation from a point source placed above a layered structure.

A Green's function that satisfies (7.85) for a position **r** in the upper semi-infinite layer is the homogeneous medium, or direct, Green's function (7.93). The appropriate boundary conditions here are that for positions $y > y'$, the electromagnetic wave must propagate into the upper half-plane, and for positions $y < -d$, the wave must propagate into the lower half-plane, and along the x-axis, the wave must always propagate away from **r'**. Such a Green's function can be obtained for $y > 0$ by considering each plane wave in the expansion (7.93) that propagates towards the layered structure and adding the appropriate Fresnel reflection term. In a similar way, it is possible to construct the transmission through the layered geometry from the Fresnel transmission coefficients for each incident plane wave. Thus, for positions $y > 0$, this leads to

$$g(\mathbf{r},\mathbf{r}') = -\frac{i}{2\pi} \int\limits_0^{+\infty} \frac{\cos(k_x[x-x'])\left(e^{-ik_{y1}|y-y'|} + r^{(TE)}(k_x)e^{-ik_{y1}(y+y')}\right)}{k_{y1}}\,dk_x, \qquad (7.94)$$

where $r^{(TE)}(k_x)$ is the Fresnel reflection coefficient for TE-polarized light being incident on the layered structure. Note that in Figure 7.12, we have shown two paths from **r'** to **r**, namely a direct path and an indirect path via reflection from the layered structure. The direct part is governed by the free-space Green's function in Equation (7.93), which we will in the following refer to as $g^{(d)}$, and the remaining part of Equation (7.94) will be referred to as $g^{(i)}$.

In the case when $y < -d$,

$$g(\mathbf{r},\mathbf{r}') = -\frac{i}{2\pi} \int\limits_0^{+\infty} \frac{\cos(k_x[x-x'])\left(e^{-i(k_{y1}y' - k_{y2}(y+d))}t^{(TE)}(k_x)\right)}{k_{y1}}\,dk_x, \qquad (7.95)$$

where $t^{(TE)}$ is the Fresnel transmission coefficient for TE polarization through the layered structure. Here, $k_{yi} = \sqrt{k_0^2 n_i^2 - k_x^2}$ with $\mathrm{Im}(k_{yi}) \leq 0$.

7.6.3   2D TM-Polarized Scalar Green's Function for a Layered Structure

For TM-polarized light, it is possible to a large extent to follow the same procedure as in the previous section. In this case, it is the magnetic field which is oriented along the z-axis and propagates only in the xy-plane, and with fields and geometry being invariant along the z-axis, that is $\mathbf{H}(\mathbf{r}) = \hat{z}H(x, y)$ and $\varepsilon(\mathbf{r}) = \varepsilon(x, y)$. In that case, the wave equation for the magnetic field reduces to

$$\left(-\frac{\nabla \varepsilon_{ref}(\mathbf{r}) \cdot \nabla}{\varepsilon_{ref}(\mathbf{r})} + \nabla^2 + k_0^2 \varepsilon_{ref}(\mathbf{r}) \right) H(\mathbf{r}) = 0. \tag{7.96}$$

When solving for the magnetic field using Green's function methods, the relevant Green's function must satisfy the same boundary conditions across interfaces as the magnetic field and be governed by the operator in the magnetic field wave equation. Green's function must thus satisfy

$$\left(-\frac{\nabla \varepsilon_{ref}(\mathbf{r}) \cdot \nabla}{\varepsilon_{ref}(\mathbf{r})} + \nabla^2 + k_0^2 \varepsilon_{ref}(\mathbf{r}) \right) g(\mathbf{r}, \mathbf{r}') = -\delta(\mathbf{r}, \mathbf{r}'). \tag{7.97}$$

For a layered structure consisting of layers with constant (position-independent) dielectric constant within each layer, the first term will disappear when $\mathbf{r}$ is away from interfaces, and we are left with an equation that is similar to the one we considered in the previous section for the TE-polarized case. However, the extra term is required in order to take care of the magnetic field boundary conditions across interfaces, which are different from those of the TE-polarized case. However, except for the different boundary conditions, the procedure is the same as in the previous section resulting in Green's function

$$g(\mathbf{r}, \mathbf{r}') = -\frac{i}{2\pi} \int_0^{+\infty} \frac{\cos(k_x[x - x']) \left(e^{-ik_{y1}|y-y'|} + r^{(TM)}(k_x)e^{-ik_{y1}(y+y')} \right)}{k_{y1}} dk_x, \tag{7.98}$$

when $y, y' > 0$, and Green's function

$$g(\mathbf{r}, \mathbf{r}') = -\frac{i}{2\pi} \int_0^{+\infty} \frac{\cos(k_x[x - x']) \left(e^{-i(k_{y1}y' - k_{y2}(y+d))} t^{(TM)}(k_x) \right)}{k_{y1}} dk_x, \tag{7.99}$$

when $y' > 0$ and $y < -d$, where $r^{(TM)}(k_x)$ and $t^{(TM)}(k_x)$ are the Fresnel reflection and transmission coefficients, respectively, for TM-polarized light being incident on the layered structure.

7.6.4   Fresnel Reflection and Transmission Coefficients for a Few Simple Geometries

In this section, the reflection and transmission coefficients will be given for a single interface and for a film on a substrate. In the case of a single interface, and the requirement of continuity of tangential electric and magnetic field components across the interface, that is E and $\frac{\partial E}{\partial y}$ must be continuous across interfaces for TE polarization, and H and $\frac{1}{\varepsilon}\frac{\partial H}{\partial y}$ must be continuous across interfaces for TM polarization, we find

$$r_{12}^{(TE)}(k_x) = \frac{k_{y1} - k_{y2}}{k_{y1} + k_{y2}}, \tag{7.100}$$

$$t_{12}^{(TE)}(k_x) = \frac{2k_{y1}}{k_{y1} + k_{y2}}, \tag{7.101}$$

$$r_{12}^{(TM)}(k_x) = \frac{k_{y1}\varepsilon_2 - k_{y2}\varepsilon_1}{k_{y1}\varepsilon_2 + k_{y2}\varepsilon_1}, \tag{7.102}$$

$$t_{12}^{(TM)}(k_x) = \frac{2k_{y1}\varepsilon_2}{k_{y1}\varepsilon_2 + k_{y2}\varepsilon_1}. \tag{7.103}$$

The first two of these expressions related directly to the electric field coincide with expressions presented in Chapter 6 in the case of normal incidence. From TM polarization, we can notice that if medium 1 is a dielectric, that is $\varepsilon_1 > 0$, and medium 2 is a metal, where we neglect the losses so that ε_2 is purely real and negative, and if furthermore $\varepsilon_2 < -\varepsilon_1$, the metal–dielectric interface will support surface plasmon polaritons (SPPs) [23] with a real in-plane wave number given by $k_{spp} = k_0\sqrt{\frac{\varepsilon_1\varepsilon_2}{\varepsilon_1+\varepsilon_2}}$. Exactly when $k_x = k_{spp}$, the denominators of Equations (7.102) and (7.103) are zero, that is the Fresnel reflection and transmission coefficients are singular right at the in-plane wave number of guided or bound modes supported by the layered geometry. In this regard, the metal–dielectric interface is particularly simple since this geometry only supports one bound mode, that is the SPP, being bound to and propagating along the interface. The presence of such singularities must be dealt with carefully in any numerical implementation of the so-called Sommerfeld integrals in (7.94), (7.95), (7.98), and (7.99). If the metal is lossy such that k_{spp} is not a real number, the singularity will be moved off the real axis, and in principle, the integration along the real axis can be carried out straightforwardly. For a dielectric–dielectric interface, that is where $\varepsilon_1 > 0$ and $\varepsilon_2 > 0$, the single interface geometry does not support any bound modes, and there will not be any singularities in the Fresnel reflection and transmission coefficients.

In the case of a film on a substrate, such that we have a three-layer structure with dielectric constants ε_1, ε_{L1}, and ε_2 (see Figure 7.12) and layer thickness d, the reflection and transmission coefficients become

$$r^{(TE)}(k_x) = \frac{r_{1,L1}^{(TE)} + r_{L1,2}^{(TE)}e^{-i2k_{y,L1}d}}{1 + r_{1,L1}^{(TE)}r_{L1,2}^{(TE)}e^{-i2k_{y,L1}d}}, \tag{7.104}$$

$$t^{(TE)}(k_x) = \frac{t_{1,L1}^{(TE)}t_{L1,2}^{(TE)}e^{-ik_{y,L1}d}}{1 + r_{1,L1}^{(TE)}r_{L1,2}^{(TE)}e^{-i2k_{y,L1}d}}, \tag{7.105}$$

$$r^{(TM)}(k_x) = \frac{r_{1,L1}^{(TM)} + r_{L1,2}^{(TM)}e^{-i2k_{y,L1}d}}{1 + r_{1,L1}^{(TM)}r_{L1,2}^{(TM)}e^{-i2k_{y,L1}d}}, \tag{7.106}$$

$$t^{(TM)}(k_x) = \frac{t_{1,L1}^{(TM)}t_{L1,2}^{(TM)}e^{-ik_{y,L1}d}}{1 + r_{1,L1}^{(TM)}r_{L1,2}^{(TM)}e^{-i2k_{y,L1}d}}. \tag{7.107}$$

In this case, for example if $n_1 = n_2$ and $n_{L1} > n_1$, then the dielectric film will support one or more modes with wave numbers corresponding to the poles of the expressions (7.104)–(7.107). Reflection and transmission coefficients for multilayer geometries can for example be obtained by following the transfer-matrix formalism as presented in Ref. [24].

7.6.5 CALCULATING THE SOMMERFELD INTEGRAL

The integral expressions (7.94), (7.95), (7.98), and (7.99) are known as Sommerfeld integrals [3,4]. For the first part of the integrals (7.94) and (7.98), we might as well just use the analytic expression for $g^{(d)}$ given in for example Equation (7.6). For the rest of the integral, we can observe that as long as $y + y' > 0$, the factor $e^{-ik_{y1}(y+y')}$ will decrease exponentially with increasing k_x for large values of k_x, and thus, the integral will converge, and the maximum values of k_x that one needs to consider will decrease with increasing $y + y'$. The integral is, however, slowly convergent in the limit $y + y' = 0$. In that case, convergence is still ensured due to the oscillating integrand $\cos(k_x[x - x'])$ multiplied with the monotonically but slowly decreasing factor $1/k_{y1}$ (for large k_x).

Consider the indirect part of Green's function in Equation (7.98):

$$g^{(i)}(\mathbf{r}, \mathbf{r}') = -\frac{i}{2\pi} \int_0^{+\infty} \frac{\cos(k_x[x - x'])r^{(TM)}(k_x)e^{-ik_{y1}(y+y')}}{k_{y1}} dk_x, \quad y > 0. \qquad (7.108)$$

Here, we note that in a numerical calculation of $g^{(i)}$, we must carefully consider that the denominator in $1/k_{y1}$ is zero when $k_x = k_0 n_1$ and that the reflection coefficient $r^{(TM)}$ may also have poles on or near the real axis. The factor $1/k_{y1}$ can be dealt with for the first part of the integral from $k_x = 0$ to $k_x = k_0 n_1$ by making the variable change

$$k_x = k_0 n_1 \cos \alpha, \qquad (7.109)$$

$$k_{y1} = k_0 n_1 \sin \alpha, \qquad (7.110)$$

leading to

$$\int_0^{k_0 n_1} \frac{\cos(k_x[x - x'])r^{(TM)}(k_x)e^{-ik_{y1}(y+y')}}{k_{y1}} dk_x$$

$$= \int_0^{\pi/2} \cos(k_0 n_1 \cos \alpha[x - x'])r^{(TM)}(\alpha)e^{-ik_0 n_1 \sin \alpha(y+y')} d\alpha, \qquad (7.111)$$

in which case, the k_{y1} in $dk_x = -k_{y1}d\alpha$ cancelled the k_{y1} in the denominator of (7.108). In a similar way, we can, for example consider the part of the integral from $k_x = k_0 n_1$ to $k_x = 2k_0 n_1$ with $k_x = \sqrt{k_0^2 n_1^2 + \beta^2}$, in which case, $k_{y1} = -i\beta$ and again $dk_x = \frac{ik_{y1}}{k_x}d\beta$ contains a factor k_{y1} that will cancel out with the k_{y1} in the denominator of (7.108) leaving a well-behaved integrand.

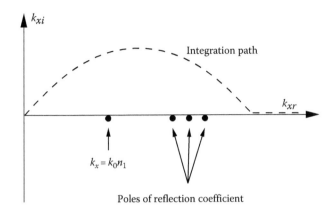

FIGURE 7.13 Schematic of integration path going into the complex plane to avoid singularities in the integrand.

When there are poles in the reflection or transmission coefficients, we will suggest another integration strategy, namely a change of the integration path into the complex plane, such that the integrand will not be taken at values near where $k_{y1} = 0$ (at $k_x = k_0 n_1$) nor near where the reflection coefficient $r^{(TM)}(k_x)$ is singular due to poles related to bound modes in the layered geometry. A schematic of this integration strategy is shown in Figure 7.13, showing an integration path into the complex plane of values for k_x. Here, k_{xr} is the real part of k_x, and k_{xi} is the imaginary part of k_x. Note that in this figure, it appears as if the poles of $r^{(TM)}$, and the point where $k_{y1} = 0$, are slightly offset into the lower half-plane. This is deliberate and can be justified by the limiting procedure with $\varepsilon \to 0$ in Equation (7.92), that is

$$k_{y1} = \lim_{\varepsilon \to 0} \sqrt{k_0^2 n_1^2 - k_x^2} - i\varepsilon. \tag{7.112}$$

In a similar way, any poles on the real axis are in fact approaching the real axis infinitesimally in a similar way such that they still are in principle located in the lower half-plane. Thus, in between the real axis and the integration path drawn in Figure 7.13, there are no poles. Furthermore, the integrand is analytic in this region, since $k_{yi}(k_x) = \sqrt{k_0^2 n_i^2 - k_x^2}$ with $\mathrm{Im}(k_{yi}) \le 0$ will be continuous in the upper-right quadrant, while any branch cuts will be placed elsewhere, such that the residue theorem can be applied. For a path in the upper-right quadrant of the complex plane, the residue theorem can be used to obtain

$$\oint \frac{\cos(k_x[x - x']) \, r^{(TM)}(k_x) \, e^{-ik_{y1}(y+y')}}{k_{y1}} dk_x = 0, \tag{7.113}$$

which means that we can replace the integral along the real axis from $k_x = 0$ to $k_x = k_0 n$ with an integral along another curve starting and ending at the same points. We can, for example choose as integration path a half ellipse going into the complex plane. The ellipse should be chosen such that it keeps a reasonable distance to the poles, and n must be chosen such that $k_0 n$ is larger than all pole values on the real axis. In that case, numerical integration is fairly unproblematic.

Note that one should not go too far into the complex plane since then $\cos(k_x[x-x'])$ will assume very large magnitudes and more so with increasing magnitude of $x-x'$. The integral can be carried out by dividing the integration path into a number of segments and applying Gauss and Gauss–Kronrod quadrature [22], and then if higher accuracy is required, the number of segments that the integration path is divided into can be increased until the difference between using the two quadratures is below a suitable threshold.

7.6.6 FAR-FIELD APPROXIMATION

In this section, we will consider the approximate behaviour of Green's function for a source position at a very large distance from the scatterer and layered geometry. We will assume that the dielectric constant ε_1 is purely real and positive, such that light will not be absorbed in this medium. Consider an observation point in cylindrical coordinates given by

$$\mathbf{r} = \hat{\mathbf{r}} r = \hat{\mathbf{x}} x + \hat{\mathbf{y}} y = r\left(\hat{\mathbf{x}} \cos\theta + \hat{\mathbf{y}} \sin\theta\right), \qquad (7.114)$$

where the distance from the origin r is large, that is $k_0\sqrt{\varepsilon_1} r \gg 1$ and where $\theta \in {]}0; \pi[$, that is we consider an observation point in the upper half-plane. It is then clear from the exact expression for Green's function that, since, in the far field, $|y-y'| = (y-y')$ and $y+y'$ will also be large distances, then the part of the integral (Equation 7.98) where k_{y1} is complex can be neglected, that is

$$g^{ff}(\mathbf{r}, \mathbf{r}') \approx -\frac{i}{2\pi} \int_0^{k_0 n_1} \frac{\cos(k_x[x-x']) \left(e^{-ik_{y1}|y-y'|} + r^{(TM)}(k_x)e^{-ik_{y1}(y+y')}\right)}{k_{y1}} dk_x, \qquad (7.115)$$

We can then again make the variable change in Equations (7.109) and (7.110), and use $\cos(k_x[x-x']) = (e^{ik_x(x-x')} + e^{-ik_x(x-x')})/2$, such that Equation (7.115) becomes

$$g^{ff}(\mathbf{r}, \mathbf{r}') \approx -\frac{i}{4\pi} \int_{\alpha=0}^{\pi} e^{-ik_0 n_1 r(\cos\alpha\cos\theta + \sin\alpha\sin\theta)} f(\alpha, x', y') d\alpha$$

$$= -\frac{i}{4\pi} e^{-ik_0 n_1 r} \int_{\alpha=0}^{\pi} e^{ik_0 n_1 r(1-\cos(\alpha-\theta))} f(\alpha, x', y') d\alpha, \qquad (7.116)$$

where $f(\alpha, x', y') = e^{ik_0 n_1 \cos\alpha x'} \left(e^{ik_0 n_1 \sin\alpha y'} + r^{(TM)}(\alpha)e^{-ik_0 n_1 \sin\alpha y'}\right)$. Here, we should notice that when r is very large, the term $e^{ik_0 n_1 r(1-\cos(\alpha-\theta))}$ will oscillate very fast with α except when $\alpha \approx \theta$, whereas the rest of the integrand, f, will be comparably slowly varying with α. This means that in the limit of very large r, the only part of the integral that will contribute is a small range of values for α around θ, in which case,

$$1 - \cos(\alpha - \theta) \approx \frac{1}{2}(\alpha - \theta)^2. \qquad (7.117)$$

We can thus replace $e^{ik_0 n_1 r(1-\cos(\alpha-\theta))}$ with $e^{i\frac{1}{2}k_0 n_1 r(\alpha-\theta)^2}$ and further extend the integration range from $-\infty$ to $+\infty$ since the added integration does not contribute anything due to the fast oscillatory behaviour of $e^{i\frac{1}{2}k_0 n_1 r(\alpha-\theta)^2}$ with α when $\alpha \neq \theta$. Furthermore, since there will only be a contribution for $\alpha \approx \theta$, we can insert $\alpha = \theta$ into the slowly varying term and place this outside the integral, that is

$$g^{ff}(\mathbf{r},\mathbf{r}') \approx \frac{-i}{4\pi} e^{-ik_0 n_1 r} f(\theta,x',y') \int_{-\infty}^{+\infty} e^{-(-i\frac{1}{2}k_0 n_1 r)u^2}\, du$$

$$= \frac{-i}{4\pi} e^{-ik_0 n_1 r} e^{ik_0 n_1 \cos\theta x'} \left(e^{ik_0 n_1 \sin\theta y'} + r^{(TM)}(\theta)e^{-ik_0 n_1 \sin\theta y'}\right) \sqrt{\frac{2\pi}{-ik_0 n_1 r}}$$

$$= \frac{e^{-ik_0 n_1 r}}{\sqrt{r}} \frac{e^{-i\frac{\pi}{4}}}{4} \sqrt{\frac{2}{\pi k_0 n_1}} e^{ik_0 n_1 \cos\theta x'} \left(e^{ik_0 n_1 \sin\theta y'} + r^{(TM)}(\theta)e^{-ik_0 n_1 \sin\theta y'}\right). \quad (7.118)$$

In the case when there is no reflection from the layered structure, this expression becomes identical with Equation (7.17).

In the case when the medium after the layered structure with dielectric constant ε_2 is also lossless, that is ε_2 is real and positive, we can in a similar way obtain a far-field expression, where here we will again use $\mathbf{r} = \hat{\mathbf{r}}r = \hat{\mathbf{x}}x + \hat{\mathbf{y}}y = r\left(\hat{\mathbf{x}}\cos\theta + \hat{\mathbf{y}}\sin\theta\right)$ but with $\theta \in]-\pi; 0[$ such that we consider a position deep into the semi-infinite medium below the layered geometry. Also, in this case, when $-(y+d)$ is a very large distance into the substrate along the y axis, we can neglect the part of the integral of Equation (7.99) where k_{y2} is imaginary, that is

$$g(\mathbf{r},\mathbf{r}') \approx -\frac{i}{2\pi} \int_{0}^{k_0 n_2} \frac{\cos(k_x[x-x'])\left(e^{-i(k_{y1}y'-k_{y2}(y+d))}t^{(TM)}(k_x)\right)}{k_{y1}}\, dk_x. \quad (7.119)$$

We will make a similar change of variables

$$k_x = k_0 n_2 \cos\alpha, \quad (7.120)$$

$$k_{y2} = k_0 n_2 \sin\alpha, \quad (7.121)$$

resulting in $dk_x = -k_{y2}d\alpha$ and again use $\cos(k_x[x-x']) = (e^{ik_x(x-x')} + e^{-ik_x(x-x')})/2$, such that Equation (7.119) becomes

$$g^{ff}(\mathbf{r},\mathbf{r}') \approx -\frac{i}{4\pi} \int_{\alpha=0}^{\pi} e^{-ik_0 n_2 r(\cos\alpha\cos\theta - \sin\alpha\sin\theta)} f(\alpha,x',y')\, d\alpha$$

$$= -\frac{i}{4\pi} e^{-ik_0 n_2 r} \int_{\alpha=0}^{\pi} e^{ik_0 n_2 r(1-\cos(\alpha+\theta))} f(\alpha,x',y')\, d\alpha, \quad (7.122)$$

where here $f(\alpha,x',y') = \frac{k_{y2}}{k_{y1}} e^{ik_0 n_2 \cos\alpha x'} e^{-ik_{y1}y'} e^{ik_0 n_2 \sin\alpha d} t^{(TM)}(\alpha)$. Here, we should notice that when r is very large, the term $e^{ik_0 n_2 r(1-\cos(\alpha+\theta))}$ will oscillate very fast with α except when $\alpha \approx -\theta$. Notice the sign difference in comparison with the case of reflection. Again the rest of the integrand, f, will be comparably slowly varying with α.

For values of α in the vicinity of $-\theta$, we can use the approximation

$$1 - \cos(\alpha + \theta) \approx \frac{1}{2}(\alpha - [-\theta])^2. \tag{7.123}$$

We can thus replace $e^{ik_0 n_2 r(1-\cos(\alpha+\theta))}$ with $e^{i\frac{1}{2}k_0 n_2 r(\alpha-[-\theta])^2}$ and, as mentioned earlier, extend the integration range from $-\infty$ to $+\infty$. Using similar arguments as mentioned earlier, we obtain

$$g^{\mathrm{ff}}(\mathbf{r}, \mathbf{r}') \approx -\frac{i}{4\pi} e^{-ik_0 n_2 r} f(-\theta, x', y') \int\limits_{-\infty}^{+\infty} e^{-(-i\frac{1}{2}k_0 n_2 r)x^2} dx$$

$$= \frac{e^{-ik_0 n_2 r}}{\sqrt{r}} \frac{1}{4} e^{-i\pi/4} \sqrt{\frac{2}{\pi k_0 n_2}} \frac{k_{y2}}{k_{y1}} e^{ik_0 n_2 \cos\theta x'} e^{-ik_{y1} y'} e^{-ik_0 n_2 \sin\theta d} t^{(TM)}(-\theta). \tag{7.124}$$

In the case when $\varepsilon_1 = \varepsilon_2 = \varepsilon_{L1} = \varepsilon_{L2} = \dots$ such that $t^{(TM)} = e^{ik_0 n_2 \sin\theta d}$ ($t^{(TM)} = 1$ for $d = 0$), again this expression becomes identical with Equation (7.17).

In the case of TE-polarized incident light instead, it is sufficient to replace $r^{(TM)}$ and $t^{(TM)}$ in equations (7.118) and (7.124) with the corresponding $r^{(TE)}$ and $t^{(TE)}$.

7.6.7 EXCITATION OF BOUND WAVEGUIDE MODES

In this section, we will consider an approximation to the part of Green's function governing the excitation of bound modes. When the observation point $\mathbf{r}$ is far from the scatterer but close to the surface of the layered geometry, and propagation losses of bound modes are not excessively high, then the scattered field at those positions will be dominated by the modes that are bound to and propagating along the layered geometry. The simplest geometry that we can consider that supports bound modes is the case of a planar metal–dielectric interface, since such a reference geometry only supports one type of bound modes, namely the SPP that we already referred to in Section 7.6.4 as giving rise to a pole in the Fresnel reflection coefficient for TM polarization. The case of SPP waves is particularly simple here because the dispersion equation of SPPs can be expressed analytically in closed form. For structures with more interfaces such as a thin metal film, this is no longer the case [25]. The contribution to Green's function due to SPP waves for a metal–dielectric interface geometry has been obtained with the mode-expansion method in Ref. [17] for the 2D case and in Ref. [26] for the 3D geometry. In both cases, analytic expressions were obtained for the far-field approximation to this part of Green's function. Here, we will follow the appendix of Ref. [17] to discuss the excitation of bound waveguide modes.

We will consider the case of a reference medium consisting of a single metal–dielectric interface and TM-polarized light. In this case, we wish to find solutions for the magnetic field that satisfies the equation

$$\Theta H(\mathbf{r}) = 0, \tag{7.125}$$

where

$$\Theta = -\frac{\nabla \varepsilon(\mathbf{r}) \cdot \nabla}{\varepsilon(\mathbf{r})} + \nabla^2 + k_0^2 \varepsilon(\mathbf{r}).$$ (7.126)

Thus, Green's function of interest must satisfy the equation

$$\Theta g(\mathbf{r}, \mathbf{r}') = -\delta(\mathbf{r} - \mathbf{r}').$$ (7.127)

Here, however, we prefer at first to introduce the related Green's function G_g being a solution to

$$\Theta_g G_g(\mathbf{r}, \mathbf{r}') = -\delta(\mathbf{r} - \mathbf{r}'),$$ (7.128)

where

$$\Theta_g = -\frac{\nabla \varepsilon(\mathbf{r}) \cdot \nabla}{\varepsilon(\mathbf{r})^2} + \frac{\nabla^2}{\varepsilon(\mathbf{r})} + k_0^2.$$ (7.129)

This is because in the case when $\varepsilon(\mathbf{r})$ is real, the operator Θ_g will be Hermitian, in which case, a complete orthogonal set of eigenmodes g_i exists for the operator given by

$$\Theta_g g_i(\mathbf{r}) = \lambda_i g_i(\mathbf{r}),$$ (7.130)

where λ_i is the corresponding eigenvalue. In that case, the total Green's function can in principle be constructed based on all eigenfunctions as

$$G_g(\mathbf{r}, \mathbf{r}') = -\sum_i \frac{g_i(\mathbf{r})[g_i(\mathbf{r}')]^*}{N_i \lambda_i},$$ (7.131)

where $\sum$ must be interpreted as a summation for discrete eigenvalues and an integration for a continuous spectrum of eigenvalues. The normalization factor is given by

$$N_i \delta_{ij} = \int [g_i(\mathbf{r})]^* g_j(\mathbf{r}) d^2 r,$$ (7.132)

where δ_{ij} must be interpreted as the Kronecker delta function for the discrete spectrum of eigenfunctions and the Dirac delta function for the continuous spectrum. Green's function $g(\mathbf{r}, \mathbf{r}')$ can then finally be obtained from the relation

$$g(\mathbf{r}, \mathbf{r}') = \frac{G_g(\mathbf{r}, \mathbf{r}')}{\varepsilon(\mathbf{r}')}.$$ (7.133)

The eigenmodes that are bound to the metal dielectric interface are given by

$$g_{k_x}(\mathbf{r}) = e^{-ik_x x} e^{-|k_x|\sqrt{\varepsilon_1/(-\varepsilon_2)}y}, \quad y > 0,$$ (7.134)

$$g_{k_x}(\mathbf{r}) = e^{-ik_x x} e^{|k_x|\sqrt{(-\varepsilon_2)/\varepsilon_1}y}, \quad y < 0,$$ (7.135)

where the corresponding eigenvalues are given by

$$\lambda_{k_x} = k_0^2 - k_x^2 \frac{\varepsilon_1 + \varepsilon_2}{\varepsilon_1 \varepsilon_2}. \tag{7.136}$$

Notice that $g_{k_x}(\mathbf{r})$ and $\frac{1}{\varepsilon(\mathbf{r})} \frac{\partial g_{k_x}(\mathbf{r})}{\partial y}$ are continuous across the interface such that the electromagnetics boundary conditions are satisfied.

The normalization factor can be obtained from Equation (7.132) as

$$N_{k_x} = \frac{\pi}{|k_x|} \sqrt{\frac{\varepsilon_1}{-\varepsilon_2}} \left(1 - \frac{\varepsilon_2}{\varepsilon_1} \right). \tag{7.137}$$

Thus, when $y, y' > 0$, the SPP contribution of Green's function becomes

$$g^{\mathrm{spp}}(\mathbf{r}, \mathbf{r}') = -\frac{\varepsilon_1}{\varepsilon_1 - \varepsilon_2} \frac{\varepsilon_1 \varepsilon_2}{\pi (\varepsilon_1 + \varepsilon_2)} \int_{k_x = -\infty}^{\infty} \frac{|k_x| e^{-ik_x(x-x')} e^{-i|k_x|\sqrt{\frac{\varepsilon_1}{-\varepsilon_2}}(y+y')}}{\left(k_{\mathrm{spp}}^2 - k_x^2 - i0^+ \right) \sqrt{\frac{\varepsilon_1}{-\varepsilon_2}}} dk_x, \tag{7.138}$$

where $k_{\mathrm{spp}} = k_0 \sqrt{\varepsilon_1 \varepsilon_2 / (\varepsilon_1 + \varepsilon_2)}$. Here, $i0^+$ is shorthand notation for adding a small positive imaginary number and observing the limit as this number approaches zero. If we, in a similar way, construct the part of Green's function related to modes that are not bound to the interface, and add all contributions to Green's function together, we obtain the total Green's function (7.94) [17]. The evaluation of the integral (7.138) is probably just as difficult to carry out as the Sommerfeld integral.

In the far field, however, the expression (7.138) can be simplified into an expression in closed form. When $|x - x'|$ is very large, as will be the case in the far field, the term $\exp(-ik_x(x - x'))$ will be a rapidly oscillating function with k_x changing along the real axis, and thus, the parts of the integration path where the remaining part of the integrand is slowly varying will asymptotically not contribute to the integral. Consider now a part of the integration path along the real axis going past one of the poles of the integrand located either slightly into the upper or the lower half-plane. If we extend this part of the integral into being along a closed curve going into the complex plane to the side where $\exp(-ik_x(x - x'))$ will vanish as k_x becomes complex due to the large $(x - x')$, then this extension practically does not change the integral value. However, it allows us to use the residue theorem, in which case, the integral will give the pole contribution from one of the poles only, namely the one being enclosed inside the integration path. This leads finally to the asymptotic expression

$$g^{\mathrm{spp,ff}}(\mathbf{r}, \mathbf{r}') = -i \frac{\varepsilon_1 \varepsilon_2}{\varepsilon_1 + \varepsilon_2} \frac{1}{\sqrt{\frac{\varepsilon_1}{-\varepsilon_2}} [\varepsilon_1 - \varepsilon_2]} e^{-ik_{\mathrm{spp}}|x-x'|} e^{-k_{\mathrm{spp}} \sqrt{\frac{\varepsilon_1}{-\varepsilon_2}}(y+y')}. \tag{7.139}$$

It is actually possible to use a similar argument to derive the expression (7.139) directly from Equation (7.108) by considering the integral from $k_0 n_1$ to ∞. For small values of y and y' but with $y > 0$ and $y' > 0$, this part of the integral cannot be neglected, not even in the far field. In that case, it will be the poles of the Fresnel reflection coefficient that give rise to closed form terms to the far-field Green's function governing bound modes. There will be one term for each bound mode.

Note that it is possible to explicitly separate the total field into the excitation of each type of mode by applying different components of Green's functions related to the different modes of light. This can be used to straightforwardly calculate the amount of light being scattered into for example out-of-plane propagating modes and bound modes.

7.7 CONSTRUCTION OF THE PERIODIC GREEN'S FUNCTION

The 1D periodic Green's function for a homogeneous reference medium must satisfy the equation

$$\left(\nabla^2 + k_0^2 \varepsilon_{\text{ref}}\right) g(\mathbf{r}, \mathbf{r}') = -\delta(\mathbf{r}, \mathbf{r}'),$$

(7.140)

the radiating boundary condition in the direction where the structure is not periodic, and the periodic Bloch boundary condition $g(x + \Lambda, y; x', y') = g(x, y; x', y') \exp(-ik_x\Lambda)$, where k_x is the Bloch wave number, in the direction where the structure *is* periodic.

Green's function can be obtained with the mode-expansion method, where we consider a complete set of eigenmodes to the left-hand-side operator in Equation (7.140) that must also satisfy the periodic boundary condition, that is

$$\left(\nabla^2 + k_0^2 \varepsilon_{\text{ref}}\right) E_\lambda(\mathbf{r}) = \lambda E_\lambda(\mathbf{r}),$$

(7.141)

and $E_\lambda(x + \Lambda, y) = E_\lambda(x, y) \exp(-ik_x\Lambda)$. This leads to the eigenfunctions

$$E_{n,k_y}(\mathbf{r}) = e^{-i(k_x+nG)x - ik_y y},$$

(7.142)

where the eigenvalue is given by $\lambda_{n,k_y} = k_0^2 n_{\text{ref}}^2 - (k_x + nG)^2 - k_y^2$ and $G = 2\pi/\Lambda$.
The normalization condition is given by

$$\int_{x=0}^{\Lambda} \int_{y=-\infty}^{\infty} E_{n,k_y}(\mathbf{r}) \left(E_{n',k_y'}(\mathbf{r})\right)^* dA = \Lambda \delta_{n,n'} 2\pi\delta(k_y' - k_y) = N_{n,k_y}\delta_{n,n'}\delta(k_y' - k_y), \quad (7.143)$$

where $N_{n,k_y} = (2\pi)\Lambda$.

A Green's function satisfying Equation (7.140) and the periodic boundary condition can now be constructed with the mode-expansion method, that is

$$g(\mathbf{r}, \mathbf{r}') = -\sum_n \int_{k_y} \frac{E_{n,k_y}(\mathbf{r})\left(E_{n,k_y}(\mathbf{r}')\right)^*}{N_{n,k_y}\lambda_{n,k_y}} dk_y$$

$$= -\sum_n \int_{k_y} \frac{e^{-i(k_x+nG)(x-x')}e^{-ik_y(y-y')}}{(2\pi)\Lambda(k_0^2 n_{\text{ref}}^2 - (k_x + nG)^2 - k_y^2)} dk_y.$$

(7.144)

In order to get Green's function that satisfies the radiating wave boundary condition, we add a small imaginary part (0^+) in the denominator, that is

$$g(\mathbf{r}, \mathbf{r}') = -\sum_n \int_{k_y} \frac{e^{-i(k_x+nG)(x-x')} e^{-ik_y(y-y')}}{(2\pi)\Lambda(k_0^2 n_{\text{ref}}^2 - (k_x + nG)^2 - k_y^2 - i0^+)} dk_y$$

$$= \sum_n \int_{k_y} \frac{e^{-i(k_x+nG)(x-x')} e^{-ik_y(y-y')}}{(2\pi)\Lambda(k_y - k_{y,n} + i0^+)(k_y + k_{y,n} - i0^+)} dk_y, \qquad (7.145)$$

where again the only effect of the small imaginary part is to add a homogeneous solution to Equation (7.140), and thus, Equation (7.140) is still satisfied, but now the resulting solution will satisfy the radiating boundary condition. Here, $k_{y,n} = \sqrt{k_0^2 n_{\text{ref}}^2 - (k_x + nG)^2}$ and $\text{Im}(k_{y,n}) \leq 0$. Thus, we notice that there are two poles located in the upper and lower half-plane of complex k_y, respectively. By using the residue theorem similar to Section 7.6.1, we find

$$g(\mathbf{r}, \mathbf{r}') = \frac{-i}{4\pi} \sum_n \frac{e^{-i(k_x+nG)(x-x')} e^{-ik_{y,n}|y-y'|}}{k_{y,n}} \cdot G. \qquad (7.146)$$

It is clear that for a source point located at some value of y', the y-behaviour of the solution corresponds to a solution propagating away from the source point either into the upper or lower half-plane relative to the source plane.

7.7.1 1D PERIODIC SCALAR GREEN'S FUNCTION FOR A LAYERED STRUCTURE

From the expression (7.146) for the free-space periodic Green's function, we can apply similar principles as in Sections 7.6.2 and 7.6.3 to obtain the corresponding Green's function which is periodic in one direction (x) and layered in the other direction (y). For TE polarization, this results in

$$g(\mathbf{r}, \mathbf{r}') = \frac{-i}{4\pi} \sum_n \frac{e^{-i(k_x+nG)(x-x')} \left(e^{-ik_{y1,n}|y-y'|} + r^{(TE)}(k_x + nG)e^{-ik_{y1,n}(y+y')} \right)}{k_{y1,n}} G \qquad (7.147)$$

when $y, y' > 0$, and

$$g(\mathbf{r}, \mathbf{r}') = \frac{-i}{4\pi} \sum_n \frac{e^{-i(k_x+nG)(x-x')} e^{-ik_{y1,n}y'} e^{+ik_{y2,n}(y+d)} t^{(TE)}(k_x + nG)}{k_{y1,n}} G \qquad (7.148)$$

when $y' > 0$ and $y < -d$. Similar expressions can be obtained for TM polarization by replacing $r^{(TE)}$ and $t^{(TE)}$ with $r^{(TM)}$ and $t^{(TM)}$, respectively. Here, $k_{yi,n} = \sqrt{k_0^2 \varepsilon_i - (k_x + nG)^2}$ with $\text{Im}(k_{yi,n}) \leq 0$.

7.8 REFLECTION FROM A PERIODIC SURFACE MICROSTRUCTURE

In this section, we will consider a geometry which is both periodic in one dimension and can be considered as a modification to a layered structure. The structure we

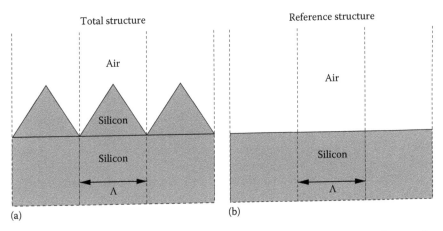

FIGURE 7.14   Schematic of (a) total structure and (b) reference structure for a silicon surface being periodically microstructured with a triangular surface geometry.

will study is a silicon surface with a periodic triangular surface microstructure (see Figure 7.14). The reference structure is a layered geometry with reference dielectric constant given by

$$\varepsilon_{\text{ref}}(\mathbf{r}) = \begin{cases} \varepsilon_{air} & \text{if } y > 0 \\ \varepsilon_{Si} & \text{if } y < 0 \end{cases}$$

We are interested in finding a solution to the equation

$$\left(\nabla^2 + k_0^2 \varepsilon(\mathbf{r})\right) E(\mathbf{r}) = 0, \tag{7.149}$$

which is in the form of a given incident wave and a scattered or reflected wave propagating away from the microstructure on the silicon surface. In the case when the surface microstructure is absent, the electric field $E_0(\mathbf{r})$ must be a solution to the equation

$$\left(\nabla^2 + k_0^2 \varepsilon_{\text{ref}}(\mathbf{r})\right) E_0(\mathbf{r}) = 0. \tag{7.150}$$

If we are interested in calculating reflection and transmission in the case when a plane wave of the form $E_{0i}(\mathbf{r}) = e^{-ik_{sx}x}e^{ik_{y1}y}$ is incident on the surface microstructure, then in order to obtain an appropriate solution E_0, we have to add a reflected field E_{0r} when $y > 0$ and construct a transmitted field E_{0t} when $y < 0$. Thus, in this case, the reference field E_0 becomes

$$E_0(\mathbf{r}) = \begin{cases} e^{-ik_{sx}x}\left(e^{ik_{y1}y} + r^{(TE)}(k_x)e^{-ik_{y1}y}\right) & \text{if } y > 0 \\ e^{-ik_{sx}x}e^{ik_{y2}y}t^{(TE)}(k_x) & \text{if } y < 0 \end{cases}$$

where $r^{(TE)}$ and $t^{(TE)}$ are appropriate Fresnel reflection and transmission coefficients for a single interface, $k_{y1} = \sqrt{k_0^2\varepsilon_{air} - k_x^2}$ and $k_{y2} = \sqrt{k_0^2\varepsilon_{Si} - k_x^2}$ with $\text{Im}(k_{yi}) \le 0$.

We will apply the periodic Green's function for a layered structure from Section 7.7.1. Considering first only one period of the structure, this Green's function satisfies the equation

$$\left(\nabla^2 + k_0^2 \varepsilon_{\text{ref}}(\mathbf{r})\right) g(\mathbf{r}, \mathbf{r}') = -\delta(\mathbf{r} - \mathbf{r}'). \tag{7.151}$$

Furthermore, Green's function satisfies the Bloch boundary condition $g(x + \Lambda, y; x', y') = g(x, y; x', y')e^{-ik_x\Lambda}$ and the radiating boundary condition that in the substrate $(y < 0)$, scattered light propagates along the negative y-axis and for a position $y > y'$, the scattered light propagates along the positive y-axis. The Bloch boundary condition means that for general positions $\mathbf{r}$ and $\mathbf{r}'$ not necessarily located inside the same period, Green's function will in fact satisfy the equation

$$\left(\nabla^2 + k_0^2 \varepsilon_{\text{ref}}(\mathbf{r})\right) g(\mathbf{r}, \mathbf{r}') = -\sum_n \delta(\mathbf{r} - \mathbf{r}' + \hat{\mathbf{x}}n\Lambda)e^{-ik_x n\Lambda}. \tag{7.152}$$

We can rewrite the wave equations (7.149) and (7.150) into the expression

$$\left(\nabla^2 + k_0^2 \varepsilon_{\text{ref}}(\mathbf{r})\right)(E(\mathbf{r}) - E_0(\mathbf{r})) = -k_0^2 \left(\varepsilon(\mathbf{r}) - \varepsilon_{\text{ref}}(\mathbf{r})\right) E(\mathbf{r}), \tag{7.153}$$

and due to the properties of Green's function, we find that

$$E(\mathbf{r}) = E_0(\mathbf{r}) + \int_{x'=-\Lambda/2}^{\Lambda/2} \int_{y'} g(\mathbf{r}, \mathbf{r}')k_0^2 \left(\varepsilon(\mathbf{r}') - \varepsilon_{\text{ref}}(\mathbf{r}')\right) E(\mathbf{r}')d^2r' \tag{7.154}$$

is a particular solution to the wave equation (7.149) which is in the form of the sum of a given reference field and a term propagating away from the surface microstructure and which also satisfies the Bloch boundary condition $E(x + \Lambda, y) = E(x, y)e^{-ik_x\Lambda}$ because both g and E_0 satisfy this condition. It is also worth noticing that the integration is over only one period of the periodic structure. In the case when $\mathbf{r}$ is located inside the period that is integrated over in Equation (7.154), then it follows that this expression is a solution to Equation (7.153) by insertion in the left-hand side and using that Green's function satisfies Equation (7.151). If we instead consider a position $\mathbf{r} \rightarrow \mathbf{r} + \hat{\mathbf{x}}n\Lambda$, it follows from Equation (7.152) that we will find the same result as mentioned earlier except for a phase factor $\exp(-ik_x n\Lambda)$, and thus, the total field satisfies the required Bloch boundary condition.

For any position where $y < 0$, and any other position when $y > 0$ outside the silicon, we find that $\varepsilon(\mathbf{r}) - \varepsilon_{\text{ref}}(\mathbf{r}) = 0$, and the total field can be obtained at any of those positions by an overlap integral over the field inside the silicon triangle that has been added to the reference structure and Green's function. The numerical problem at first is then reduced to calculating the total field in the region where $\varepsilon(\mathbf{r}) - \varepsilon_{\text{ref}}(\mathbf{r}) \neq 0$. This is completely equivalent to Section 7.3 except that here $\varepsilon_{\text{ref}}(\mathbf{r})$ depends on the position, which means that even though there is a substrate $(y < 0)$, this region does not have to be included in the numerical problem.

Also similar to Section 7.3, we can discretize the region where $\varepsilon(\mathbf{r}) - \varepsilon_{\text{ref}}(\mathbf{r}) \neq 0$ into small square-shaped elements and assume that the total field E and reference field

E_0 inside each element are constant, that is $E_i = E(\mathbf{r}_i)$ and $E_{0,i} = E_0(\mathbf{r}_i)$, respectively, where $\mathbf{r}_i$ is the centre position of element i, and assign only one dielectric constant to each element given by $\varepsilon_i = \varepsilon(\mathbf{r}_i)$ and $\varepsilon_{\text{ref},i} = \varepsilon_{\text{ref}}(\mathbf{r}_i)$ for the total and reference geometry, respectively. In that case, the discretized integral equation (7.154) can be written as

$$E_i = E_{0,i} + \sum_j g_{ij} k_0^2 \left(\varepsilon_j - \varepsilon_{\text{ref},j} \right) E_j A_j, \tag{7.155}$$

where A_j is the area of element j and

$$g_{ij} = \frac{1}{A_j} \int_j g(\mathbf{r}_i, \mathbf{r}') dA'. \tag{7.156}$$

Here, we should note that in the region where the reference geometry has been modified, Green's function consists of a direct and an indirect contribution of the form $g(\mathbf{r}, \mathbf{r}') = g^{(d)}(x - x'; y - y') + g^{(i)}(x - x', y + y')$, where the direct part has a singularity, while the indirect part does not, since we will not encounter a situation with $y + y' = 0$ when sampling at the centre of discretization elements.

Thus, as the first approximation to the matrix elements g_{ij}, we can use

$$g_{ij} = g(\mathbf{r}_i, \mathbf{r}_j), \quad i \neq j \tag{7.157}$$

$$g_{ii} = g^{(i)}(\mathbf{r}_i, \mathbf{r}_i) + \frac{1}{A_i} \int_i g^{(d)}(\mathbf{r}_i, \mathbf{r}') dA'. \tag{7.158}$$

Since the singular part of the direct Green's function is similar to the free-space Green's function when the distance is small, we can, as the first approximation, use the result (7.15) for $g_{ii}^{(d)}$, that is

$$g_{ii} \approx g^{(i)}(\mathbf{r}_i, \mathbf{r}_i) + \frac{1}{2i(ka)^2} \left(kaH_1^{(2)}(ka) - i\frac{2}{\pi} \right), \tag{7.159}$$

where $k = k_0 n_{\text{ref},i}$ and $\pi a^2 = A_i$. This approximation can be acceptable when the cell width is much smaller than the period. It is possible to improve this slightly by noting that it is only the real part of Green's function which is singular. Then we can keep the real part of this expression but use the imaginary part of Green's function calculated for a short distance between $\mathbf{r}$ and $\mathbf{r}'$ such as the distance between neighbour elements. If i and i' are neighbour elements, this amounts to

$$g_{ii} \approx \text{Re} \left(g^{(i)}(\mathbf{r}_i, \mathbf{r}_i) + \frac{1}{2i(ka)^2} \left(kaH_1^{(2)}(ka) - i\frac{2}{\pi} \right) \right) + i\text{Im}(g_{ii'}). \tag{7.160}$$

7.8.1 CALCULATING REFLECTION AND TRANSMISSION

When considering reflection of light from the microstructured surface, it will be preferable to consider the reflected light at a large distance from the surface, that is for a large value of y ($k_0 n_{air} y \gg 1$). If we further consider the case with $y' > 0$, then we can use the following far-field approximation to Green's function (7.147) that includes only those terms that are not evanescent (exponentially decreasing) along the y-axis, that is

$$g^{ff}(\mathbf{r}, \mathbf{r}') = \frac{-i}{4\pi} \sum_{n, \, \mathrm{Im}(k_{y1,n})=0} \frac{e^{-i(k_x+nG)(x-x')} \left(e^{-ik_{y1,n}|y-y'|} + r^{(TE)}(k_x + nG)e^{-ik_{y1,n}(y+y')} \right)}{k_{y1,n}} G.$$

(7.161)

We should also remember that even without the microstructuring of the surface, there will be a reflection given by $E_{0r}(\mathbf{r})$. In the far field, the total reflected field can be written as

$$E_r^{ff}(\mathbf{r}) = \sum_{n, \, \mathrm{Im}(k_{y1,n})=0} e^{-i(k_x+nG)x} e^{-ik_{y1,n}y} A_n,$$

(7.162)

where

$$A_n = \delta_{n,0} r^{(TE)}(k_x) + \int\limits_{x'=-\Lambda/2}^{x'=\Lambda/2} \int\limits_{y'} f_n(\mathbf{r}')k_0^2 \left(\varepsilon(\mathbf{r}') - \varepsilon_{ref}(\mathbf{r}') \right) E(\mathbf{r}')dx' dy'$$

(7.163)

with

$$f_n(\mathbf{r}') = -\frac{i}{4\pi} \frac{G}{k_{y1,n}} e^{i(k_x+nG)x'} \left(e^{ik_{y1,n}y'} + r^{(TE)}(k_x + nG)e^{-ik_{y1,n}y'} \right).$$

(7.164)

The time-averaged incident power per period is given by

$$P_i = \frac{1}{2} \mathrm{Re} \left(\int\limits_{x=-\Lambda/2}^{x=\Lambda/2} \mathbf{E}_{0i} \times \mathbf{H}_{0i}^* \cdot (-\hat{\mathbf{y}})dx \right) = \frac{1}{2} \frac{k_{y1,0}\Lambda}{\omega\mu_0} |E_{0,i}|^2,$$

(7.165)

and the corresponding total reflected power is given by

$$P_r = \frac{1}{2} \mathrm{Re} \left(\int\limits_{x=-\Lambda/2}^{x=\Lambda/2} \mathbf{E}_r^{ff} \times \mathbf{H}_r^{ff*} \cdot (\hat{\mathbf{y}})dx \right) = \sum_{n, \, \mathrm{Im}(k_{y1,n})=0} P_{r,n},$$

(7.166)

where the power reflected into the nth diffraction order is given by

$$P_{r,n} = \frac{1}{2} \frac{k_{y1,n}\Lambda}{\omega\mu_0} |A_n|^2.$$

(7.167)

Thus, the reflectance for the nth diffraction order is given by

$$R_n = \frac{P_{r,n}}{P_i} = \frac{|A_n|^2}{|E_{0i}|^2} \frac{k_{y1,n}}{k_{y1,0}}.$$

(7.168)

The total transmission, and transmission related to different transmission diffraction orders, can be calculated with similar principles. If we use the approximation that we neglect the imaginary part of the silicon refractive index, then in that case, we obtain the far-field approximation to Green's function by keeping only those terms n where $k_{y2,n}$ is real ($y' > 0$, $y < 0$), that is

$$g(\mathbf{r},\mathbf{r}') = \frac{-i}{4\pi} \sum_{n,\, \mathrm{Im}(k_{y2,n})=0} \frac{e^{-i(k_x+nG)(x-x')} e^{-ik_{y1,n}y'} e^{+ik_{y2,n}y} t^{(TE)}(k_x+nG)}{k_{y1,n}} G \qquad (7.169)$$

Thus, the transmitted far field ($k_0 n_{Si} y \ll -1$) is given by

$$E_t^{ff}(\mathbf{r}) = \sum_{n,\, \mathrm{Im}(k_{y2,n})=0} e^{-i(k_x+nG)x} e^{ik_{y2,n}y} B_n, \qquad (7.170)$$

where

$$B_n = t^{(TE)}(k_x)\delta_{n,0} + \int_{x'=-\Lambda/2}^{x'=\Lambda/2} \int_{y'} h_n(\mathbf{r}')k_0^2 \left(\varepsilon(\mathbf{r}') - \varepsilon_{\mathrm{ref}}(\mathbf{r}')\right) E(\mathbf{r}') dx' dy'. \qquad (7.171)$$

with

$$h_n(\mathbf{r}') = -\frac{i}{4\pi} \frac{G}{k_{y1,n}} t^{(TE)}(k_x+nG)e^{i(k_x+nG)x'} e^{-ik_{y1,n}y'} \qquad (7.172)$$

The transmitted power per period can be written as

$$P_t = \frac{1}{2}\mathrm{Re}\left(\int_{x=-\Lambda/2}^{x=\Lambda/2} E_t^{ff} \times H_t^{ff*} \cdot (-\hat{y})dx\right) = \sum_{n,\, \mathrm{Im}(k_{y2,n})=0} P_{t,n}, \qquad (7.173)$$

where

$$P_{t,n} = \frac{1}{2}\frac{k_{y2,n}\Lambda}{\omega\mu_0}|B_n|^2. \qquad (7.174)$$

The corresponding transmittance is given by

$$T = \sum_{n,\, \mathrm{Im}(k_{y2,n})=0} \frac{P_{t,n}}{P_i}, \qquad (7.175)$$

and the part of the transmittance being transmitted into the nth diffraction order is given by

$$T_n = \frac{P_{t,n}}{P_i}. \qquad (7.176)$$

As an example, we will consider a TE-polarized plane-wave incident on a periodic triangular surface geometry similar to Figure 7.14 with angle of incidence 40°, 55°, or 70°. We will use the period $\Lambda = 600$ nm and height of triangles $h = 413$ nm. The calculated zero-order reflectance versus wavelength for wavelengths in the range $\lambda \in [300; 900$ nm], and calculated using 60×42 square-shaped discretization

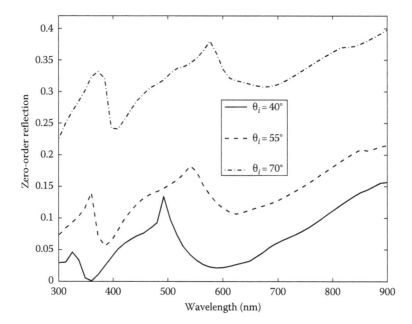

FIGURE 7.15 Zero-order reflection for light incident on a silicon surface being microstructured with a periodic array of silicon triangles with period 600 nm and height 413 nm.

elements, is shown in Figure 7.15. Note that the refractive index of silicon depends appreciably on the wavelength for this range, and we used the values (or linear interpolation between the values) tabulated in Ref. [27]. An alternative method for calculating the same result could be to use the GFSIEM with periodic Green's functions. In that case, it is actually not necessary to use Green's function for a layered geometry. It is sufficient to use two Green's functions (7.146) corresponding to different refractive indices. Some examples of this approach can be found in Refs. [28–31]. We have checked that we also obtain the same result as in Figure 7.15 with the periodic GFSIEM.

7.9 ITERATIVE SOLUTION SCHEME TAKING ADVANTAGE OF THE FAST FOURIER TRANSFORM

In this section, we will consider an iterative approach for solving the discretized integral equation in the GFAIEM. The advantage of such an approach lies in the possible vast reduction of both calculation time and memory requirements. Consider a linear system of equations with N unknowns expressed on matrix form as

$$\overline{\overline{C}}\,\overline{x} = \overline{y}, \tag{7.177}$$

where
$\overline{\overline{C}}$ is an $N \times N$ matrix
$\overline{y}$ is a given vector with N elements
$\overline{x}$ is the vector that we want to find such that Equation (7.177) is satisfied

Solving the linear system of equations with the usual matrix inversion will require a calculation time that scales as N^3. The calculation time for a matrix–vector multiplication on the other hand scales as N^2. An alternative is to use an iterative algorithm where we start out with a guess for $\bar{x}$ and improve the guess in each of a series of steps such that Equation (7.177) is increasingly better satisfied. Then if the number of steps required to find the solution $\bar{x}$ with an acceptable precision is not too large, this can be much faster than usual matrix inversion. For this strategy to be fast, it is very important that the matrix–vector product can be calculated very fast. In some cases, we may notice from an equation such as Equation (7.12) with Green's function of the form $g(\mathbf{r}, \mathbf{r}') = g(\mathbf{r} - \mathbf{r}')$ that it is in the form of a convolution integral, that is

$$E(\mathbf{r}) = E_0(\mathbf{r}) + \int g(\mathbf{r} - \mathbf{r}')k_0^2 \left(\varepsilon(\mathbf{r}') - \varepsilon_{\text{ref}} \right) E(\mathbf{r}')dA'. \qquad (7.178)$$

In the case when we have discretized a square-shaped region containing our geometry into elements of the same size and shape, the aforementioned equation becomes a discrete convolution, which for an appropriate number of discretization elements along each axis on the form $N_x = 2^{n_x}$ and $N_y = 2^{n_y}$, where n_x and n_y are integers, can be calculated very fast by using the FFT algorithm. First, each of $g(\mathbf{r} - \mathbf{r}')$ and $f(\mathbf{r}) = k_0^2 (\varepsilon(\mathbf{r}) - \varepsilon_{\text{ref}}) E(\mathbf{r})$ should be Fourier transformed, then they should be multiplied in reciprocal space, and then the convolution is obtained by Fourier transforming the product back to real space. The time required for the FFT scales as $N \ln N$, and the multiplication in reciprocal space scales as N. Thus, the time for the convolution scales as $N \ln N$ instead of N^2 being the scaling for usual matrix–vector products. Furthermore, a lot of information is repeated several times in the matrix $\overline{\overline{C}}$. As a consequence of the form of Green's function $g(\mathbf{r}, \mathbf{r}') = g(\mathbf{r} - \mathbf{r}')$, there will be many pairs of positions $\mathbf{r}$ and $\mathbf{r}'$ with the same $\mathbf{r} - \mathbf{r}'$ and thus giving rise to the same value of matrix elements for all those pairs in $\overline{\overline{C}}$.

When we calculate the matrix–vector product with the FFT approach, we only need to store a small part of the matrix $\overline{\overline{C}}$, and instead of storage requirements scaling as N^2, here storage requirements will only scale as N. This is very important since it is often not the calculation time but the available computer memory that limits how large a structure that can be modelled. The drawback is that in some cases, a large number of iterative steps are required before sufficient accuracy is obtained. It might be the case that the approach is very fast for one particular structure, while it is not fast at all for another structure because many more iterative steps are required. Here, we will consider a variant of the conjugate gradient algorithm that was previously applied in astrophysics by B. T. Draine [7,32] formulated for a linear system of equations (7.177) on matrix form.

Conjugate Gradient Algorithm [7,32]

In this algorithm, we start out with an initial guess $\mathbf{x}_0$ and set

$$\bar{z} \equiv \overline{\overline{C}}^{\dagger} \bar{y}, \qquad (7.179)$$

$$\overline{g}_0 = \overline{z} - \overline{\overline{C}}^\dagger \overline{\overline{C}} \overline{x}_0, \tag{7.180}$$

$$\overline{p}_0 = \overline{g}_0, \tag{7.181}$$

$$\overline{w}_0 = \overline{\overline{C}} \overline{x}_0, \tag{7.182}$$

$$\overline{v}_0 = \overline{\overline{C}} \overline{p}_0, \tag{7.183}$$

where † represents Hermitian conjugation, that is $[\overline{\overline{C}}^\dagger]_{jk} = [\overline{\overline{C}}^*]_{kj}$ or $\overline{\overline{C}}^\dagger = \overline{\overline{C}}^{T*}$, corresponding to both transposition and conjugation. The estimate $\overline{x}_i$ is now iteratively improved ($i = 0, 1, 2, \ldots$) using the following relations:

$$\alpha_i = \frac{\overline{g}_i^\dagger \overline{g}_i}{\overline{v}_i^\dagger \overline{v}_i}, \tag{7.184}$$

$$\overline{x}_{i+1} = \overline{x}_i + \alpha_i \overline{p}_i, \tag{7.185}$$

$$\overline{w}_{i+1} = \overline{w}_i + \alpha_i \overline{v}_i, \tag{7.186}$$

$$\overline{g}_{i+1} = \overline{z} - \overline{\overline{C}}^\dagger \overline{w}_{i+1}, \tag{7.187}$$

$$\beta_i = \frac{\overline{g}_{i+1}^\dagger \overline{g}_{i+1}}{\overline{g}_i^\dagger \overline{g}_i}, \tag{7.188}$$

$$\overline{p}_{i+1} = \overline{g}_{i+1} + \beta_i \overline{p}_i, \tag{7.189}$$

$$\overline{v}_{i+1} = \overline{\overline{C}} \overline{g}_{i+1} + \beta_i \overline{v}_i. \tag{7.190}$$

The aforementioned expressions for $\overline{v}_i$ and $\overline{w}_i$ are equivalent to

$$\overline{v}_i \equiv \overline{\overline{C}} \overline{p}_i, \tag{7.191}$$

$$\overline{w}_i \equiv \overline{\overline{C}} \overline{x}_i. \tag{7.192}$$

In order to avoid that a rounding off error accumulates, the latter two expressions can be used, for example for every 10 iterations.

When $\overline{x}_i$ is close to the actual solution $\overline{x}$, then $\overline{w}_i$ and $\overline{y}$ must be nearly the same vector resulting in the error measure as follows:

$$\text{error} = 1 - \frac{(\overline{w}_i^\dagger \overline{y})(\overline{y}^\dagger \overline{w}_i)}{(\overline{w}_i^\dagger \overline{w}_i)(\overline{y}^\dagger \overline{y})}. \tag{7.193}$$

We can, for example choose to iterate until the *error* is smaller than 10^{-8}.

7.9.1 2D DISCRETE CONVOLUTION

If we are not going to solve the linear system of equations by using matrix inversion, then it is also not necessary to arrange the discrete field values in a vector. In that case, we can arrange the field values in a matrix $\overline{\overline{E}}$ with matrix elements $[\overline{\overline{E}}]_{i_x,i_y} = E_{i_x,i_y} = E(\mathbf{r}_{i_x,i_y})$, where the latter $E(\mathbf{r}_{i_x,i_y})$ is the physical field value at sampling position $\mathbf{r}_{i_x,i_y} = \hat{\mathbf{x}}(x_0 + \Delta x(i_x - \frac{1}{2})) + \hat{\mathbf{y}}(y_0 + \Delta y(i_y - \frac{1}{2}))$. Here, $\mathbf{r}_{i_x,i_y}$ with $i_x = 1, 2, \ldots, N_x$ and $i_y = 1, 2, \ldots, N_y$ represents the centre position of the elements. The discretized region is assumed to be contained within a rectangular region with lower-left corner given by (x_0, y_0). For a scatterer in free space, Green's function translational symmetry means that we can formulate the linear system of equations in the following way:

$$E_{i_x,i_y} = E_{0,i_x,i_y} + \sum_{j_x,j_y} g_{i_x-j_x,i_y-j_y} k_0^2 \left(\varepsilon_{j_x,j_y} - \varepsilon_{\text{ref}}\right) E_{j_x,j_y} \Delta A, \tag{7.194}$$

where we have assumed that all elements have the same size ΔA. Notice that Equation (7.194) is a 2D discrete convolution.

Consider also the matrix $\overline{\overline{G}}$ where the matrix elements are given by $[\overline{\overline{G}}]_{N_x+i_x,N_y+i_y} = G_{i_x+N_x,i_y+N_y} = g_{i_x,i_y}$, where since here i_x and i_y can assume negative values in Equation (7.194), we have added N_x and N_y such that only indices larger than or equal to one will be considered in the matrix. The range of indices for $\overline{\overline{G}}$ is larger than what is the case for $\overline{\overline{E}}$. In order to work with matrices of the same size, the matrix $\overline{\overline{E}}$ can be padded with zeroes resulting in the matrix $\overline{\overline{E}}'$, in which case, the discrete convolution given by

$$H_{i_x,i_y} = \sum_{j_x,j_y} G_{i_x-j_x+N_x,i_y-j_y+N_y} E_{j_x,j_y}, \tag{7.195}$$

can be calculated fast with first an FFT of $\overline{\overline{E}}'$ and $\overline{\overline{G}}$, a product in reciprocal space, and then an inverse fast Fourier transform (IFFT), that is the matrix $\overline{\overline{H}}$ can be obtained by

$$\overline{\overline{H}} = \left[IFFT \left(FFT(\overline{\overline{G}}) . * FFT(\overline{\overline{E}}') \right) \right]_{\text{Submatrix of } N_x \times N_y \text{ elements}}, \tag{7.196}$$

where " $. *$ " means that elements at the same positions in the two matrices are multiplied together to form the new element at that position.

If we are going to use the conjugate gradient iterative scheme from the previous section but for data organized in matrices rather than vectors, then the operation $\overline{h} = \overline{\overline{C}}\,\overline{\overline{E}}$ must be understood as

$$h_{i_x,i_y} = E_{i_x,i_y} - \sum_{j_x,j_y} g_{i_x-j_x,i_y-j_y} k_0^2 \left(\varepsilon_{j_x,j_y} - \varepsilon_{\text{ref}}\right) E_{j_x,j_y} \Delta A, \tag{7.197}$$

and $\overline{f} = \overline{\overline{C}}^{\dagger} \overline{\overline{E}}$ must be understood as

$$f_{i_x,i_y} = E_{i_x,i_y} - \sum_{j_x,j_y} \left(\varepsilon_{i_x,i_y} - \varepsilon_{\text{ref}}\right)^* g^*_{j_x-i_x,j_y-i_y} k_0^2 E_{j_x,j_y} \Delta A. \tag{7.198}$$

Note that, for example for the GFAIEM for a layered structure in Section 7.8, we will not immediately have an integral equation being a convolution as in

Equation (7.178). However, for $y, y' > 0$, Green's function can be split into two terms such that $g(x, y; x', y') = g^{(1)}(x - x', y - y') + g^{(2)}(x - x', y + y')$, and in that case, the integral equation can be written as

$$E(\mathbf{r}) = E_0(\mathbf{r}) + \int g^{(1)}(x - x', y - y') k_0^2 \left(\varepsilon(\mathbf{r}') - \varepsilon_{\text{ref}} \right) E(\mathbf{r}') dA'$$
$$+ \int g^{(2)}(x - x', y + y') k_0^2 \left(\varepsilon(\mathbf{r}') - \varepsilon_{\text{ref}} \right) E(\mathbf{r}') dA' \qquad (7.199)$$

such that there are two convolution integrals that each can be handled fast with the FFT technique. Using that approach, we have reproduced the results in Figure 7.15.

As an example of a large scattering problem that can be dealt with by the iterative method, we will consider a finite photonic crystal with an array of square-shaped cylinders placed on a square lattice with period Λ and length and width of the cylinders given by $a = \Lambda/3$. We will also consider a background dielectric constant $\varepsilon_1 = 1$ and cylinder dielectric constant $\varepsilon_2 = 12$ (silicon). This kind of geometry is known as a 2D photonic crystal [33]. A schematic of the fully periodic photonic crystal and corresponding band diagram for our choice of parameters are shown in Figure 7.16.

From the band diagram, we will expect that the photonic crystal will efficiently reflect light with a wavelength in the bandgap, while this will not be the case for light with a frequency that is not in the bandgap. Here, we will consider a plane-wave incident on a finite array of 9×9 cylinders and calculate the field for a wavelength in the bandgap ($\lambda = \Lambda/0.35$) and a wavelength outside the bandgap (e.g. $\lambda = \Lambda/0.2$). The field distributions calculated using the method presented in this section are shown in Figure 7.17. Observe that in one case, $\lambda = \Lambda/0.2$, the light can penetrate into the photonic crystal, whereas in the other case (e.g. $\lambda = \Lambda/0.35$), it cannot, since there

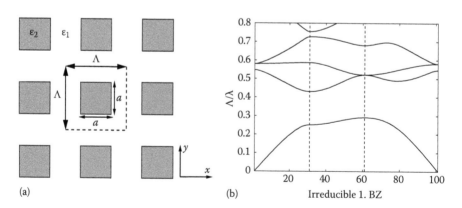

(a) (b) Irreducible 1. BZ

FIGURE 7.16 (a) Schematic of a periodic array of square cylinders with dielectric constant ε_2 and side length a arranged on a square lattice with period Λ in a material with dielectric constant ε_1 and (b) the corresponding band diagram when $\varepsilon_1 = 1$, $\varepsilon_2 = 12$, and $a = \Lambda/3$ for in-plane propagating TE-polarized light calculated with the plane-wave expansion method. (From Joannopoulos, J.D. et al., *Photonic Crystals: Molding the Flow of Light*, Princeton University Press, Princeton, NJ, 2008. With permission.)

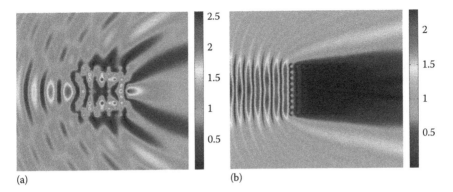

FIGURE 7.17 (**See colour insert.**) Magnitude of electric field for a plane TE-polarized wave incident on a photonic crystal with 9×9 square rods of size $a = \Lambda/3$, where Λ is the period and where the rod dielectric constant is $\varepsilon_2 = 12$ and the background dielectric constant is $\varepsilon_1 = 1$. The results correspond to using the frequency given by either (a) $\Lambda/\lambda = 0.2$ or (b) $\Lambda/\lambda = 0.35$.

are no propagating modes of light in the photonic crystal for that wavelength. For the calculation we have used 18 discretization elements per period. This is favourable since it is divisible by 3, and the rod width $a = \Lambda/3$. Thus, we end up with a linear system of equations with $162 \times 162 = 26,244$ unknowns when including only the photonic crystal region in the calculation at first. Instead of constructing an additional routine for obtaining the field outside the region with the array of cylinders, we actually just extended the window of calculation such that the array of cylinders is the same, but the window of calculation is increased by a factor 3 along each direction. This means that we in principle solved a linear system of equations with $(162 * 3) \times (162 * 3) = 236,196$ unknowns, which directly gave us the field distribution for the entire window shown in Figure 7.17.

7.10 FURTHER READING

For a deeper study of the GFIEM for electromagnetics scattering problems, the reader is referred to Refs. [11,18,34] and references therein. For a deeper study of the theory of Green's functions, the reader is referred to Refs. [1,2,35]. Scattering from a periodic 2D array and the related Green's function has been studied in Ref. [36].

In this chapter, we have not considered GFSIEMs for 3D objects. However, this is also possible, and that technique has been used for micro- and nanostructures in several recent papers [37–41] and earlier, for example in [42–44]. For a study of more advanced discretization schemes, the reader can find inspiration in, for example Ref. [45]. These references by no means represent a complete list but can get you started if you intend to do research in this area.

EXERCISES

7.1 Implement the GFAIEM from Section 7.3 in matrix form in for example MATLAB® and use standard matrix inversion to calculate the field distributions shown in Figure 7.2. Calculate the far-field radiation patterns shown in Figure 7.3.

7.2 Implement the GFVIEM from Section 7.4 in for example MATLAB and calculate the field distributions shown in Figure 7.5.

7.3 Derive Green's function surface integral equations similar to Section 7.5 for a situation where it is the electric field that is polarized along the z-axis and where it is assumed that the fields and the geometry are invariant along the z-axis.

7.4 Prove Equation (7.57).

7.5 Use the GFSIEM to calculate the field and normal derivative on the outer boundary of a silver cylinder of diameter $D = 500$ nm located in air ($\varepsilon_1 = 1$) when a plane wave of wavelength $\lambda = 700$ nm propagating along the x-axis ($H_0(\mathbf{r}) = e^{-ik_0 x}$) is incident on the cylinder, that is reproduce Figure 7.8 using different numbers of discretization elements and different subsectioning of each element.

7.6 Use the result from Exercise 7.5 to calculate the differential scattering cross section shown in Figure 7.9.

7.7 Observe the behaviour of expression (7.146) in the limit of very small $x - x'$ and $y - y'$ and show that the behaviour will be similar to that of the non-periodic Green's function in expression (7.93). Use this to prove that (7.146) satisfies Equation (7.140).

7.8 Find a practical numerical approach to calculate (7.146) when $|y - y'| = 0$ such that convergence with the number of terms included in the sum is improved. Hint: Subtract a sum that can be analytically calculated such that the summand of the remaining sum will asymptotically decrease as $1/n^2$.

7.9 Construct a numerical program that tabulates (7.146) for values of $x - x'$ and $|y - y'|$ on a grid. Use the case of period $\Lambda = 500$ nm and $k_x = k_0 \sin \theta$. Choose, for example a spacing between gridpoints of $\Delta x = \Delta y = 20$ nm and tabulate for $x - x' \in [0; 500$ nm$]$ and $|y - y'| \in [0; 500$ nm$]$. When $x - x' = |y - y'| = 0$, use an approximation to the average value over the cell. The imaginary part can be approximated with an actual calculation for a small but not zero value of $x - x'$ and $|y - y'|$, whereas the real part can be approximated using the real part of Equation (7.15). For negative and other values of $x - x'$ outside the tabulated interval, Green's function can be obtained by exploiting the Bloch boundary condition or symmetries.

7.10 Construct a numerical program that tabulates the indirect part of (7.147) for values of $x - x'$ and $y + y'$ on a grid. Use the case of period $\Lambda = 500$ nm and $k_x = k_0 \sin \theta$. Choose, for example a spacing between grid points of $\Delta x = \Delta y = 20$ nm and tabulate for $x - x' \in [0; 500$ nm$]$ and $y + y' \in [0; 500$ nm$]$. Note that in a practical calculation with the GFAIEM, where, for example square elements are sampled at the centre of the elements, we will always have that $y + y' > 0$, so we will not be troubled with any singularity in the indirect part of Green's function.

7.11 Implement the GFAIEM for a layered structure and calculate reflection for a periodic array of silicon triangles on a silicon surface similar to Figure 7.14.

7.12 Modify the program developed in Exercise 7.1 such that instead of usual matrix inversion, the solution for the field inside the scatterer is obtained by the iterative algorithm described in Section 7.9.

7.13 Modify the program from Exercise 7.12 further such that the matrix–vector products are carried out by using the FFT. Note that in this case, it is possible to arrange the field values in a matrix instead of a vector and carry out the 2D FFT on the matrix.

7.14 After having solved Exercises 7.12 and 7.13, you are now ready to model large structures since the required computer memory only scales as N and not N^2, and the calculation time scales much better than N^3. Use the program from Exercise 7.13 to calculate scattering of light by a finite photonic crystal similar to Figure 7.17 for a frequency inside and outside the bandgap of the photonic crystal.

REFERENCES

1. E. Economou, *Green's Functions in Quantum Physics*, 1st edn. Berlin, Germany: Springer-Verlag, 1979.
2. P. M. Morse and H. Feshbach, *Methods of Theoretical Physics*. New York: McGraw-Hill, 1953.
3. L. Novotny, B. Hecht, and D. W. Pohl, Interference of locally excited surface plasmons, *J. Appl. Phys.*, 81, 1798 (1997).
4. L. Novotny, Allowed and forbidden light in near-field optics. I. A single dipolar light source, *J. Opt. Soc. Am. A*, 14, 91 (1997).
5. S. Lipschutz, M. R. Spiegel, and J. Liu, *Schaum's Outline of Mathematical Handbook of Formulas and Tables*, 4th edn. New York: McGraw-Hill, 2013.
6. R. F. Harrington, *Field Computation by Moment Methods*. New York: Macmillan, 1968.
7. B. T. Draine, The discrete-dipole approximation and its application to interstellar graphite grains, *Astrophys. J.*, 333, 848–872 (1988).
8. E. M. Purcell and C. R. Pennypacker, Scattering and absorption of light by nonspherical dielectric grains (in the interstellar medium), *Astrophys. J.*, 186, 705 (1973).
9. G. H. Goedecke and S. G. O'Brien, Scattering by irregular inhomogeneous particles via the digitized Green's function algorithm, *Appl. Opt.*, 27, 2431 (1988).
10. M. F. Iskander, H. Y. Chen, and J. E. Penner, Optical scattering and absorption by branched chains of aerosols, *Appl. Opt.*, 28, 3083 (1989).
11. T. Søndergaard, Modeling of plasmonic nanostructures: Green's function integral equation methods, *Phys. Stat. Sol. (b)*, 244, 3448 (2007).
12. J. P. Kottmann and O. J. F. Martin, Accurate solution of the volume integral equation for high-permittivity scatterers, *IEEE Trans. Antennas Propag.*, 48, 1719 (2000).
13. J. Van Bladel, *Singular Electromagnetic Fields and Sources*. Piscataway, NJ: IEEE Press, 1991.
14. O. J. F. Martin, C. Girard, and A. Dereux, Generalized field propagator for electromagnetic scattering and light confinement, *Phys. Rev. Lett.*, 74, 526 (1995).
15. A. D. Yaghjian, Electric dyadic Greens-functions in the source region, *Proc. IEEE*, 68, 248 (1980).
16. J. Nachamkin, Integrating the dyadic Green's function near sources, *IEEE Trans. Antennas Propag.*, 38, 919 (1990).

17. J. Jung and T. Søndergaard, Green's function surface integral equation method for theoretical analysis of scatterers close to a metal interface, *Phys. Rev. B*, 77, 245310 (2008).

18. J. Jin, *The Finite Element Method in Electromagnetics*, 2nd edn. New York: Wiley, 2002.

19. W. C. Chew, *Waves and Fields in Inhomogeneous Media*. Piscataway, NJ: IEEE Press, 1995.

20. D. W. Prather, M. S. Mirotznik, and J. N. Mait, Boundary integral methods applied to the analysis of diffractive optical elements, *J. Opt. Soc. Am. A*, 14, 34 (1997).

21. P. B. Johnson and R. W. Christy, Optical constants of the noble metals, *Phys. Rev. B*, 6, 4370 (1972).

22. E. Kreyszig, *Advanced Engineering Mathematics*, 9th edn. Singapore: John Wiley & Sons, 2006.

23. H. Raether, *Surface Plasmons on Smooth and Rough Surfaces and on Gratings*. New York: Springer-Verlag, 1986.

24. M. V. Klein, and T. E. Furtak, *Optics*, 2nd edn. New York: John Wiley & Sons, 1986.

25. V. Siahpoush, T. Søndergaard, and J. Jung, Green's function approach to investigate the excitation of surface plasmon polaritons in a nanometer-thin metal film, *Phys. Rev. B*, 85, 075305 (2012).

26. T. Søndergaard and S.I. Bozhevolnyi, Surface plasmon polariton scattering by a small particle placed near a metal surface: An analytical study, *Phys. Rev. B*, 69, 045422 (2004).

27. E. D. Palik, *Handbook of Optical Constants of Solids*. Boston, MA: Academic Press, 1985. The data is also available at http://refractiveindex.info.

28. T. Søndergaard, J. Gadegaard, P. K. Kristensen, T. K. Jensen, T. G. Pedersen, and K. Pedersen, Guidelines for 1D-periodic surface microstructures for antireflective lenses, *Opt. Express*, 18, 26245 (2010).

29. T. Søndergaard, S. I. Bozhevolnyi, J. Beermann, S. M. Novikov, E. Devaux, and T. W. Ebbesen, Extraordinary optical transmission with tapered slits: Effect of higher diffraction and slit resonance orders, *J. Opt. Soc. Am. B*, 29, 130 (2012).

30. T. Søndergaard, S. M. Novikov, T. Holmgaard, R. L. Eriksen, J. Beermann, Z. Han, K. Pedersen, and S. I. Bozhevolnyi, Plasmonic black gold by adiabatic nanofocusing and absorption of light in ultra-sharp convex grooves, *Nat. Commun.*, 3, 969 (2012). doi:10.1038/ncomms1976.

31. T. Søndergaard, and S. I. Bozhevolnyi, Surface-plasmon polariton resonances in triangular-groove metal gratings, *Phys. Rev. B*, 80, 195407 (2009).

32. M. Petravic and G. Kuo-Petravic, An ILUCG algorithm which minimizes in the euclidean norm, *J. Comput. Phys.*, 32, 263 (1979).

33. J. D. Joannopoulos, S. G. Johnson, J. N. Winn, and R. D. Meade, *Photonic Crystals: Molding the Flow of Light*. Princeton, NJ: Princeton University Press, 2008.

34. L. Novotny and B. Hecht, *Principles of Nano-Optics*, 2nd edn. Cambridge, U.K.: Cambridge University Press, 2012.

35. C.-T. Tai, *Dyadic Green's Functions in Electromagnetic Theory*. London, U.K.: Intext Educational Publishers, 1971.

36. G. Kobidze, B. Shanker, and D. P. Nyquist, Efficient integral-equation-based method for accurate analysis of scattering from periodically arranged nanostructures, *Phys. Rev. E*, 72, 056702 (2005).

37. A. M. Kern and O. J. F. Martin, Surface integral formulation for 3D simulations of plasmonic and high permittivity nanostructures, *J. Opt. Soc. Am. B*, 26, 732–740 (2009).

38. U. Hohenester and J. Krenn, Surface plasmon resonances of single and coupled metallic nanoparticles: A boundary integral method approach, *Phys. Rev. B*, 72, 195429 (2005).

39. B. Gallinet and O. J. F. Martin, Scattering on plasmonic nanostructures arrays modeled with a surface integral formulation, *Photon. Nanostruct.*, 8, 278 (2010).

40. B. Gallinet, A. M. Kern, and O. J. F. Martin, Accurate and versatile modeling of electromagnetic scattering on periodic nanostructures with a surface integral approach, *J. Opt. Soc. Am. A*, 27, 2261 (2010).

41. R. Rodríguez-Oliveros and J. A. Sánchez-Gil, Localized surface-plasmon resonances on single and coupled nanoparticles through surface integral equations for flexible surfaces, *Opt. Express*, 19, 12208 (2011).

42. N. Marly, D. De Zutter, and H. F. Pues, A surface integral equation approach to the scattering and absorption of doubly periodic lossy structures, *IEEE Trans. Electrom. Compat.*, 36, 14 (1994).

43. N. Marly, B. Baekelandt, D. De Zutter, and H. F. Pues, Integral equation modeling of the scattering and absorption of multilayered doubly-periodic lossy structures, *IEEE Trans. Antennas Prop.*, 43, 1281 (1995).

44. A. F. Peterson, S. L. Ray, and R. Mittra, *Computational Methods for Electromagnetics*. New York: IEEE Press, 1998.

45. A. F. Peterson, *Mapped Vector Basis Functions for Electromagnetic Integral Equations*. San Rafael, CA: Morgan & Claypool Publishers, 2005.

8 Finite Element Method

The finite element method (FEM) is based on a *variational formulation* of Maxwell's equations that involves integral expressions on the computational domain. Unlike, for example the finite-difference method which approximates Maxwell's equations directly (via approximation of the differential operators), the FEM leaves Maxwell's equations completely intact but approximates the solution space in which one tries to find a reasonable approximation to the exact solution. This solution space is obtained by subdividing the computational domain into small patches and by providing a number of polynomials on each patch for the approximation of the solution. The patches together with the local polynomials defined on them are called *finite elements*. The most common examples of finite elements are triangles and rectangles in 2D and tetrahedrons and cuboids in 3D together with constant, linear, quadratic, and cubic polynomials. These locally defined polynomial spaces have to be pieced together to ensure tangential continuity of the electric and magnetic field across the boundaries of neighbouring patches. Once these local approximations with proper continuity conditions have been defined, these are inserted into the variational equation. This results in a linear system whose solution is a piecewise polynomial approximation of the exact solution. Hence, the two basic ingredients in each FEM for Maxwell's equations are

1. A variational formulation of Maxwell's equations
2. A suitable construction of finite elements based on polynomials defined on local geometric patches to transform the variational formulation into a discrete, algebraic problem

For each of the problems discussed in the following, namely time-harmonic Helmholtz and Maxwell's scattering problems, we will provide both the variational form and examples of finite elements. Unlike the finite-difference method, where we get a pointwise approximation of the field, we get throughout the computational domain a piecewise polynomial approximation. Plugged into the right computational framework, the essential properties of the FEM are as follows:

- Complex geometrical shapes can be treated without geometrical approximations, for example roundings can be well and easily approximated.
- The finite element mesh can easily be adapted to the behaviour of the solution, for example to singularities at corners.
- High-order approximations are available and ensure fast convergence.

Several excellent books treating the finite element analysis of Maxwell's equations are available; very recommendable are the books of Peter Monk [5] and Leszek Demkowicz [3].

8.1 INTRODUCTION: HELMHOLTZ EQUATION IN 1D

For the introduction of the finite element concept, we start with the most simple approximation of the time-harmonic Maxwell's equations – the scalar Helmholtz equation in 1D. We will solve the equation inside a homogeneous, that is spatially invariant, medium. We let our computational domain Ω be the interval $(0, L)$. The source field is a plane wave coming from the left. A possible problem formulation reads: Find $u(x)$, $x \in \Omega = (0, L)$ such that

$$\partial_x^2 u(x) + k^2 u(x) = 0 \tag{8.1}$$

$$u(0) = e^{ik0} = 1 \tag{8.2}$$

$$\partial_x u(L) = iku(L). \tag{8.3}$$

The situation is illustrated in Figure 8.1. The source field is the plane wave $u_{\text{source}} = \exp(ikx)$. The interior solution $u(x)$ at the left boundary $x = 0$ should equal the source field, since there is no back scattering from the interior. The boundary condition on the right boundary, $\partial_x u = iku$, is the so-called Sommerfeld radiation condition in 1D. We will discuss radiation conditions in more detail in Section 8.3.1. In 1D, it becomes immediately clear that every right propagating plane wave must satisfy this condition. Hence, we have a well-described problem: the Helmholtz equation plus two boundary conditions – first, a so-called Dirichlet boundary condition with given data on the left boundary and second, a so-called *Dirichlet-to-Neumann* (DtN) boundary condition on the right boundary. We will come back to this in (8.5) and more generally in Section 8.2.1. Other choices of boundary conditions must be used for general formulations of time-harmonic scattering problems in higher space dimensions, but for the illustration of the concept in 1D, these are sufficient.

8.1.1 VARIATIONAL FORMULATION

Figure 8.2 gives a road map for turning the Helmholtz equation into a solvable discrete problem. Currently, we are at the top position: the Helmholtz equation. First, we want to construct a variational equation. To this end, we multiply (8.1) with the complex conjugate of a so-called test function $v(x)$ and integrate over the domain Ω to arrive at

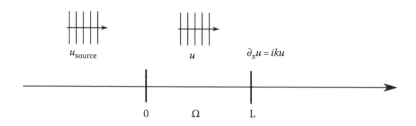

FIGURE 8.1 The 1D Helmholtz problem with given data on the left boundary.

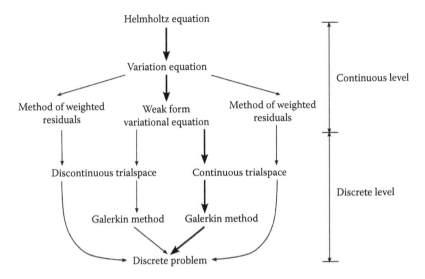

FIGURE 8.2 Methods to obtain the discrete problem. We follow the bold arrows.

$$\int_0^L v^*(x) \left(\partial_x^2 u(x) + k^2 u(x) \right) dx = 0,$$

$$u(0) = 1,$$

$$\partial_x u(L) = iku(L), \tag{8.4}$$

where nothing has been specified so far about $v(x)$, except that we must be allowed to compute the integral. This formulation cannot be equivalent to (8.1) through (8.3), since very special choices of v would equate the integral to zero even for functions, which do not satisfy (8.1). Hence, in order to have the chance to get something equivalent to (8.1), Equation (8.4) must hold true for many different functions $v(x)$. We call the class of functions from which we take $v(x)$ the *test space*. In contrast, the space of functions containing the desired solution $u(x)$ is called the *trial* or *ansatz space*. In general, both spaces can be different. It is subject to the convergence theory of finite elements to define precisely the test and trial spaces and to conclude convergence rates. For our purpose here, it is sufficient to define these spaces in a way that makes the construction of the numerical algorithms reasonable. In the first attempt, we require $\int_0^L v^* v \, dx < \infty$, which is the class of square-integrable functions, the $L^2(0, L)$ functions. We will make the choice of the space more precise in the following but only in a framework that introduces usual FEM notions and gives a first orientation.* A comprehensive treatment can be found, for example in Ref. [5]. The functions $u(x)$ should be twice differentiable since the second derivative is contained in (8.1). Hence, we seek them in the class $C^2(0, L)$. So in a first attempt to construct a variational equation, we formulate the following: Find a function $u(x) \in C^2(0, L)$ such that

---

* For a collection of typical notions of FEM theory, see Section 8.2.6.

$$\int_0^L v^*(x) \left(\partial_x^2 u(x) + k^2 u(x) \right) dx = 0 \text{ for all } v \in L^2(0, L) \tag{8.5}$$

$$u(0) = 1 \tag{8.6}$$

$$\partial_x u(L) = iku(L), \tag{8.7}$$

where the boundary conditions appear as constraints to the variational equation in the first line. Equation (8.7) relates so-called *Dirichlet* data $u(L)$ to so-called *Neumann* data $\partial_x u(L)$. The operator which mediates this action is called DtN operator (see Section 8.2.1 for more details).

Note: The method of weighted residuals would start here by incorporating the boundary conditions either into the right-hand side or by a suitable restriction to the test and ansatz spaces. We will not follow this path here.

8.1.2 WEAK FORM

Next, we step further down in Figure 8.2 to obtain the variational formulation in weak form. We perform an integration by parts on the left-hand side of (8.5):

$$\int_0^L v^*(x) \left(\partial_x^2 u(x) + k^2 u(x) \right) dx$$

$$= \int_0^L \left(-\partial_x v^*(x) \partial_x u(x) + k^2 v^*(x) u(x) \right) dx + v^*(x) \partial_x u(x)|_0^L. \tag{8.8}$$

It is convenient to abbreviate the negative of the integral including the boundary terms by

$$a(v, u) := \int_0^L \left(\partial_x v^*(x) \partial_x u(x) - k^2 v^*(x) u(x) \right) dx - v^*(x) \partial_x u(x)|_0^L. \tag{8.9}$$

The integral in (8.9) is the core of the weak form. The term is called a *sesquilinear form* because it is anti-linear in its first argument (the v, due to complex conjugation) and linear in its second argument (the u). Hence, $a(v, u)$ maps two complex functions via integration to a complex number. In addition to the integration by parts, we would like to get rid of the use of the boundary conditions (8.6) and (8.7) in terms of explicit constraints. This is easily accomplished for the DtN condition (8.7). We just set

$$v^*(x) \partial_x u(x)|^L = v^*(L) \partial_x u(L) = v^*(L) iku(L).$$

This specialization modifies the sesquilinear form (8.9) to

$$a(v, u) := \int_0^L \left(\partial_x v^*(x) \partial_x u(x) - k^2 v^*(x) u(x) \right) dx - v^*(L) iku(L) + v^*(0) \partial_x u(0). \tag{8.10}$$

Now we can rewrite our first variational form (8.5). Without further specification, let us denote the test space by V and the trial space by U. Then the variational equation employing the weak form reads

$$\text{Find } u \in U : \ a(v, u) = 0 \text{ for all } v \in V$$
$$u(0) = 1.$$

Equation (8.7) is now incorporated into the sesquilinear form $a(v, u)$ and does not appear as an extra constraint. To consider the left boundary condition (8.6) describing the incoming field, we do a trick. Consider an almost arbitrary function $g(x)$ subtracted from $u(x)$ such that the difference is a new function $\tilde{u}(x)$ having zero boundary conditions on the left boundary:

$$a(v, u - g + g) = 0$$
$$a\left(v, \underbrace{u - g}_{\tilde{u}} \right) = -a\,(v, g)\,. \tag{8.11}$$

Hence, $\tilde{u} := u - g$. Let us require

$$g(0) = u(0) = 1 \tag{8.12}$$

and denote the artificially right-hand side created this way as

$$b(v) := -a\,(v, g)\,.$$

Due to the subtraction $u - g$, our trial space for $\tilde{u}$ is smaller than the one for u; it is restricted to functions having zero values on the left boundary.

Let us put things together. Dropping the tilde on top of u to simplify notation, the variational formulation in weak form reads

$$\text{Find } u \in U : \ a(v, u) = b(v) \text{ for all } v \in V.$$

8.1.3 GALERKIN METHOD

The next step leads to the Galerkin formulation. This requires a consideration of the test and trial spaces. We do it in a simplified manner. First, the trial space must be chosen such that $u(0) = 0$. Without cancelling information from the equations, we can enforce the same property from the test space $v(0) = 0$. Further, due to the symmetry in the formulation, both spaces U and V must have the same properties to allow for a proper differentiation and integration; hence, we may choose $V = U$. This choice is called the Galerkin formulation, in short notation:

$$\text{Find } u \in V : \ a(v, u) = -a(v, g) \text{ for all } v \in V. \tag{8.13}$$

Here, the sesquilinear form $a(v, u)$ is given as

$$a(v, u) := \int_0^L \left(\partial_x v^*(x) \partial_x u(x) - k^2 v^*(x) u(x) \right) dx - v^*(L) i k u(L).$$

Introductory Scattering Problem in 1D

Let the source field be a plane wave $\exp(ikx)$. The variational form of the 1D Helmholtz equation reads

$$\text{Find } u \in V: \quad a(v, u) = -a(v, g) \text{ for all } v \in V$$

with

$$a(v, u) := \int_0^L \left(\partial_x v^*(x) \partial_x u(x) - k^2 v^*(x) u(x) \right) dx - v^*(L) i k u(L) \tag{8.14}$$

and an almost arbitrary continuous function $g(x)$ fixed on the left boundary by

$$g(0) = 1.$$

This is a simplification of (8.10) due to the choice $v(0) = 0$. Such kinds of variational formulations are the workhorses in FEMs.

8.1.4 DISCRETE PROBLEM

Discretization in the context of a Galerkin method means the restriction of the infinite-dimensional space V to a finite-dimensional space $V_h \subset V$, which is a subset of the large space. The index h of V_h indicates that this space is constructed with respect to a subdivision of the computational domain into patches with a typical geometrical size of h. Therefore, we indicate usually the discrete approximation of u and v by u_h and v_h. Thus, the discrete problem reads very similar to the continuous one (8.13):

$$\text{Find } u_h \in V_h: \ a(v_h, u_h) = -a(v_h, g) \text{ for all } v_h \in V_h. \tag{8.15}$$

8.1.5 LINEAR FINITE ELEMENTS

We recall the way how we derive the discrete problem from the continuous one (8.13) — just restrict the infinite-dimensional test and trial space V to a finite-dimensional space V_h:

$$\text{Find } u \in V: \ a(v, u) = b(v) \quad \text{for all } v \in V$$

$$\downarrow$$

$$\text{Find } u_h \in V_h: \ a(v_h, u_h) = b(v_h) \quad \text{for all } v_h \in V_h. \tag{8.16}$$

The space V_h is not uniquely predetermined. The different types of the Galerkin methods differ in the construction of the elements of V_h. Let V_h be spanned by N linearly independent functions $v_j(x)$, $j = 1, \ldots, N$. We represent the discrete approximation of the exact solution u by $u_h(x) = \sum_{j=1}^{N} u_j v_j(x)$, insert this into (8.16)

$$a\left(v_h, \sum_{j=1}^{N} u_j v_j(x)\right) = b(v_h),$$

and get the linear system

$$i = 1, \ldots, N : \quad \sum_{j=1}^{N} a_{ij} u_j = b_i \quad \text{with } a_{ij} = a\left(v_i, v_j\right) \quad \text{and } b_i = b\left(v_i\right).$$

For a convenient computation of the matrix and vector components a_{ij} and b_i, it is useful to decompose the form $a(v, u)$ of the variational formulation (8.14) into the so-called *stiffness* and *mass* parts $s(v, u)$ and $m(v, u)$, respectively:

$$a(v, u) = s(v, u) - m(v, u)$$

$$s(v, u) := \int_0^L \partial_x v^* \partial_x u$$

$$m(v, u) := \int_0^L k^2 v^* u.$$

Note: The notions stiffness and mass come from mechanical systems and have no meaning in electrodynamics. But as we will see, both parts have very different mathematical properties.

We want to construct the matrix $\left(a_{ij}\right)$ as sparse matrix, that is a matrix with predominantly zero entries, which is a very useful numerical property. To this end, we construct test functions v_j that exist only on small patches. The most common finite elements are the *piecewise linear hat* functions shown in the left part of Figure 8.3:

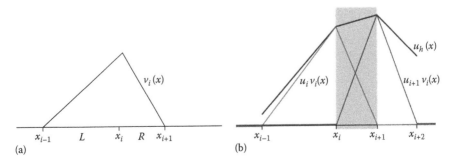

FIGURE 8.3 (a) Piecewise linear test function with left subdomain L and right subdomain R. (b) Piecewise linear function u_h composed from trial functions $v_i(x)$ and $v_{i+1}(x)$.

$$
v_i(x) = \begin{cases} \dfrac{x - x_{i-1}}{h_{i-1}} & \text{for} \quad x \in [x_{i-1}, x_i] \\[2mm] -\dfrac{x - x_{i+1}}{h_i} & \text{for} \quad x \in [x_i, x_{i+1}] \\[2mm] 0 & \text{elsewhere,} \end{cases}
\tag{8.17}
$$

where we introduced the lengths of the subdomains with $h_{i-1} = x_i - x_{i-1}$ and $h_i = x_{i+1} - x_i$. Let us compute the stiffness matrix based on the hat functions. Using (8.17), we get

$$
\begin{aligned}
s_{ij} &= s\left(v_i, v_j\right) \\
&= \int_0^L \partial_x v_i^* \partial_x v_j \, dx \\
&= \int_{x_{i-1}}^{x_{i+1}} \partial_x v_i^* \partial_x v_j \, dx \\
&= \int_{x_{i-1}}^{x_i} \partial_x v_i^* \partial_x v_j \, dx + \int_{x_i}^{x_{i+1}} \partial_x v_i^* \partial_x v_j \, dx.
\end{aligned}
$$

If we fix the row i for a moment, that is the test function v_i, we see that there exist contributions to $\left(s_{ij}\right)$ only for the $j = i-1,\, i,\, i+1$. Let us decompose s_{ij} further with respect to contributions which come from the left segment (x_{i-1}, x_i), indicated as L, and from the right segment (x_i, x_{i+1}), indicated as R,

$$
\begin{aligned}
s_{i,i-1}\big|_L &= \int_{x_{i-1}}^{x_i} \partial_x v_i^* \partial_x v_{i-1} \\
&= (x_i - x_{i-1}) \frac{-1}{(x_i - x_{i-1})^2} \\
&= -\frac{1}{h_{i-1}},
\end{aligned}
$$

$$
\begin{aligned}
s_{i,i-1}\big|_R &= \int_{x_i}^{x_{i+1}} \partial_x v_i^* \partial_x v_{i-1} \\
&= \int_{x_i}^{x_{i+1}} \partial_x v_i^* \cdot 0 \\
&= 0
\end{aligned}
$$

and similarly

$$
s_{i,i}\big|_L = \frac{1}{h_{i-1}} \qquad\qquad s_{i,i}\big|_R = \frac{1}{h_i}
$$

$$S_{i,i+1}\big|_L = \int\limits_{\underbrace{x_{i-1}}_{=0}}^{x_i} \dots \qquad\qquad S_{i,i+1}\big|_R = -\frac{1}{h_i}.$$

Let us write the ith row of the stiffness matrix:

$$s\left(v_i,\ v_{i-1}u_{i-1} + v_iu_i + v_{i+1}u_{i+1}\right)$$

$$= \underbrace{\frac{1}{h_{i-1}}\begin{pmatrix} -1 & 1 \end{pmatrix}\begin{pmatrix} u_{i-1} \\ u_i \end{pmatrix}}_{\text{segment } i-1,i} + \underbrace{\frac{1}{h_i}\begin{pmatrix} 1 & -1 \end{pmatrix}\begin{pmatrix} u_i \\ u_{i+1} \end{pmatrix}}_{\text{segment } i,i+1}.$$

The same way, we get the next row shifting $i \to i+1$:

$$s\left(v_{i+1},\ v_iu_i + v_{i+1}u_{i+1} + v_{i+2}u_{i+2}\right)$$

$$= \underbrace{\frac{1}{h_i}\begin{pmatrix} -1 & 1 \end{pmatrix}\begin{pmatrix} u_i \\ u_{i+1} \end{pmatrix}}_{\text{segment } i,i+1} + \underbrace{\frac{1}{h_{i+1}}\begin{pmatrix} 1 & -1 \end{pmatrix}\begin{pmatrix} u_{i+1} \\ u_{i+2} \end{pmatrix}}_{\text{segment } i+1,i+2}.$$

Now we collect the contributions of the individual segments, for example segment $i, i+1$, given by local 2×2 matrices:

$$S^{loc}_{\{i,i+1\}} = \frac{1}{h}\begin{pmatrix} -1 & 1 \\ 1 & -1 \end{pmatrix}$$

$$h = x_{i+1} - x_i.$$

These local matrices are then summed up to a global matrix S as explained in the following.

8.1.6 DOMAIN MAPPING

In higher space dimensions, it simplifies the computation of the local matrices, if the local patches (triangle, quadrilaterals, etc.), which are in 1D just segments, are mapped to simply shaped reference domains. The situation in 1D is depicted in Figure 8.4. The reference domain is the interval $[0, 1]$ with a normalized coordinate $\hat{x} \in [0, 1]$. If we use hat functions $\hat{v}_1$ and $\hat{v}_2$ on the unit interval, exactly as defined in (8.17), that is we attribute the inner vertex of $\hat{v}_1$ to $\hat{x} = 0$ and the inner vertex of $\hat{v}_2$ to $\hat{x} = 1$, we can formalize the interval mapping as

$$[0, 1] \to [x_i, x_j]$$
$$\hat{x} \mapsto x = x_i\hat{v}_1\left(\hat{x}\right) + x_j\hat{v}_2\left(\hat{x}\right). \tag{8.18}$$

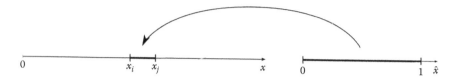

FIGURE 8.4 Mapping from the reference interval to the original interval.

8.1.7 ASSEMBLY PROCESS

Assembly is the process of summing up the global system matrix (a_{ij}) and the right-hand side (b_i) from local matrices. Algorithm 8.1 shows the scheme. It would not be necessary to formalize this procedure in 1D since it is obvious how to do it. But since the same basic scheme can be used in all dimensions, it is convenient to discuss it first by means of the 1D example. The idea is to compute the contributions from the interval $[x_i, x_j]$ (which of course would be the interval $[x_i, x_{i+1}]$ for normal counting of the indices) based on computations on the unit interval. A proper transformation between reference and original domain yields the desired results for the original domain. A particular problem is how to assign the vertices in the reference domain to those in the original domain. Let the vertices of segment (x_i, x_j); this means vertices i and j, be assigned to the vertices with numbers 1 and 2 in the reference domain. The relation to the original domain can be expressed via permutation matrices P. These matrices have in each row exactly one entry of one and are zero otherwise. A 2×1 vector $(a, b)^T$ maps from the $\hat{x}$-system to the x-system by

$$
\begin{array}{c}
i \\
\\
\\
\\
j \\
\\
\end{array}
\underbrace{\begin{pmatrix} 0 & 0 \\ & \vdots \\ 1 & \\ & \vdots \\ & 1 \\ & \vdots \\ 0 & 0 \end{pmatrix}}_{P^T}
\begin{pmatrix} a \\ b \end{pmatrix}
=
\begin{pmatrix} 0 \\ \vdots \\ a \\ \vdots \\ b \\ \vdots \\ 0 \end{pmatrix}.
\tag{8.19}
$$

Hence, the small vector distributes its entries according to the positions in the x-system. Accordingly, a small 2×2 matrix S^{loc} contributes to the global $N \times N$ matrix via $P^T S^{\text{loc}} P$. Hence, the global matrix S results from

$$
S = \sum_{i=1}^{N+1} P^T S^{\text{loc}}_{\{i-1,i\}} P.
\tag{8.20}
$$

The permutation matrix P connects the locally ordered quantities to the globally ordered quantities. It is often also called *connectivity matrix* C. The distribution to the global matrix is in practice not realized via matrix multiplications as in our short notation, since this would be too expensive, but by direct index mapping.

8.1.8 ALGORITHM: PLANE-WAVE PROPAGATION

The realization of the 1D FEM algorithm with given data on the left boundary and a transparent boundary condition in terms of a DtN operator on the right condition is summarized in Algorithm 8.1. In Figure 8.5, we display the result with $N = 10$ segments.

Algorithm 8.1 Semi-Transparent Plane-Wave Propagation in 1D (DtN)

Input: gridpoints $\{x_1, \ldots, x_N\}$

$S = 0; M = 0$

$g = 1$ % constant vector, $g(0) = u(0)$, see (8.12)

compute list of all segments $L = \{(x_i, x_j), \ldots\}$

$$S_{\text{loc}} = \begin{pmatrix} 1 & -1 \\ -1 & 1 \end{pmatrix}; \quad M_{\text{loc}} = \begin{pmatrix} \frac{1}{3} & \frac{1}{6} \\ \frac{1}{6} & \frac{1}{3} \end{pmatrix}$$

for $l \in L$ **do**

    $P = 0$ % reset permutation matrix, see (8.19)

    $P(1, i) = 1$ % assign local index 1 to global index i

    $P(2, j) = 1$ % assign local index 2 to global index j

    $h = x_j - x_i$

    $S = S + P^T \frac{1}{h} S_{\text{loc}} P; \quad M = M + P^T h k^2 M_{\text{loc}} P$

end for

$A = S - M$

$b = -A \cdot g$

                           %Correct for the zero boundary condition left

$A(1, :) = [\,]$ %Delete first row

$A(:, 1) = [\,]$ %Delete first column

$b(1) = [\,]$ %Delete first entry

                           %Correct for the right boundary condition

$A(\text{end}, \text{end}) = A(\text{end}, \text{end}) - ik$

$u = A \backslash b$

$u = u + g$

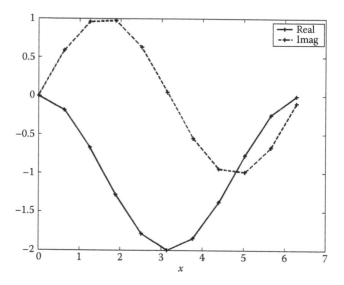

FIGURE 8.5 Real and imaginary part of $\tilde{u}$ with $N = 10$ segments over one wavelength.

8.2 GENERAL SCATTERING PROBLEM IN 1D

Figure 8.6 sketches a typical scattering problem. We have a source field living in the exterior Ω_{ext} that satisfies Maxwell's equations there. The source field enters the interior domain Ω_{int}, which typically contains some objects. The objects (not drawn in Figure 8.6) cause scattered fields which leave Ω_{int} after a while. We will collect the necessary equations to describe the situation. Additionally, we supply a general solution representation in each domain based on the different fundamental solutions ϕ. The goal is to count the number of unknowns in the entire description of the scattering problem:

$$
\begin{aligned}
\Omega_{int}: &\quad u'' + k^2(x)u = 0 &\quad u = c_1\phi_1 + c_2\phi_2 \\
\Omega_{ext}: &\quad u''_{source} + k^2 u_{source} = 0 &\quad u_{source}\ \text{given} \\
\text{left}\ \Omega_{ext}: &\quad u''_{s,left} + k^2 u_{s,left} = 0 &\quad u = c_3\phi_3 + c_4\phi_4 \\
\text{right}\ \Omega_{ext}: &\quad u''_{s,right} + k^2 u_{s,right} = 0 &\quad u = c_5\phi_5 + c_6\phi_6.
\end{aligned}
\tag{8.21}
$$

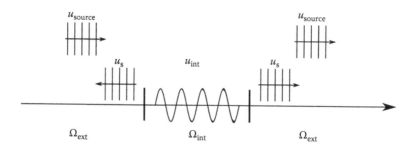

FIGURE 8.6 Sketch of a scattering problem.

We have three domains: left exterior domain, interior domain, and right exterior domain. In each, we have two fundamental solutions whose superposition gives the general solution in the domain. Hence, we have altogether six constants $c_1, \ldots, c_6$ to determine. If we take the continuity of the fields and the continuity of the first derivative of the fields across the left and right boundary, we have four conditions:

$$\text{Continuity of } u = u_{\text{source}} + u_s, \quad x \in \{0, L\} \tag{8.22}$$

$$\text{Continuity of } \partial_n u = \partial_n (u_{\text{source}} + u_s), \quad x \in \{0, L\}. \tag{8.23}$$

The two missing conditions are supplied by the following radiation conditions:

$$\text{Radiation condition (left and right): } \partial_n u_s - iku_s = 0, \tag{8.24}$$

where ∂_n denotes the normal derivative, that is $\partial_n = -\partial_x$ on the left and $\partial_n = \partial_x$ on the right boundary.

Equations (8.21) through (8.24) define the 1D scattering problem. The construction remains essentially the same in higher space dimensions. However, Sommerfeld radiation condition in d dimensions, $d \geq 1$, gets more complicated:

$$\lim_{r \to \infty} r^{\frac{d-1}{2}} (\partial_r u_s - iku_s) = 0 \text{ uniformly in all directions.} \tag{8.25}$$

8.2.1 VARIATIONAL FORMULATION IN 1D WITH DtN OPERATOR

Next, we repeat the procedure from the previous section to derive a variational formulation. First, we multiply the Helmholtz equation describing the field in the interior domain (8.21) with a test function $v \in V$ and integrate by parts:

$$\int_0^L \partial_x v^* \partial_x u - v^* \partial_x u \Big|_0^L - \int_0^L k^2 v^* u = 0. \tag{8.26}$$

In the exterior domain, it holds for the scattered fields

$$\partial_x^2 u_s + k^2 u_s = 0, \ x \in \mathbb{R} \setminus [0, L]$$

with the fundamental solutions $\phi_\pm = e^{\pm ikx}$ whose possible superpositions give all possible scattered fields u_s.

8.2.1.1 DtN Operator

Via the radiation condition, we select for the right exterior domain solution

$$u_s(x) = u_0 e^{ikx};$$

hence,

$$\partial_x u_s(x) = iku_0 e^{ikx}$$

$$= iku_s(x).$$

We call the expression

$$\partial_x u_s(L) = i k u_s(L)$$

the DtN operator. Now we have

$$
\begin{aligned}
\partial_x u\Big|_{x=L} &= \partial_x \left(u_{\text{source}} + u_s\right)_{x=L} \\
&= \left(\partial_x u_{\text{source}} + ik\,(u_s)\right)_{x=L} \\
&= \left(\partial_x u_{\text{source}} + ik\,(u - u_{\text{source}})\right)_{x=L} \\
&= \left(\partial_x u_{\text{source}} - iku_{\text{source}} + iku\right)_{x=L}.
\end{aligned}
$$

Here, we use all of our conditions: the continuity of the field derivative (first line), the DtN operator (second line), and the continuity of the field (third line). If the source field is a plane wave coming from the right, this yields, with ∂_n the normal derivative with respect to the boundary, $\partial_n u - iku = -2iku_{\text{source}}$. In contrast, for an outgoing field, that is for a plane wave travelling to the right, we find $\partial_n u - iku = 0$. Here, u is taken at the boundary $x = L$. The same holds true for the left boundary; hence, we have at the boundaries $\Gamma = \{0, L\}$

$$\partial_n u\Big|_{x \in \Gamma} = \left(\partial_n u_{in} - iku_{in} + iku\right)_{x \in \Gamma}.$$

We plug this into (8.26) to obtain

$$\int_0^L \partial_x v^* \partial_x u - v^*(L)\partial_n u\Big|_L - v^*(0)\partial_n u\Big|_0 - \int_0^L k^2 v^* u = 0.$$

$$\int_0^L \partial_x v^* \partial_x u - \int_0^L k^2 v^* u = v^*(L)\partial_n u\Big|_L + v^*(0)\partial_n u\Big|_0.$$

Finally, we find

$$\int_0^L \partial_x v^* \partial_x u - \int_0^L k^2 v^* u - ikv^*(L)u(L) - ikv^*(0)u(0)$$

$$= v^*(L)\left(\partial_n u_{in} - iku_{in}\right)_{x=L} + v^*(0)\left(\partial_n u_{in} - iku_{in}\right)_{x=0}.$$

Again we introduce a sesquilinear form $a(v, u)$ and an anti-linear form $b(v)$ to abbreviate the last equation which yields the 1D scattering problem with the DtN operator as a boundary condition.

Scattering Problem in 1D with DtN Operator

Find $u \in V$: $a(v, u) = b(v)$ for all $v \in V$,

with a sesquilinear form $a(v, u)$ given by (8.27) and anti-linear form $b(v)$ given by (8.28)

$$a(v, u) = \int_0^L \partial_x v^* \partial_x u - \int_0^L k^2 v^* u - ikv^*(L)u(L) - ikv^*(0)u(0) \qquad (8.27)$$

$$b(v) = v^*(L)\left(\partial_n u_{\text{source}} - iku_{\text{source}}\right)_{x=L} + v^*(0)\left(\partial_n u_{\text{source}} - iku_{\text{source}}\right)_{x=0}. \qquad (8.28)$$

8.2.2   VARIATIONAL FORMULATION IN 1D WITH PERFECTLY MATCHED LAYERS

The task of the general radiation condition (8.25) or its 1D specialization (8.24) is to select the outgoing part of the scattered field. In the previous section we employed the DtN-operator for that purpose since it is the easiest way to select the right scattering field in 1D. An alternative way is the use of the perfectly matched layer (PML) technique. It has two remarkable advantages:

1. It can be used in a similar way in higher space dimensions.
2. It can be plugged into existing codes provided that the codes allow for sufficient general materials and right-hand sides.

The idea in 1D is the following. Let us consider the right exterior domain $x > L$. The general solution of the scattered field is

$$u_s = c_- e^{-ikx} + c_+ e^{ikx}, \quad x > L.$$

We have to discriminate between the incoming version of the scattered field $c_- \exp(-ikx)$, which we need to drop, and the outgoing scattered field $c_+ \exp(ikx)$. Here, x is the real-space coordinate. Now we allow x to be a path in the complex plane, not just on the real axis. We choose the path

$$x(\tau) := L + (1 + i\sigma)\tau$$

$$\sigma > 0, \text{ real, fixed}$$

$$\tau > 0, \text{ real, path parameter,}$$

which yields the following mapping

$$c_- \exp(-ikx) \mapsto c_- \exp(-ikL) \exp(-ik\tau) \exp(\sigma k\tau)$$

$$c_+ \exp(ikx) \mapsto c_+ \exp(ikL) \exp(ik\tau) \exp(-\sigma k\tau).$$

The first expression, the incoming field, blows up for $\tau \to \infty$, due to the factor $\exp(\sigma k \tau)$, whereas the second one decreases exponentially with $\exp(-\sigma k \tau)$. So for τ large enough, we have a clear discrimination: nearly the entire field value comes from the left travelling scattered field. Hence, a proper formulation of the Sommerfeld radiation condition in terms of the complexified path $x(\tau)$ is

$$u(x(\tau)) \longrightarrow 0 \text{ as } \tau \longrightarrow \infty.$$

Let us stay in the right exterior domain and introduce the complex distance variable z

$$z(\tau) = \tau + i\sigma\tau.$$

We consider the complexification of the space variable within the Helmholtz equation:

$$\partial_x^2 u + k^2 u = 0$$

$$\downarrow$$

$$\partial_z^2 u + k^2 u = 0$$

$$\downarrow$$

$$\frac{1}{(1+i\sigma)^2} \frac{\partial^2 u}{\partial \tau^2} + k^2 u = 0.$$

The first step is decisive. If we are allowed to generalize the real derivative to a complex derivative, that is if the derivative is independent of the direction in the complex plane, we can use the linear path transformation (or others) to obtain the last equation. After this preparation, we return to the variational formulation of the scattering problem.

The description in the interior domain remains as mentioned earlier:

$$x \in \Omega_{\text{int}} : \quad \int_0^L \partial_x v^* \partial_x u - v^* \partial_x u \Big|_0^L - \int_0^L k^2 v^* u = 0. \tag{8.29}$$

The scattering in the right exterior domain $\Omega_{\text{ext}}, \Re x > L$, with complexified path, becomes

$$\int_L^\infty d\tau \frac{(1+i\sigma)}{(1+i\sigma)^2} \partial_\tau v^* \partial_\tau u_s - \frac{v^*}{1+i\sigma} \partial_\tau u_s \Big|_L^\infty - \int_L^\infty d\tau \, (1+i\sigma) k^2 v^* u_s = 0$$

and in the left exterior domain $\Omega_{\text{ext}}, \Re x < 0$, also with complexified path,

$$\int_{-\infty}^0 d\tau \frac{(1+i\sigma)}{(1+i\sigma)^2} \partial_\tau v^* \partial_\tau u_s - \frac{v^*}{1+i\sigma} \partial_\tau u_s \Big|_{-\infty}^0 - \int_{-\infty}^0 d\tau \, (1+i\sigma) k^2 v^* u_s = 0.$$

Continuity at the right boundary requires

$$u(L) = u_{\text{source}}(L) + u_s(L). \tag{8.30}$$

$$\partial_x u|_{x=L} = \partial_x u_{\text{source}}|_{x=L} + \frac{1}{1+i\sigma} \, \partial_\tau u_s|_{x=L}. \tag{8.31}$$

We rewrite (8.29) based on these conditions

$$\int_0^L \partial_x v^* \partial_x u - v^* \partial_x u_s \Big|_0^L - \int_0^L k^2 v^* u = v^* \partial_x u_{\text{source}} \Big|_0^L$$

together with the equations of the scattered field

$$\int_L^\infty d\tau \frac{(1+i\sigma)}{(1+i\sigma)^2} \partial_\tau v^* \partial_\tau u_s - \frac{v^*}{1+i\sigma} \partial_\tau u_s \Big|_L^\infty - \int_L^\infty (1+i\sigma) k^2 v^* u_s = 0$$

$$\int_{-\infty}^0 d\tau \frac{(1+i\sigma)}{(1+i\sigma)^2} \partial_\tau v^* \partial_\tau u_s - \frac{v^*}{1+i\sigma} \partial_\tau u_s \Big|_{-\infty}^0 - \int_{-\infty}^0 (1+i\sigma) k^2 v^* u_s = 0.$$

We add these three equations and all boundary terms cancel out. This is due to the directional independence of $\partial_x u_s = \frac{1}{1+i\sigma} \partial_\tau u_s$ and the vanishing scattering terms at infinity. We find

$$\int_0^L \partial_x v^* \partial_x u - \int_0^L k^2 v^* u + \int_L^\infty d\tau \frac{(1+i\sigma)}{(1+i\sigma)^2} \partial_\tau v^* \partial_\tau u_s - \int_L^\infty (1+i\sigma) k^2 v^* u_s$$

$$+ \int_{-\infty}^0 d\tau \frac{(1+i\sigma)}{(1+i\sigma)^2} \partial_\tau v^* \partial_\tau u_s - \int_{-\infty}^0 (1+i\sigma) k^2 v^* u_s = v^* \partial_x u_{\text{source}} \Big|_0^L. \tag{8.32}$$

To simplify the notation, we introduce two sesquilinear forms, one for the interior domain and one for the exterior domain:

$$a_{\text{int}}(v, u) := \int_0^L \partial_x v^* \partial_x u - \int_0^L k^2 v^* u$$

$$a_{\text{ext}}(v, u_s) := \int_L^\infty d\tau \frac{(1+i\sigma)}{(1+i\sigma)^2} \partial_\tau v^* \partial_\tau u_s - \int_L^\infty d\tau (1+i\sigma) k^2 v^* u_s$$

$$+ \int_{-\infty}^0 d\tau \frac{(1+i\sigma)}{(1+i\sigma)^2} \partial_\tau v^* \partial_\tau u_s - \int_{-\infty}^0 d\tau (1+i\sigma) k^2 v^* u_s.$$

Hence, (8.32) reads

$$a_{\text{int}}(v, u) + a_{\text{ext}}(v, u_s) = v^* \partial_x u_{\text{source}} \Big|_0^L. \tag{8.33}$$

It is tempting to construct an integral expression which goes from $-\infty$ to ∞ but note that the interior field u and the scattered fields u_s jump across the boundaries.

8.2.2.1 Completion to a Continuous Function

Now we use the same trick as used in (8.11) transforming Dirichlet data to zero boundary data. We introduce an auxiliary function g with

$$u_s = \underbrace{u_s + g}_{w} - g$$

$$g(L) = u_{\text{source}}(L)$$

$$g(\infty) = 0 \tag{8.34}$$

and a function $w := u_s + g$ living in the exterior domains and being continuous with the function u in the interior domain. Therefore, (8.33) changes to

$$a_{\text{int}}(v, u) + a_{\text{ext}}(v, w) = v^* \partial_x u_{\text{source}} \Big|_0^L + a_{\text{ext}}(v, g). \tag{8.35}$$

Since u lives in the interior domain and w in the exterior domain, we may rename w to u for $x \in \mathbb{R} \setminus [0, L]$ and define a single sesquilinear form

$$a(v, u) := a_{\text{int}}(v, u) + a_{\text{ext}}(v, u).$$

As the only difference in the definition of $a_{\text{int}}(v, u)$ and $a_{\text{ext}}(v, u)$ is the chosen spatial (real or complex) path, we may use the equivalent definition

$$a(v, u) := \int_\gamma \partial_z v^* \partial_z u \, dz - \int_\gamma k^2 v^* u \, dz, \tag{8.36}$$

(see Figure 8.7 for a path $\gamma(\tau)$).

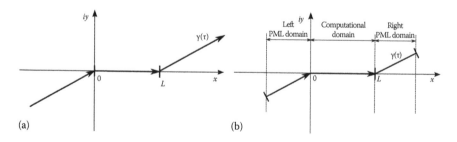

FIGURE 8.7 Complex path defining the PML. (a) PML on the unbounded domain. (b) Truncation to PML layers with finite thickness.

Scattering Problem in 1D with PML

Let the source field data $u_{source}(x)$ and $\partial_n u_{source}(x)$ at the boundary $x \in \{0, L\}$ be given. The variational form of the 1D Helmholtz equation reads as follows:

Find $u \in V : a(v, u) = v^* \, \partial_n u_{source}|_{x=0} + v^* \, \partial_n u_{source}|_{x=L} + a_{ext}(v, g)$ for all $v \in V$

with

$$a(v, u) = \int_\gamma \partial_z v^* \partial_z u \, dz - \int_\gamma k^2 v^* u \, dz$$

$$a_{ext}(v, u) = \int_{\gamma_{ext}} \partial_z v^* \partial_z u \, dz - \int_{\gamma_{ext}} k^2 v^* u \, dz$$

where $a_{ext}(v, g)$ is computed only on the exterior part of the path γ and g is taken from (8.34).

8.2.3   Discretization

The discretization of both kinds of variational equations, the type with DtN operator and the type with PML, follows the same scheme as applied for the semi-transparent Helmholtz equation in Section 8.1.

First, we discretize the computational domain. In the case of the formulation involving the DtN operator as transparent boundary condition, we just have to decompose the interval $(0, L)$ into subintervals as in Section 8.1. In the case of the formulation with the PML as non-reflecting layer, we first have to truncate the semi-infinite exterior domains (see Figure 8.7), and second, we have to decompose the truncated domain into intervals. In both cases, we now step segment-wise through the computational domain (plus eventually the PML layers) and compute the contribution of each interval by recomputing the matrix entries in the form of local matrices on a unit interval and transform the result to the physical domain. In particular, the PML can be embedded in a natural way into this scheme. Figure 8.8 illustrates the transformation of the real unit interval to the complex PML interval as it is needed to assemble the global matrices.

This transformation is what we in fact exploit within the algorithmic realization. We define the complexification of the real x-coordinate to the complex coordinate x^c by

$$x^c = \begin{cases} x(1 + i\sigma), & x < 0 \\ x, & 0 \le x < L \\ L + (x - L)(1 + i\sigma), & x \ge L, \end{cases} \tag{8.37}$$

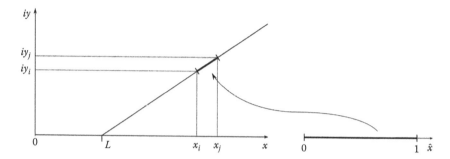

FIGURE 8.8 Transformation from the unit interval to the complex-valued PML interval.

which yields the differentials

$$dx^c = \begin{cases} dx, & 0 < x < L \\ (1 + i\sigma)\, dx, & x \notin (0, L). \end{cases} \tag{8.38}$$

In particular, the unit interval $(0, 1)$ in the reference domain maps to the complexified physical interval by

$$(0, 1) \mapsto \begin{cases} x_j - x_i, & 0 < x < L \\ (1 + i\sigma)\,(x_j - x_i), & x \notin (0, L). \end{cases} \tag{8.39}$$

The entire scattering procedure is summarized in the pseudo-code Algorithm 8.2.

8.2.4 A POSTERIORI ERROR ESTIMATION

Figure 8.5 displays the discrete approximation of a plane wave as solution of (8.15) with $N = 10$ segments of equal length. We want to have information about the accuracy of this approximation. In Figure 8.9, we repeat the simulation with two different grids, $N = 5$ and $N = 10$, with the corresponding grids displayed at the bottom.

The goal of the a posteriori error estimation is to find reliable statements about

- The global error of an approximate solution in comparison to the exact solution
- The local contributions to the global error in order to control adaptive mesh refinement

There are a number of different ways to derive an a posteriori error estimation. In their core, these are different ways to obtain information, if possible in a cheap way, that have higher accuracy than the approximate solution at hand. We sketch here an approach called *goal-oriented* error estimation (cf. [8,11]), which is based on the pioneering work of Rannacher and coworkers [1]. Goal orientation means that we

---

Algorithm 8.2 Scattering in 1D (PML)

Input: grid $\{x_1, \ldots, x_N\}$
   $u_{\text{source}}(0), \partial_n u_{\text{source}}(0), u_{\text{source}}(L), \partial_n u_{\text{source}}(L)$

$S = 0; \ M = 0; \ SB = 0; \ MB = 0$
$g = 0$
compute list of all segments $L = \{(x_i, x_j), \ldots\}$

$$S_{\text{loc}} = \begin{pmatrix} 1 & -1 \\ -1 & 1 \end{pmatrix}; \quad M_{\text{loc}} = \begin{pmatrix} \dfrac{1}{3} & \dfrac{1}{6} \\ \dfrac{1}{6} & \dfrac{1}{3} \end{pmatrix}$$

for $l \in L$ **do**
   $P = 0$ % reset permutation matrix, see (8.19)
   $P(1, i) = 1$ % assign local index 1 to global index i
   $P(2, j) = 1$ % assign local index 2 to global index j

$$h = \begin{cases} (x_j - x_i)(1 + i\sigma) & \text{if } (x_i, x_j) \text{ within the PML layers} \\ (x_j - x_i) & \text{if } 0 \le (x_i, x_j) \le L \end{cases}$$

$$S = S + P^T \frac{1}{h} S_{\text{loc}} P; \quad M = M + P^T h k^2 M_{\text{loc}} P$$

if (x_i, x_j) within the PML layers **then**

$$SB = SB + P^T \frac{1}{h} S_{\text{loc}} P; \quad MB = MB + P^T h k^2 M_{\text{loc}} P$$

   end if
end for

$$A = S - M$$
$$AB = SB - MB$$

Set $g = u_{\text{source}}$ at the two boundary vertices, zero elsewhere
$b = \partial_n u_{\text{source}}(0) + \partial_n u_{\text{source}}(L) + AB \cdot g$
$u = A \backslash b$
$u = u + g$

---

first define a quantity of interest, which we want to compute to a sufficient accuracy. For a general formulation of the approach, it is convenient to define the quantity of interest in terms of a functional J

$$J : V \to \mathbb{C}$$

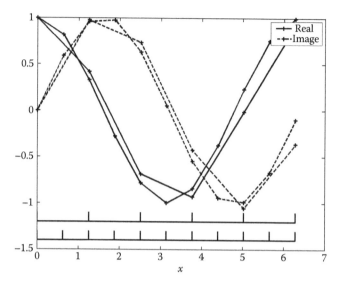

FIGURE 8.9 Plane-wave approximation u_H with $N = 5$ and u_h with $N = 10$ segments.

mapping the space of trial functions in a certain way to complex numbers. The example for a functional we want to consider here is

$$J(v) = v(L), \tag{8.40}$$

hence, the functional returns the value of the function v at position $x = L$.

Further, it simplifies the notation, especially when switching between the discrete and the continuous problem, if we denote the functional evaluated for a given function v by angle brackets (the ':=' means *is defined by*):

$$\langle J, v \rangle := J(v).$$

In the discrete case, the bracket notation can be identified with the Euclidean scalar product

$$\langle a, b \rangle := \sum_{i=1}^{N} a_i^* b_i.$$

Using this notion, the discrete variational problem (8.15) reads as follows:

$$\text{Find } u_h \in V: \quad \langle v_h, A_h u_h \rangle = -\langle v_h, A_h g \rangle \quad \text{for all } v_h \in V_h. \tag{8.41}$$

Together with the discrete variational problem (8.41), we consider the discrete dual problem:

$$\text{Find } z_h \in V: \quad \langle z_h, A_h v_h \rangle = \langle J, v_h \rangle \quad \text{for all } v_h \in V_h, \tag{8.42}$$

which has a structure similar to the primal problem (8.41) except that the test functions appear on the right-hand side of the inner product and that the right-hand side of the equation is somewhat arbitrarily constructed by the functional J. In order to elucidate the meaning of (8.42), we consider its continuous counterpart:

$$\text{Find } z \in V : a\,(z, v) = \langle J, v \rangle \quad \text{ for all } v \in V,$$

or using the definition of $a\,(z, v)$,

Find $z \in V$:

$$\int_0^L \partial_x z^* \partial_x v - \int_0^L k^2 z^* v - ikz^*(L)v(L) - ikz^*(0)v(0) = \langle J, v \rangle \quad \text{ for all } v \in V.$$

Let us undo the integration by parts

$$-\int_0^L \partial_x^2 z^* v + v\partial_x z^* |_0^L - \int_0^L k^2 z^* v - ikz^*(L)v(L) - ikz^*(0)v(0) = \langle J, v \rangle .$$

Collecting the corresponding boundary terms yields

$$-\int_0^L \partial_x^2 z^* v - \int_0^L k^2 z^* v + v(L)\,(\partial_x z^*(L) - ikz^*(L)) - v(0)\,(\partial_x z^*(0) + ikz^*(0)) = \langle J, v \rangle .$$

The continuous form of the functional J, which should return the value at position L of its argument function, is

$$\langle J, v \rangle = \int_0^L \delta\,(x - L)\, v dx.$$

If we choose the boundary conditions

$$\partial_n z^*(L) = ikz^*(L)$$
$$\partial_n z^*(0) = -ikz^*(0)$$

and compare the functions under the integral, we find

$$\partial_x^2 z + k^2 z = -\delta\,(x - L) .$$

Hence, the solution of the dual equation is *Green's function*

$$z(x) = \frac{e^{-ik|x-L|}}{2ik}$$

obeying the boundary conditions for incoming fields

$$\partial_n z(L) = -ikz(L)$$
$$\partial_n z(0) = ikz(0).$$

For the purpose of this introduction, it is sufficient to compare two different discrete solutions: a coarse function $u_H \in V_H$ defined on a coarse grid and a function $u_h \in V_h$ defined on a fine grid, where the finer grid originates from the coarser one by a subdivision of its segments. Hence, for the finite element spaces, it holds that $V_H \subset V_h$. We consider the coarse grid solution u_H as given and ask which segments should be refined in order to provide a good error reduction in terms of the functional (8.40) without refining all segments. Suppose the fine grid solution would also be available. We subtract the dual equation (8.42) computed for the fine grid solution from the one computed for the coarse grid solution and obtain

$$\langle z_h, A_h u_h - A_h u_H \rangle = \langle J, u_h \rangle - \langle J, u_H \rangle.$$

Here, the quantity

$$\langle v_h, A_h u_h - A_h u_H \rangle =: \rho(v_h)$$

is called the residual functional (of the coarse grid solution with respect to the fine grid solution) which we also denote as

$$\rho(v_h) =: \langle v_h, \rho \rangle.$$

The expression

$$\langle z_h, \rho \rangle = \langle J, u_h \rangle - \langle J, u_H \rangle \tag{8.43}$$

tells us that the difference in the functional value, that is $u_h(\text{end}) - u_H(\text{end})$, can be computed via the inner product of the fine grid dual solution and the residual. This is the core insight used for the construction of local error estimators.

8.2.4.1 Galerkin Orthogonality

We want to extend result (8.43) in a certain way. Suppose a fine grid solution u_h with respect to an arbitrary right-hand side f has been computed

$$\langle v_h, A_h u_h \rangle = \langle v_h, f \rangle \quad \text{for all } v_h \in V_h. \tag{8.44}$$

Since $V_H \subset V_h$, the following is a special case of (8.44):

$$\langle v_H, A_h u_h \rangle = \langle v_H, f \rangle.$$

On the other hand, the coarse grid solution follows from

$$\langle v_H, A_h u_H \rangle = \langle v_H, f \rangle \quad \text{for all } v_H \in V_H.$$

Subtraction of the last two equations yields the so-called Galerkin orthogonality

$$\langle v_H, A_h (u_h - u_H) \rangle = 0 \quad \text{for all } v_H \in V_H \tag{8.45}$$

which says that the difference between the fine and the coarse solution is A-orthogonal to each element of the coarse space. Using this property, we generalize (8.43) by subtracting an arbitrary element v_H from z_h to get the final global result

$$\langle z_h - v_H, \rho \rangle = \langle J, u_h \rangle - \langle J, u_H \rangle. \tag{8.46}$$

8.2.4.2 A Different Viewpoint

We summarize the results so far from a slightly different viewpoint. From the Galerkin orthogonality (8.45), we know

$$\langle v_H, A_h (u_h - u_H) \rangle = 0$$

for all v_H defined on the coarse grid. If we take the test function from the larger space V_h defined on the fine grid, we have in general a non-zero residual $\rho(v_h)$:

$$\langle v_h, A_h (u_h - u_H) \rangle = \rho(v_h).$$

For a special choice of v_h, namely $v_h = z_h$, we get a special residual, namely $\rho(z_h) = J(u_h) - J(u_H)$. We want to ensure to have exactly this special kind of residual small, not any other residual possible. Hence, the dual function z_h (up to an arbitrary shift v_H) can be seen as a weight in computing the expression $\langle z_h - v_H, A_h (u_h - u_H) \rangle$. From the deviation $u_h - u_H$ on the entire domain, only the part which projects onto z_h plays a role.

8.2.4.3 Error Localization and Error Indicator

The test function v_H in (8.46) may be chosen arbitrarily from V_H. If we want to make the expression as small as possible, we have to chose v_H close to z_h but defined in V_H. Therefore, we may use $v_H = z_H$, the dual solution on the coarse grid, and need to compute

$$\langle z_h - z_H, A_h (u_h - u_H) \rangle.$$

This, however, is a global statement, and no statements about the local contributions to the global error are given directly. To localize the contribution from each segment of the fine grid, we use the representation of the global matrix A_h as the sum of local matrices defined on each segment as shown in (8.20):

$$A_h = \sum_{i=1}^{N+1} P^T A_{h_i}^{\mathrm{loc}} P.$$

With $z_h - z_H$ and $u_h - u_H$ restricted to the segments $\{i - 1, i\}$, we obtain

$$\langle z_h - z_H, A_h (u_h - u_H)\rangle = \sum_{i=1}^{N+1} \langle (z_h - z_H)_{\{i-1,i\}}, A^{\mathrm{loc}}_{h\{i-1,i\}} (u_h - u_H)_{\{i-1,i\}}\rangle.$$

Here, we can identify the contribution of each fine grid segment

$$\eta_{\{i-1,i\}} = \left|\langle (z_h - z_H)_{\{i-1,i\}}, A^{\mathrm{loc}}_{h\{i-1,i\}} (u_h - u_H)_{\{i-1,i\}}\rangle\right| \tag{8.47}$$

and the sum of two neighbouring fine grid segments gives the indicator we need to control the adaptive mesh refinement. An algorithmic realization of this error localization is given in Algorithm 8.4. The procedure *local_assemble* returns the stiffness and mass matrices in localized, which means unconnected, form and a matrix I which condenses the local matrices to the desired global matrices, with N as the number of segments:

$$\text{Sparsity of } \left(S_{\mathrm{loc}} \text{ or } k^2 M_{\mathrm{loc}}\right) = \begin{pmatrix} \blacksquare & \blacksquare & & & & \\ \blacksquare & \blacksquare & & & & \\ & & \blacksquare & \blacksquare & & \\ & & \blacksquare & \blacksquare & & \\ & & & & \ddots & \\ & & & & & \blacksquare & \blacksquare \\ & & & & & \blacksquare & \blacksquare \end{pmatrix} \in \mathbb{C}^{2N \times 2N}$$

and with

$$I = \begin{pmatrix} 1 & & & \\ & 1 & & \\ & 1 & 1 & \\ & & 1 & \ddots \\ & & & & 1 \\ & & & & 1 \\ & & & & & 1 \end{pmatrix} \in \mathbb{Z}^{(2N-2) \times N}.$$

The condensations

$$S = I^T S_{\mathrm{loc}} I$$

$$r = I^T r_{\mathrm{loc}}$$

follow.

8.2.5 Adaptive Mesh Refinement

The computation of (8.47) is done on the fine mesh x_h, which is a uniform refinement of the coarse mesh x_H. Since we perform any computation on x_h, we need a nodal representation of a coarse grid function v_H on the fine mesh x_h, that is a *prolongation (interpolation) operator P. P* transfers nodal values from the coarse to the fine mesh via linear interpolation.

Algorithm 8.3 Adaptive Semi-Transparent Plane-Wave Propagation in 1D

Input: initial mesh $x = \{x_0 \ldots x_n\}$

%compute "exact" (up to TOL) functional value J_{ex}
$J_{ex} = J_{exact}(x, TOL)$

$[A, rhs]$ = assemble(x)
$u = A \backslash rhs$
Jf = goal_functional(x)   %functional for the current mesh
$J = Jf \cdot u$ %functional value for the current solution
while $|J - J_{ex}| <$ eps **do**
    x =refine_mesh(x, u)   %refine mesh (see Algorithm 8.4)
    $[A, rhs]$ = assemble(x)   %assemble new system
    $u = A \backslash rhs$ %new solution
    Jf =goal_functional(x)   %new functional
    $J = Jf \cdot u$ %new functional value
end while

$$P = \begin{pmatrix} 1 & & & & & \\ \frac{1}{2} & \frac{1}{2} & & & & \\ & 1 & & & & \\ & \frac{1}{2} & \frac{1}{2} & & & \\ & & 2 & 2 & & \\ & & & \ddots & & \\ & & & & 1 & \\ & & & & \frac{1}{2} & \frac{1}{2} \\ & & & & 2 & 2 \\ & & & & & 1 \end{pmatrix}.$$

The entire procedure for an adaptive grid refinement is sketched in the pseudo-codes (Algorithms 8.3 and 8.4).

Example: Transmission through a Metallic Layer

We consider the transmission through a layer of thickness $\lambda/4$ with refractive index squared $n^2 = 1 + 8i$ embedded in air. The results of the computation are shown in Figure 8.10. The metallic layer causes a damping of the field; hence, the accuracy of the solution at the end of the domain at $x = 2\pi$ depends essentially on the accuracy of the field computation inside the metallic layer. This is reflected by the higher density of plus signs there.

Note: The adaptive strategy has the benefit of the same accuracy for fewer unknowns required compared to a uniform discretization. On the other hand, adaptive techniques have certain computational overheads. In typical 1D cases, adaptivity

---

Algorithm 8.4 Adaptive Mesh Refinement in 1D

---

function $x_r = $ **refine_mesh**(x, u)
%Coarse grid quantities
$[A_H, rhs] = $ assemble(x)
$J_{fH} = $ goal_functional(x)
$z_H = A_H^H \backslash J_{fH}^H$

%Compute fine grid quantities
$P = $ prolong(length(x))   %prolongation matrix
$x_h = $ refine_uniformly(x)
$[A_h, rhs] = $ assemble$(x_h,)$
$J_{fh} = $ goal_functional(x)
$z_h = A_h^H \backslash J_{fh}^H$
$u_h = A_h \backslash rhs$

$[A, rhs, I] = $ assemble_local(x_h) %local assembly
$rho = (A * I * (P * u_H - u_h))$ %elementwise residal
$ind = $ indicator$(zh - P * zH, rho, P, I)$ %interval indicator
$x_r = $ refine_adaptively(x, ind) %refine as indicated
return x_r

---

does not yield a net advantage in computation time. In higher dimensions, this may change drastically. In examples with localized phenomena, like field singularities or plasmonic fields, adaptivity guarantees effective meshes and usually much shorter computation times compared to uniform approaches.

8.2.6 FEM NOTIONS: ELEMENT SUPPORT, BASIS FUNCTIONS, SHAPE FUNCTIONS, FINITE ELEMENTS, AND FINITE ELEMENT SPACES

For the understanding of the 1D FEM treatment of scattering problems, it is not necessary to provide the typical FEM notions right from the beginning. However, introducing them helps to prepare the transition to spatial dimensions 2 and 3. Figure 8.11, referring to Figure 8.3, shows again our 1D situation with linear finite elements.

Element support: The computational domain is geometrically decomposed into simple subdomains, in our case into line segments. These elementary patches are typically denoted as K or K_j.

Basis functions: The basis functions are often denoted as $\phi_i(x)$. They exist on the entire computational domain and a weighted superposition of them gives the finite element solution $u_h(x)$. Usually, the basis function has to fulfil certain continuity conditions. In our example case, we only require them to be continuous.

Shape functions: The part of a given basis function $\phi(x)$ living on a patch K is called a shape function $N(x)$. The shape function exists only on K. In Figure 8.11, the patch K knows two shape functions: $N_1(x)$ and $N_2(x)$.

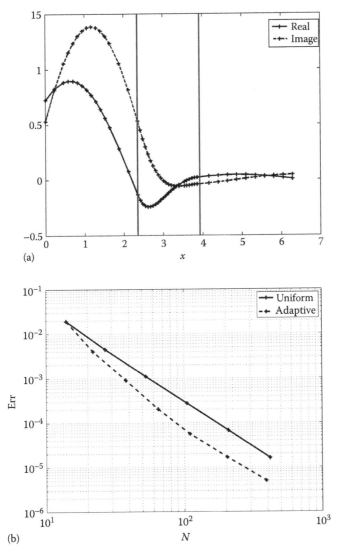

FIGURE 8.10 Adaptively discretized field transmission through a metallic layer. (a) Field amplitude. The plus signs indicate the discretization and the vertical lines the position of the metallic layer. (b) Convergence of the uniform and the adaptive discretization. N is the number of unknowns and err $= |u(2\pi) - u_h(2\pi)|$.

Finite elements: Usually, we consider the patch K together with its shape functions as a finite element. In a slightly more general definition, we consider a finite element as a patch K plus the polynomial approximation space related to K (in our case so far, the space of linear polynomials) plus the so-called DOFs, which are recipes telling us how to construct the shape functions.

Finite element spaces: We used V to denote the space which contains the exact solution u and V_h to denote the discrete finite element space which contains

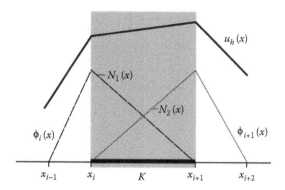

FIGURE 8.11 Global basis functions $\phi_i(x)$, $\phi_{i+1}(x)$, local finite element support K, and the local shape functions $N_1(x)$, $N_2(x)$ as projection of the global basis functions to K; compare Figure 8.3.

the discrete solution u_h. In the Helmholtz case, typically the so-called Sobolev spaces $H^1(\Omega)$ and $H_0^1(\Omega)$ are used, where the superscript 1 refers to the smoothness of the functions (one time weak differentiable) and the subscript 0 to zero boundary conditions. In the Maxwell's case, the spaces $H(\mathbf{curl}, \Omega)$ are used. For the understanding of the algorithmic ideas, the corresponding definitions are not important. Again we refer to Ref. [5].

8.3 MATHEMATICAL BACKGROUND: MAXWELL AND HELMHOLTZ SCATTERING PROBLEMS AND THEIR VARIATIONAL FORMS

Whereas many of the other methods described in this book could be derived directly from Maxwell's equations, the application of the FEM requires some preparation. We have to transform the differential form of Maxwell's equations into a variational form and to find a way that allows the inclusion of source fields, interior fields, and scattered fields into the variational framework. In the following section, we generalize the 1D variational form of scattering problems discussed in Section 8.2 to 2D and 3D problems.

8.3.1 Maxwell's Scattering Problem

Figure 8.12 shows the situation of a general scattering problem. This is the analogue of Figure 8.6 in the 1D case. We have a source field $\mathbf{E}_{\text{source}}$ travelling through the unbounded space, being itself a solution to Maxwell's equations. If no scatterer would be present, this would be the only field we have. But if it hits a scattering object, some parts of the field will be refracted into the object (in case it has a finite refractive index), some parts will be diffracted around the object, and some parts will be reflected. The refracted and diffracted field, that is the field outside the scatterer different from the source field, is called the scattered field $\mathbf{E}_{\text{scattered}}$ or in short $\mathbf{E}_s$. We want to compute the field inside a computational domain Ω which we also call the

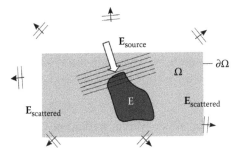

FIGURE 8.12 Scattering problem in 2D and 3D.

interior domain, in contrast to the exterior domain surrounding Ω. Further, we denote the boundary of the computational domain with $\partial\Omega$.

We formalize the situation as follows. Let the geometry of the scattering problem be given by a position-dependent permittivity $\varepsilon(\mathbf{r})$ defined in the entire real, 3D space, that is $\mathbf{r} \in \mathbb{R}^3$. Further, let us know the source field $\mathbf{E}_{source}(\mathbf{r})$, with $\mathbf{r}$ in the exterior domain, $\mathbf{r} \in \mathbb{R}^3 \backslash \bar{\Omega}$. Then the general scattering problem in 3D is defined by the following four conditions:

1. Interior problem

$$\frac{1}{\varepsilon}\nabla \times \frac{1}{\mu}\nabla \times \mathbf{E} - \omega^2\mathbf{E} = 0 \quad \text{for } \mathbf{r} \in \Omega.$$

General Maxwell's equations in the interior domain has to be solved.

2. Exterior problem

$$\left.\begin{array}{l} \dfrac{1}{\varepsilon}\nabla \times \dfrac{1}{\mu}\nabla \times \mathbf{E}_{source} - \omega^2\mathbf{E}_{source} = 0 \\[2ex] \dfrac{1}{\varepsilon}\nabla \times \dfrac{1}{\mu}\nabla \times \mathbf{E}_s - \omega^2\mathbf{E}_s = 0 \end{array}\right\} \quad \text{for } \mathbf{r} \in \mathbb{R}^3 \backslash \bar{\Omega}.$$

The fields in the exterior domain are a superposition of the source and the scattered fields $\mathbf{E}_{source}$ and $\mathbf{E}_s$. The source field and the scattered field must obey Maxwell's equations.

3. Continuity of the tangential data along the boundary of the computational domain

$$(\mathbf{E} - (\mathbf{E}_{source} + \mathbf{E}_s)) \times \mathbf{n} = 0 \quad \text{for } \mathbf{r} \in \partial\Omega$$

$$(\nabla \times (\mathbf{E} - (\mathbf{E}_{source} + \mathbf{E}_s))) \times \mathbf{n} = 0 \quad \text{for } \mathbf{r} \in \partial\Omega.$$

4. The Silver–Müller radiation condition for the scattered field

$$\lim_{r \to \infty} r\left((\nabla \times \mathbf{E}_s) \times \mathbf{r}^0 - i|\mathbf{k}|\mathbf{E}_s\right) = 0 \tag{8.48}$$

holds uniformly in all directions $\mathbf{r}$, where $\mathbf{r}^0$ denotes the radial vector of unit length.

Note: Unlike the other chapters, we reserve in the FEM chapter the hat symbol $\frown$ on top of a quantity exclusively to identify quantities that belong to a reference coordinate system.

8.3.1.1 Discussion of the Silver–Müller Radiation Condition

The radiation condition is needed to define a uniquely solvable scattering problem. The following discussion illuminates its role and also provides an understanding of the PML in Maxwell's case. First, the Silver–Müller condition is an asymptotic statement, characterizing the field if we move towards infinity. Suppose we have a local field approximation like

$$\mathbf{E}_s(\mathbf{r}) = \mathbf{E}_0(\mathbf{r})e^{ikr}, \tag{8.49}$$

which should approximate the field in a small sphere far away from the origin, where $\mathbf{E}_0(\mathbf{r})$ decays like $1/r$ but is supposed to be almost constant within this small sphere. If we replace $\mathbf{E}_s$ in (8.48) with (8.49), we get

$$r\left((\nabla \times \mathbf{E}_s) \times \mathbf{r}^0 - i\,|\mathbf{k}|\,\mathbf{E}_s\right) \xrightarrow{r \to \infty} 0$$

$$r\left((i\mathbf{k} \times \mathbf{E}_0) \times \mathbf{r}^0 - i\,|\mathbf{k}|\,\mathbf{E}_0\right) \xrightarrow{r \to \infty} 0$$

$$r\left(\mathbf{k}\left(\mathbf{r}^0 \cdot \mathbf{E}_0\right) - \mathbf{E}_0\left(\mathbf{k} \cdot \mathbf{r}^0\right) + |\mathbf{k}|\,\mathbf{E}_0\right) \xrightarrow{r \to \infty} 0$$

$$r\left((\mathbf{k} \cdot \mathbf{r}^0 - |\mathbf{k}|)\,\mathbf{E}_0 - \mathbf{k}\left(\mathbf{r}^0 \cdot \mathbf{E}_0\right)\right) \xrightarrow{r \to \infty} 0,$$

where we neglected $\nabla \times \mathbf{E}_0$ compared to $\mathbf{E}_0$ (almost constant) in the second line. From the last line, we conclude the following:

1. Since $\mathbf{E}_0 = \mathcal{O}\left(1/r\right)$, the expression in the first parenthesis decays faster than $1/r$; hence,

$$\mathbf{k} \cdot \mathbf{r}^0 - |\mathbf{k}| = o\left(1/r\right),$$

that is the wave propagates for large distances *radially*, $\mathbf{k} \parallel \mathbf{r}^0$ as $r \to \infty$.
2. Additionally, the second term has to vanish at the same rate; hence,

$$\mathbf{r}^0 \cdot \mathbf{E}_0 = o\left(1/r\right),$$

that is the field approaches asymptotically a transversal plane wave.

Once we see that the scattered part of the field $\mathbf{E}_s$ approaches asymptotically a local plane wave with phase velocity directed away from the scatterer, we can apply the PML approach via complex stretching as introduced in Section 8.2.2, since the entire discussion of mapping a propagating plane wave to an exponentially damped plane wave can be repeated.

8.3.2 SLIGHT SIMPLIFICATION: THE HELMHOLTZ SCATTERING PROBLEM

Due to its scalar nature, it is sometimes easier to study the Helmholtz scattering problem instead of Maxwell's scattering problem. We start from the Helmholtz equation for the electric field (2.51).

For continuity with the previous sections, we now return to denoting u instead of E and use the definition of the squared wavenumber:

$$k^2 := \omega^2 \varepsilon \mu. \tag{8.50}$$

Independently of the dimension and of the special field considered, one often denotes the Helmholtz equation as

$$\Delta u + k^2 u = 0. \tag{8.51}$$

In parallel to the definition of Maxwell's scattering problem in Section 8.3.1, we state the Helmholtz scattering problem in 3D:

1. Interior problem

$$\Delta u + k^2 u = 0 \quad \text{for } \mathbf{r} \in \Omega.$$

The Helmholtz equation in the interior domain has to be solved.
2. Exterior problem

$$\left. \begin{aligned} \Delta u_{\text{source}} + k^2 u_{\text{source}} &= 0 \\ \Delta u_{\text{s}} + k^2 u_{\text{s}} &= 0. \end{aligned} \right\} \quad \text{for } \mathbf{r} \in \mathbb{R}^3 \backslash \bar{\Omega}.$$

The fields in the exterior domain are a superposition of the source and the scattered fields u_{source} and u_{s}. The source field and the scattered field must obey the Helmholtz equation.
3. Continuity of the data and their normal derivative along the boundary of the computational domain

$$u - (u_{\text{source}} + u_{\text{s}}) = 0 \quad \text{for } \mathbf{r} \in \partial \Omega$$
$$(\nabla (u - (u_{\text{source}} + u_{\text{s}}))) \cdot \mathbf{n} = 0 \quad \text{for } \mathbf{r} \in \partial \Omega.$$

4. The Sommerfeld radiation condition for the scattered field

$$\lim_{r \to \infty} r (\partial_r u_{\text{s}} - i k u_{\text{s}}) = 0 \tag{8.52}$$

holds uniformly in all directions.

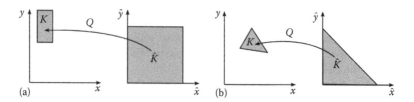

FIGURE 8.13 Two typical domain maps in 2D FEM. (a) A unit square is mapped to a rectangle. (b) The unit triangle is mapped to a general triangle.

8.3.3 TRANSFORMATION RULES

The geometric flexibility of the FEM is based in a large part on the fact that a mapping of given patches of a mesh to simply shaped reference patches greatly simplifies the entire computational process. Once we know the mapping to a standard or *reference* situation, we can reduce an essential part of the computation to the *reference* situation, which can be prepared in great detail. Figure 8.13 illustrates two typical situations in 2D, which we want to use later in the construction of the FEM algorithms. Moreover, in Section 8.2.2, Figure 8.7, we consider the PML as a linear mapping from the real physical domain to a complexified spatial domain. Hence, having transformation rules describing the change of the equations when we change local coordinate systems, we can treat the realization of FEM and PML in a unified and elegant manner.

First, we consider the mapping of the geometric quantities and use the derived rules plus suitable conservation properties to derive the transformation of the differential operators **grad**, **curl**, and **div** and subsequently the mapping of field quantities. We will see that geometric quantities and field quantities transform differently. For an easy algebraic notation, it is useful to write geometric quantities in row vectors and field quantities in column vectors (of their coefficients). Expressions like $d\mathbf{l} \cdot \mathbf{E}$ with $d\mathbf{l} = (dx, dy, dz)$ (a row vector of path components) and $\mathbf{E} = (E_x, E_y, E_z)^T$ (a column vector of field components) have then the right algebraic form and can be programmed directly as they appear in the formula.

8.3.3.1 Mapping of Geometric Quantities

Let an arbitrary domain mapping $Q : \hat{K} \rightarrow K$ given by

$$\begin{pmatrix} x \\ y \\ z \end{pmatrix} = Q\left(\hat{x}, \hat{y}, \hat{z}\right). \tag{8.53}$$

In this way, we map *points* from the reference domain to *points* in the original domain, where the points are defined by the *coefficients* $\hat{x}, \hat{y}, \hat{z}$ in the reference system and by the *coefficients* x, y, z in the original system. For our purposes, it would be more intuitive to have (8.53) in transposed form to have the result of the mapping as row vector. However, it is convention to write it in the given form. Hence, in our further elaboration, we have to transpose expressions related to Q and its derivatives.

First, we want to study how infinitesimally small path, face, and volume elements, denoted as $d\mathbf{l}$, $d\mathbf{A}$, and dV, respectively, transform. In 3D, we define $d\mathbf{l}$, $d\mathbf{A}$ by three-coefficient row vectors. A linearization of the mapping at point $(\hat{x}, \hat{y}, \hat{z})$ yields

$$\begin{pmatrix} dx \\ dy \\ dz \end{pmatrix} = \begin{pmatrix} \partial_{\hat{x}}Q_1 & \partial_{\hat{y}}Q_1 & \partial_{\hat{z}}Q_1 \\ \partial_{\hat{x}}Q_2 & \partial_{\hat{y}}Q_2 & \partial_{\hat{z}}Q_2 \\ \partial_{\hat{x}}Q_3 & \partial_{\hat{y}}Q_3 & \partial_{\hat{z}}Q_3 \end{pmatrix} \begin{pmatrix} d\hat{x} \\ d\hat{y} \\ d\hat{z} \end{pmatrix}.$$

Introducing the Jacobian

$$J(\hat{x}, \hat{y}, \hat{z}) := \begin{pmatrix} \partial_{\hat{x}}Q_1 & \partial_{\hat{y}}Q_1 & \partial_{\hat{z}}Q_1 \\ \partial_{\hat{x}}Q_2 & \partial_{\hat{y}}Q_2 & \partial_{\hat{z}}Q_2 \\ \partial_{\hat{x}}Q_3 & \partial_{\hat{y}}Q_3 & \partial_{\hat{z}}Q_3 \end{pmatrix}$$

the mapping reads

$$\begin{pmatrix} dx \\ dy \\ dz \end{pmatrix} = J(\hat{x}, \hat{y}, \hat{z}) \begin{pmatrix} d\hat{x} \\ d\hat{y} \\ d\hat{z} \end{pmatrix}. \tag{8.54}$$

Defining the path element as the row vector $d\mathbf{l} = (dx, dy, dz)$, we obtain the path element transformation

$$d\mathbf{l} = d\hat{\mathbf{l}} J^T. \tag{8.55}$$

Next, we transform the face element $d\mathbf{A} = (dA_x, dA_y, dA_z)$ spanned by two path elements 1 and 2

$$\begin{aligned} d\mathbf{A} &= \left[\begin{pmatrix} dx_1 \\ dy_1 \\ dz_1 \end{pmatrix} \times \begin{pmatrix} dx_2 \\ dy_2 \\ dz_2 \end{pmatrix} \right]^T \\ &= \left(J \begin{pmatrix} d\hat{x}_1 \\ d\hat{y}_1 \\ d\hat{z}_1 \end{pmatrix} \times J \begin{pmatrix} d\hat{x}_2 \\ d\hat{y}_2 \\ d\hat{z}_2 \end{pmatrix} \right)^T \\ &= \left(|J| J^{-T} \begin{pmatrix} d\hat{x}_1 \\ d\hat{y}_1 \\ d\hat{z}_1 \end{pmatrix} \times \begin{pmatrix} d\hat{x}_2 \\ d\hat{y}_2 \\ d\hat{z}_2 \end{pmatrix} \right)^T, \end{aligned}$$

which gives, with $|J|$ the determinant of J, the transformation rule for small faces

$$d\mathbf{A} = d\hat{\mathbf{A}} J^{-1} |J|. \tag{8.56}$$

The applied algebraic rule

$$J \begin{pmatrix} d\hat{x}_1 \\ d\hat{y}_1 \\ d\hat{z}_1 \end{pmatrix} \times J \begin{pmatrix} d\hat{x}_2 \\ d\hat{y}_2 \\ d\hat{z}_2 \end{pmatrix} = |J| J^{-T} \begin{pmatrix} d\hat{x}_1 \\ d\hat{y}_1 \\ d\hat{z}_1 \end{pmatrix} \times \begin{pmatrix} d\hat{x}_2 \\ d\hat{y}_2 \\ d\hat{z}_2 \end{pmatrix}$$

comes from the triple-product computation

$$\langle Jv_1, Jv_2 \times Jv_3 \rangle = |J| \langle v_1, v_2 \times v_3 \rangle$$
$$= |J| \langle Jv_1, J^{-T}v_2 \times v_3 \rangle$$
$$= \langle Jv_1, |J| J^{-T}v_2 \times v_3 \rangle$$

and follows now from a comparison of the second part of the inner product in the last row. Finally, we transform the volume element dV. Based on the triple product, we have

$$dV = \langle d\mathbf{l}_1^T, d\mathbf{l}_2^T \times d\mathbf{l}_3^T \rangle$$
$$= \langle Jd\hat{\mathbf{l}}_1^T, Jd\hat{\mathbf{l}}_2^T \times Jd\hat{\mathbf{l}}_3^T \rangle$$
$$= |J| \langle d\hat{\mathbf{l}}_1^T, d\hat{\mathbf{l}}_2^T \times d\hat{\mathbf{l}}_3^T \rangle$$
$$= |J| \, d\hat{V},$$

and find the transformation rule for small volumes

$$dV = |J| \, d\hat{V}. \tag{8.57}$$

8.3.3.2 Mapping of grad, curl, and div

Gradient: We use the nabla notation describing a column vector

$$\mathbf{grad}\, v(x, y, z) = \nabla_{xyz} v(x, y, z) = \begin{pmatrix} \partial_x v(x, y, z) \\ \partial_y v(x, y, z) \\ \partial_z v(x, y, z) \end{pmatrix}.$$

How does this column vector change when we deform the domain? Let an arbitrary function $v(x, y, z)$ be given which lives on K and allows for the computation of the gradient. We compute the scalar quantity $d\mathbf{l} \cdot \nabla_{xyz} v$. This is a physical quantity with the meaning of a voltage (e.g. $\mathbf{E} \cdot d\mathbf{s}$) or a current ($\mathbf{H} \cdot d\mathbf{s}$). This quantity must not change if we change the coordinate system; hence, we require

$$d\mathbf{l} \cdot \nabla_{xyz} v(x, y, z) = d\hat{\mathbf{l}} \cdot \nabla_{\hat{x}\hat{y}\hat{z}} \hat{v}\left(\hat{x}, \hat{y}, \hat{z}\right),$$

where $v(x, y, z)$ in the physical domain follows from $\hat{v}\left(\hat{x}, \hat{y}, \hat{z}\right)$ in the reference domain by $v(x, y, z) = v\left(Q\left(\hat{x}, \hat{y}, \hat{z}\right)\right) =: \left(\hat{v} \circ Q\right)\left(\hat{x}, \hat{y}, \hat{z}\right)$ based on the coordinate transform (8.53). It follows from (8.55) that

$$d\hat{\mathbf{l}} J^T \cdot \nabla_{xyz} v(x, y, z) = d\hat{\mathbf{l}} \cdot \nabla_{\hat{x}\hat{y}\hat{z}} \hat{v}\left(\hat{x}, \hat{y}, \hat{z}\right).$$

Since this should hold true for all functions v, we find the mapping of the gradient

$$\nabla_{xyz} = J^{-T}\nabla_{\hat{x}\hat{y}\hat{z}}. \tag{8.58}$$

Curl: We proceed the same way and require the invariance of

$$d\mathbf{A} \cdot \nabla_{xyz} \times \mathbf{f} = d\hat{\mathbf{A}} \cdot \nabla_{\hat{x}\hat{y}\hat{z}} \times \hat{\mathbf{f}}.$$

Via (8.56), this yields

$$d\hat{\mathbf{A}}J^{-1}\,|J| \cdot \nabla_{xyz} \times \mathbf{f} = d\hat{\mathbf{A}} \cdot \nabla_{\hat{x}\hat{y}\hat{z}} \times \hat{\mathbf{f}}$$

and we obtain

$$\nabla_{xyz} \times \mathbf{f} = \frac{J}{|J|}\nabla_{\hat{x}\hat{y}\hat{z}} \times \hat{\mathbf{f}}. \tag{8.59}$$

Divergence: We consider the invariance

$$dV\nabla_{xyz} \cdot \mathbf{f} = d\hat{V}\nabla_{\hat{x}\hat{y}\hat{z}} \cdot \hat{\mathbf{f}}.$$

Based on (8.57), this is

$$|J|\,d\hat{V}\nabla_{xyz} \cdot \mathbf{f} = d\hat{V}\nabla_{\hat{x}\hat{y}\hat{z}} \cdot \hat{\mathbf{f}}$$

and yields the transform

$$\nabla_{xyz} \cdot \mathbf{f} = \frac{1}{|J|}\nabla_{\hat{x}\hat{y}\hat{z}} \cdot \hat{\mathbf{f}} \tag{8.60}$$

8.3.3.3 Mapping of Fields

Let $\mathbf{u}$ denote a vectorial field quantity given as a column vector of three numbers, which are the coefficients of the x-, y-, and z-coordinate system and $\hat{\mathbf{u}}$ the corresponding column vector of coefficients in the $\hat{x}$-, $\hat{y}$-, and $\hat{z}$-coordinate system, where $\mathbf{u}$ and $\hat{\mathbf{u}}$ are different algebraic representations of the same, coordinate-independent physical quantity. We will also consider transformed scalar field quantities u and $\hat{u}$.

We know that the voltages $d\mathbf{l} \cdot \mathbf{E}$, or the current $d\mathbf{l} \cdot \mathbf{H}$, must be coordinate-independent quantities; hence, $\mathbf{E}$ and $\mathbf{H}$ must transform like the gradient in (8.58). Further, we know that $d\mathbf{A} \cdot \mathbf{D}$ and $d\mathbf{A} \cdot \mathbf{B}$ must be preserved under coordinate transformations. Hence, $\mathbf{D}$ and $\mathbf{B}$ must transform like the **curl** operator in (8.59). Obviously, $\mathbf{D}$ and $\mathbf{B}$ transform differently than $\mathbf{E}$ and $\mathbf{H}$. Finally, the carrier density $\rho = \mathbf{div}\,\mathbf{D}$, computed over a volume $dV\rho$, must be preserved. Hence, ρ must transform as **div** in (8.60).

The following table summarizes the transformation rules:

	Geometric quantity	Field quantity		Example
Segment	$d\mathbf{l} = d\hat{\mathbf{l}} J^T$	$\mathbf{u} = J^{-T} \hat{\mathbf{u}}$	Field strength	E, H
Face	$d\mathbf{A} = d\hat{\mathbf{A}} \, |J| \, J^{-1}$	$\mathbf{u} = \frac{J}{|J|} \hat{\mathbf{u}}$	Field density	D, B
Volume	$dV = d\hat{V} \, |J|$	$u = \frac{1}{|J|} \hat{u}$	Scalar density	ρ

8.3.3.4 Mapping of μ and ε

Rewriting Maxwell's **curl** equations from the physical to the reference domain, we get

$$\frac{J}{|J|} \nabla_{\hat{x}\hat{y}\hat{z}} \times \hat{\mathbf{H}} = i\omega\varepsilon J^{-T} \hat{\mathbf{E}}$$

$$\frac{J}{|J|} \nabla_{\hat{x}\hat{y}\hat{z}} \times \hat{\mathbf{E}} = -i\omega\mu J^{-T} \hat{\mathbf{H}},$$

which shows that the entire transformation can be traced back to a mapping of the material quantities

$$\hat{\varepsilon} := |J| \, J^{-1} \varepsilon J^{-T} \tag{8.61}$$

$$\hat{\mu} := |J| \, J^{-1} \mu J^{-T}. \tag{8.62}$$

8.3.4 PML IN 2D AND 3D

The 1D complex coordinate stretching was depicted in Figure 8.7. What is the higher-dimensional analogue? The (tensor product type) generalization of (8.37) to 2D is obviously the following (σ_x is σ_y are real constants):

$$x^c = \begin{cases} x(1 + i\sigma_x), & x < 0 \\ x, & 0 \leq x < L_x \\ L_x + (x - L_x)(1 + i\sigma_x), & x \geq L_x, \end{cases} \tag{8.63}$$

and

$$y^c = \begin{cases} y(1 + i\sigma_y), & y < 0 \\ y, & 0 \leq y < L_y \\ L_y + (y - L_y)(1 + i\sigma_y), & y \geq L_y, \end{cases} \tag{8.64}$$

where we indicate the complexified coordinate with a superscript c. This is exactly what we had in the 1D case (8.7) but with a domain mapping in 2D. The Jacobian of the mapping $(x, y) \mapsto (x^c, y^c)$ depends on the position (see Figure 8.14) and evaluates, for example to

$$J^{(2,2)} = \begin{pmatrix} 1 & \\ & 1 \end{pmatrix} \qquad J^{(3,2)} = \begin{pmatrix} 1 + i\sigma_x & \\ & 1 \end{pmatrix} \qquad J^{(3,3)} = \begin{pmatrix} 1 + i\sigma_x & \\ & 1 + i\sigma_y \end{pmatrix}.$$

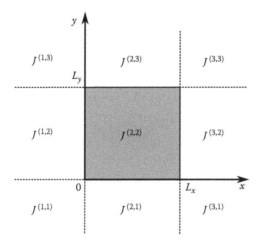

FIGURE 8.14 PML regions in 2D in lexicographic order.

The realization of the PML now consists in using the transformed permittivity (8.61) and permeability (8.62), where the Jacobians are computed from the complex stretching of the space variables (8.63) and (8.64). We proceed now as follows. We consider a subdomain of the exterior, say, $\Omega^{(2,1)} \subset \mathbb{R}^2$ from Figure 8.14 and the complex solution $u^{(2,1)} : \Omega^{(2,1)} \to \mathbb{C}$ living there. Then we use the mapping from the real to the complex coordinates to define a complex-valued spatial domain: $\Omega^{(2,1)} \to \Omega^{c,(2,1)} \subset \mathbb{C}^2$, which means $(x, y) \mapsto (x, y(1 + i\sigma_y))$. On the complexified domain $\Omega^{c,(2,1)}$, we have a complex extension of $u^{(2,1)}$ which we denote by $u^{c,(2,1)}$. The restriction of $u^{c,(2,1)}$ to the real space $\Omega^{(2,1)}$ returns the original function $u^{(2,1)}$. The key point is that our function $u^{(2,1)}(x, y)$ has a complex extension by its nature, and we point to its values by introducing complex instead of only real spatial coordinates.

8.3.5 INTEGRATION BY PARTS

One essential step in deriving the variational form of the Helmholtz equation in 1D has been the application of the integration by parts formula in (8.8). Starting with this 1D case, we summarize the integration by parts formulas we need in the following. In 1D, one might derive it like the following:

$$\int_\Omega v^* \partial_x w \, dx = \int_\Omega \partial_x \left(v^* \overset{\downarrow}{w} \right) dx \qquad\qquad \text{differentiation only wrt } w$$

$$= \int_\Omega \partial_x (v^* w) \, dx - \int_\Omega \partial_x \left(\overset{\downarrow}{v^*} w \right) dx \qquad \text{product rule}$$

$$= (v^* w)|_0^L - \int_\Omega \partial_x v^* w \, dx \qquad\qquad \text{evaluate first integral}$$

where the arrows indicate the factor on which the differential operator acts only. In 2D and 3D, we proceed the same way as follows:

$$\int_\Omega v^* \nabla \cdot \mathbf{w} = \int_\Omega \nabla \cdot \left(v^* \overset{\downarrow}{\mathbf{w}} \right) \qquad\qquad \text{differentiation only wrt } \mathbf{w}$$

$$= \int_\Omega \nabla \cdot (v^* \mathbf{w}) - \int_\Omega \nabla \cdot \left(\overset{\downarrow}{v^*} \mathbf{w} \right) \qquad \text{product rule}$$

$$= \int_{\partial\Omega} d\mathbf{s} \cdot v^* \mathbf{w} - \int_\Omega \nabla v^* \cdot \mathbf{w}. \qquad \text{Gauss's theorem} \qquad (8.65)$$

The corresponding formula with $\nabla \cdot \mathbf{w}$ replaced by $\nabla \times \mathbf{w}$, which we will need for the Maxwell's case, reads

$$\int_\Omega df\ \mathbf{v} \cdot \nabla \times \mathbf{u} = \int_\Omega df\ (\nabla \times \mathbf{v}) \cdot \mathbf{u} - \int_{\partial\Omega} d\mathbf{s} \cdot (\mathbf{v} \times \mathbf{u}). \qquad (8.66)$$

It follows from

$$\int_\Omega \mathbf{u} \cdot \nabla \times \mathbf{v} = \int_\Omega \nabla \cdot \left(\overset{\downarrow}{\mathbf{v}} \times \mathbf{u} \right) \qquad\qquad \text{cyclic shift in triple product}$$

$$= \int_\Omega \nabla \cdot (\mathbf{v} \times \mathbf{u}) - \int_\Omega \nabla \cdot \left(\mathbf{v} \times \overset{\downarrow}{\mathbf{u}} \right) \quad \text{product rule}$$

$$= \int_{\partial\Omega} d\mathbf{s} \cdot (\mathbf{v} \times \mathbf{u}) + \int_\Omega \mathbf{v} \cdot \nabla \times \mathbf{u}. \qquad \text{Gauss's theorem}$$

Note that in (8.65), the boundary term occurs with a plus sign, whereas the boundary term in (8.66) has a minus sign.

8.3.6 VARIATIONAL FORMULATION FOR THE HELMHOLTZ EQUATION WITH PML

We want to derive the variational formulation for higher spatial dimensions along the lines given for the 1D case in Section 8.2.2. For simplicity, we examine the 2D case. We follow the definition of the Helmholtz scattering problem given in Section 8.3.2.

8.3.6.1 Interior Problem

We rewrite the interior problem $\Delta u(\mathbf{r}) + k^2(\mathbf{r})u(\mathbf{r}) = 0$ in weak form, which means that we multiply with the complex conjugate of a test function $v(\mathbf{r})$ and integrate over the domain Ω:

$$\int_\Omega v^* \left(\Delta u + k^2 u \right) dx\, dy = 0.$$

We decompose the integral and use $\Delta u = \nabla \cdot \nabla u$ and the integration by parts formula (8.65) with $\mathbf{w} = \nabla u$ to obtain

$$\int_{\partial\Omega} d\mathbf{s} \cdot v^* \nabla u - \int_{\Omega} \nabla v^* \cdot \nabla u \, dx \, dy + \int_{\Omega} v^* k^2 u \, dx \, dy = 0. \tag{8.67}$$

We abbreviate $a_{\text{int}}(v,u) := \int_{\Omega} \nabla v^* \cdot \nabla u \, dx \, dy - \int_{\Omega} v^* k^2 u \, dx \, dy$ to rewrite the last equation in shorter form

$$\int_{\partial\Omega} d\mathbf{s} \cdot v^* \nabla u - a_{\text{int}}(v,u) = 0. \tag{8.68}$$

8.3.6.2 Exterior Problem for the Scattered Field

In the first step, the weak form of the Helmholtz equation with respect to the scattered field in the exterior has the same form as

$$\int_{\partial\Omega_{\text{ext}}} d\mathbf{s} \cdot v^* \nabla u_s - \int_{\Omega_{\text{ext}}} \nabla v^* \cdot \nabla u_s \, dx \, dy + \int_{\Omega_{\text{ext}}} v^* k^2 u_s \, dx \, dy = 0.$$

Further, the scattered field has to satisfy the Sommerfeld radiation condition (compare 8.3.2). We realize this via the complex stretching, that is we map the real to a complexified spatial domain, where any quantity and operation act on the complexified domain:

$$\int_{\partial\Omega_{\text{ext}}^c} d\mathbf{s}^c \cdot v^{*c} \nabla^c u_s^c - \int_{\Omega_{\text{ext}}^c} \nabla^c v^{*c} \cdot \nabla^c u_s^c \, dx^c \, dy^c + \int_{\Omega_{\text{ext}}^c} v^{*c} \left(k^2\right)^c u_s^c \, dx^c \, dy^c = 0.$$

Here, the u_s^c has the meaning of $u_s(x^c, y^c)$. Remember that in Maxwell's case, the entire mapping can be traced back to the material mappings (8.61) and (8.62). In the 2D case considered here, the Helmholtz equation is exactly Maxwell's equation for the z-component of $\mathbf{E}$; hence, the material mappings hold true in the Helmholtz 2D case. It follows that

$$\left(k^2\right)^c := \omega^2 \left(\mu^c \varepsilon^c\right)_z$$
$$= |J|^2 \, \omega^2 \varepsilon \mu \tag{8.69}$$

and the entire complex extension reduces to a mapping of k^2

$$\int_{\partial\Omega_{\text{ext}}} d\mathbf{s} \cdot v^{*c} \nabla u_s^c - \int_{\Omega_{\text{ext}}} \nabla v^{*c} \cdot \nabla u_s^c \, dx \, dy + \int_{\Omega_{\text{ext}}} v^{*c} |J|^2 k^2 u_s^c \, dx \, dy = 0,$$

where v^c and u_s^c have here the meaning of $v^c(x^c, y^c)$ and $u_s(x^c, y^c)$. Again we abbreviate $a_{\text{ext}}(v,u) := \int_{\Omega_{\text{ext}}} \nabla v^* \cdot \nabla u \, dx \, dy - \int_{\Omega_{\text{ext}}} v^* |J|^2 k^2 u \, dx \, dy$ and rewrite the last equation to

$$\int_{\partial\Omega_{\text{ext}}} d\mathbf{s} \cdot v^{*c} \nabla u_s^c - a_{\text{ext}}(v^c, u_s^c) = 0. \tag{8.70}$$

The simultaneous use of u^c (and v^c) for different functions, namely for u_s^c (x^c, y^c) and $u^c(x, y)$, can be confusing. On the other hand, a more correct notation, for example $u^c \circ \phi$ with ϕ the complex stretching $\phi : \Omega \subset \mathbb{R}^2 \to \mathbb{C}^2$ appears too clumsy. Since the correct meaning in the equation is directly indicated by the integration variables, no real confusion should be caused.

8.3.6.3 Variational Formulation on the Entire Domain

We want to combine (8.68) and (8.70) to derive the variational formulation. Together we have

$$\int_{\partial\Omega} d\mathbf{s} \cdot v^* \nabla u - a_{int}(v, u) = 0 \qquad \text{entire field in the interior}$$

$$\int_{\partial\Omega_{ext}} d\mathbf{s} \cdot v^* \nabla u_s - a_{ext}(v, u_s) = 0 \qquad \text{scattered field in the exterior,}$$

where we have dropped the superscript c in the last equation indicating that we do not compute the scattered field itself but the complex extension of it. The field u in the interior and u_s in the exterior are not continuous at the boundary since their difference is the source field. To overcome this difficulty, we proceed as in 1D (8.34). We introduce the auxiliary functions g and w

$$u_s = \underbrace{u_s + g}_{:=w} - g$$

$$g(\partial\Omega) = u_{source}(\partial\Omega)$$

$$g(\infty) = 0 \qquad\qquad (8.71)$$

and use $u_s = u - u_{source}$ at the boundary to find the pair of equations

$$\int_{\partial\Omega} d\mathbf{s} \cdot v^* \nabla u - a_{int}(v, u) = 0 \qquad \text{entire field in the interior}$$

$$\int_{\partial\Omega_{ext}} d\mathbf{s} \cdot v^* \nabla (u - u_{source}) - a_{ext}(v, w - g) = 0 \qquad \text{auxiliary } w \text{ in the exterior.}$$

We add these two equations, taking into account that the orientation of the boundary integrals concerning $\partial\Omega$ is opposite, and get

$$\int_{\partial\Omega} d\mathbf{s} \cdot v^* \nabla u_{source} - a_{int}(v, u) - a_{ext}(v, w - g) = 0. \qquad (8.72)$$

Further, sorting known and unknown quantities, we find

$$a_{int}(v, u) + a_{ext}(v, w) = \int_{\partial\Omega} d\mathbf{s} \cdot v^* \nabla u_{source} + a_{ext}(v, g),$$

which is the exact analogue to the 1D case (8.35). As in 1D, Section 8.2, we rename w in the exterior to u and define $a(\cdot, \cdot) := a_{\mathrm{int}}(\cdot, \cdot) + a_{\mathrm{ext}}(\cdot, \cdot)$. Finally, we have the Helmholtz scattering problem with PML in higher space dimensions.

Helmholtz Scattering Problem with PML

Let the source field data u_{source} and $\partial_n u_{\mathrm{source}}$ at the boundary $\partial\Omega$ of the computational domain Ω be given. The variational form of the scattering problem reads as follows:

$$\text{Find } u \in V: \quad a(v, u) = \int_{\partial\Omega} ds \cdot v^* \nabla u_{\mathrm{source}} + a_{\mathrm{ext}}(v, g) \quad \text{for all } v \in V, \quad (8.73)$$

with

$$a(\cdot, \cdot) = a_{\mathrm{int}}(\cdot, \cdot) + a_{\mathrm{ext}}(\cdot, \cdot)$$

$$a_{\mathrm{int}}(v, u) = \int_\Omega \nabla v^* \cdot \nabla u \, dx \, dy - \int_\Omega v^* k^2 u \, dx \, dy$$

$$a_{\mathrm{ext}}(v, g) = \int_{\Omega_{\mathrm{ext}}} \nabla v^* \cdot \nabla g \, dx \, dy - \int_{\Omega_{\mathrm{ext}}} v^* |J|^2 k^2 g \, dx \, dy$$

and g is an arbitrary continuous function with

$$g(\partial\Omega) = u_{\mathrm{source}}(\partial\Omega)$$

$$g(\infty) = 0.$$

Please note again that u in the interior domain is in fact the desired solution, but in the exterior domain, it is a superposition of the scattered field u_s and the auxiliary function g.

8.3.7 VARIATIONAL FORMULATION FOR MAXWELL'S EQUATIONS WITH PML

We derive the variational formulation for Maxwell's scattering problem completely parallel to the derivation for the Helmholtz scattering problem.

8.3.7.1 Interior Problem

We start from the **curl–curl** equation for the electric field

$$\nabla \times \frac{1}{\mu} \nabla \times \mathbf{E} - \omega^2 \varepsilon \mathbf{E} = 0,$$

multiply with the complex conjugate of a vectorial test function **v**, and integrate

$$\int_\Omega dx\,dy\,\mathbf{v}^* \cdot \left(\nabla \times \frac{1}{\mu}\nabla \times \mathbf{E}\right) - \omega^2 \int_\Omega dx\,dy\,\mathbf{v}^*\varepsilon\mathbf{E} = 0.$$

An application of the integration by parts formula (8.66) yields

$$\int_\Omega dx\,dy\,(\nabla \times \mathbf{v}^*) \cdot \frac{1}{\mu}(\nabla \times \mathbf{E}) - \int_{\partial\Omega} d\mathbf{s} \cdot \left(\mathbf{v}^* \times \left[\frac{1}{\mu}\nabla \times \mathbf{E}\right]\right) - \omega^2 \int_\Omega dx\,dy\,\mathbf{v}^*\varepsilon\mathbf{E} = 0.$$

We abbreviate $a_{\text{int}}(\mathbf{v},\mathbf{E}) := \int_\Omega dx\,dy\,(\nabla \times \mathbf{v}^*) \cdot (1/\mu)(\nabla \times \mathbf{E}) - \omega^2 \int_\Omega dx\,dy\,\mathbf{v}^*\varepsilon\mathbf{E}$ to rewrite the last equation more concisely

$$\int_{\partial\Omega} d\mathbf{s} \cdot \left(\mathbf{v}^* \times \left[\frac{1}{\mu}\nabla \times \mathbf{E}\right]\right) - a_{\text{int}}(\mathbf{v},\mathbf{E}) = 0. \qquad (8.74)$$

8.3.7.2 Exterior Problem for the Scattered Field

As for the Helmholtz equation, the scattered field in the exterior domain satisfies Maxwell's equations in weak form

$$\int_{\Omega_{\text{ext}}} dx\,dy\,(\nabla \times \mathbf{v}^*) \cdot \frac{1}{\mu}(\nabla \times \mathbf{E}_s) - \int_{\partial\Omega_{\text{ext}}} d\mathbf{s} \cdot \left(\mathbf{v}^* \times \left[\frac{1}{\mu}\nabla \times \mathbf{E}_s\right]\right)$$
$$- \omega^2 \int_{\Omega_{\text{ext}}} dx\,dy\,\mathbf{v}^*\varepsilon\mathbf{E}_s = 0$$

and the same formula holds true for the complex extension of all quantities

$$\int_{\Omega^c_{\text{ext}}} dx^c\,dy^c\,(\nabla^c \times \mathbf{v}^{*c}) \cdot \frac{1}{\mu^c}(\nabla^c \times \mathbf{E}^c_s) - \int_{\partial\Omega^c_{\text{ext}}} d\mathbf{s}^c \cdot \left(\mathbf{v}^{*c} \times \left[\frac{1}{\mu^c}\nabla^c \times \mathbf{E}^c_s\right]\right)$$
$$- \omega^2 \int_{\Omega^c_{\text{ext}}} dx^c\,dy^c\,\mathbf{v}^{*c}\varepsilon^c\mathbf{E}_s{}^c = 0.$$

In Maxwell's case, the entire mapping can be traced back to the material mappings (8.61) and (8.62); hence, we have to replace

$$(\mu^c)^{-1} = \frac{1}{|J|}J^T\mu^{-1}J$$
$$\varepsilon^c = |J|\,J^{-1}\varepsilon J^{-T}.$$

Therefore, the formula for the complex extended field given in original spatial coordinates is

$$\int_{\Omega_{ext}} dx\, dy\, (\nabla \times \mathbf{v}^{*c}) \cdot \frac{1}{|J|} J^T \mu^{-1} J \left(\nabla \times \mathbf{E}_s^c\right) - \omega^2 \int_{\Omega_{ext}} dx\, dy\, \mathbf{v}^{*c} \, |J|\, J^{-1} \varepsilon J^T \mathbf{E}_s^c$$

$$- \int_{\partial\Omega_{ext}} ds \cdot \left(\mathbf{v}^{*c} \times \left[\frac{1}{|J|} J^T \mu^{-1} J \nabla \times \mathbf{E}_s^c\right]\right) = 0.$$

We abbreviate

$$a_{ext}\,(\mathbf{v}, \mathbf{E}) := \int_{\Omega_{ext}} dx\, dy\, (\nabla \times \mathbf{v}^*) \cdot \frac{1}{|J|} J^T \mu^{-1} J \,(\nabla \times \mathbf{E})$$

$$- \omega^2 \int_{\Omega_{ext}} dx\, dy\, \mathbf{v}^* \, |J|\, J^{-1} \varepsilon J^{-T} \mathbf{E}$$

and rewrite the last equation to

$$\int_{\partial\Omega_{ext}} ds \cdot \left(\mathbf{v}^{*c} \times \left[\mu^{-1} \nabla \times \mathbf{E}_s^c\right]\right) - a_{ext}\,\left(\mathbf{v}^c, \mathbf{E}_s^c\right) = 0. \qquad (8.75)$$

The main difference to the Helmholtz counterpart (8.70) is the different form of the boundary integral term. This is due to the fact that the material properties in the Helmholtz case are entirely hidden within the wavenumber k, whereas in Maxwell's case, the permeability μ appears in the boundary term and must be explicitly transformed.

8.3.7.3 Variational Formulation on the Entire Domain

We combine (8.74) and (8.75)

$$\int_{\partial\Omega} ds \cdot \left(\mathbf{v}^* \times \left[\frac{1}{\mu} \nabla \times \mathbf{E}\right]\right) - a_{int}(\mathbf{v}, \mathbf{E}) = 0 \qquad \text{int. field}$$

$$\int_{\partial\Omega_{ext}} ds \cdot \left(\mathbf{v}^* \times \left[\frac{1}{|J|} J^T \mu^{-1} J \nabla \times \mathbf{E}_s\right]\right) - a_{ext}\,(\mathbf{v}, \mathbf{E}_s) = 0 \qquad \text{ext. scattered field.}$$

We introduce the auxiliary function $\mathbf{g}$ as in (8.71)

$$\mathbf{E}_s = \underbrace{\mathbf{E}_s + \mathbf{g}}_{:=\mathbf{W}} - \mathbf{g}$$

$$\mathbf{g}(\partial\Omega) = \mathbf{E}_{source}\,(\partial\Omega)$$

$$\mathbf{g}(\infty) = 0$$

to obtain

$$\int_{\partial\Omega} ds \cdot \left(\mathbf{v}^* \times \left[\frac{1}{\mu} \nabla \times \mathbf{E} \right] \right) - a_{\text{int}}(\mathbf{v}, \mathbf{E}) = 0 \qquad \text{int.}$$

$$\int_{\partial\Omega_{\text{ext}}} ds \cdot \left(\mathbf{v}^* \times \left[\frac{1}{|J|} J^T \mu^{-1} J \nabla \times (\mathbf{E} - \mathbf{E}_{\text{source}}) \right] \right) - a_{\text{ext}}(\mathbf{v}, \mathbf{W} - \mathbf{g}) = 0 \qquad \text{ext.}$$

Here, it is not immediately clear that the boundary integrals along $\partial\Omega$ have opposite signs as in the Helmholtz case where it is just indicated by the orientation of the contour. We will show this at the end of this section. Adding the two equations, we arrive at

$$\int_{\partial\Omega} ds \cdot \left(\mathbf{v}^* \times \left[\frac{1}{|J|} J^T \mu^{-1} J \nabla \times \mathbf{E}_{\text{source}} \right] \right) - a_{\text{int}}(\mathbf{v}, \mathbf{E}) - a_{\text{ext}}(\mathbf{v}, \mathbf{W} - \mathbf{g}) = 0.$$

Compare this to the corresponding result for the Helmholtz equation (8.72). Sorting known and unknown quantities, we find

$$a_{\text{int}}(\mathbf{v}, \mathbf{E}) + a_{\text{ext}}(\mathbf{v}, \mathbf{W}) = \int_{\partial\Omega} ds \cdot \left(\mathbf{v}^* \times \left[\frac{1}{|J|} J^T \mu^{-1} J \nabla \times \mathbf{E}_{\text{source}} \right] \right) + a_{\text{ext}}(\mathbf{v}, \mathbf{g}).$$

Defining again $a(\cdot, \cdot) := a_{\text{int}}(\cdot, \cdot) + a_{\text{ext}}(\cdot, \cdot)$, we end up with

---

Maxwell's Scattering Problem with PML

Let the source field data $\mathbf{E}_{\text{source}}$ and $\nabla \times \mathbf{E}_{\text{source}}$ at the boundary $\partial\Omega$ of the computational domain Ω be given. The variational form of the scattering problem reads as follows:

Find $\mathbf{E} \in V$:

$$a(\mathbf{v}, \mathbf{E}) = \int_{\partial\Omega} ds \cdot \left(\mathbf{v}^* \times \left[(\mu^c)^{-1} \nabla \times \mathbf{E}_{\text{source}} \right] \right) + a_{\text{ext}}(\mathbf{v}, \mathbf{g}) \qquad \text{for all } \mathbf{v} \in V,$$

$$(8.76)$$

with

$$a(\cdot, \cdot) = a_{\text{int}}(\cdot, \cdot) + a_{\text{ext}}(\cdot, \cdot)$$

$$a_{\text{int}}(\mathbf{v}, \mathbf{E}) = \int_{\Omega} dx\, dy\, (\nabla \times \mathbf{v}^*) \cdot \frac{1}{\mu} (\nabla \times \mathbf{E}) - \omega^2 \int_{\Omega} dx\, dy\, \varepsilon \mathbf{v}^* \mathbf{E}$$

$$a_{\text{ext}}(\mathbf{v}, \mathbf{g}) = \int_{\Omega_{\text{ext}}} dx\, dy\, (\nabla \times \mathbf{v}^*) \cdot (\mu^c)^{-1} (\nabla \times \mathbf{E})$$

$$\qquad\qquad - \omega^2 \int_{\Omega_{\text{ext}}} dx\, dy\, \varepsilon^c \mathbf{v}^* \mathbf{E}$$

and **g** is an arbitrary tangentially continuous function with

$$\mathbf{g}(\partial\Omega) = \mathbf{E}_{\text{source}}(\partial\Omega)$$

$$\mathbf{g}(\infty) = 0$$

and

$$(\mu^c)^{-1} = \frac{1}{|J|}J^T\mu^{-1}J$$

$$\varepsilon^c = |J|\,J^{-1}\varepsilon J^{-T}.$$

It remains to show that

$$\int_{\partial\Omega} ds \cdot \left(\mathbf{v}^* \times \left[\frac{1}{\mu}\nabla \times \mathbf{E}\right]\right) = \int_{\partial\Omega} ds \cdot \left(\mathbf{v}^* \times \left[\frac{1}{|J|}J^T\mu^{-1}J\nabla \times \mathbf{E}\right]\right).$$

For simplicity, we assume that **E** is continuous across $\partial\Omega$, that is we do not have any material discontinuity here. We consider the integrand (in 3D) with an arbitrary face element $d\mathbf{A}$

$$d\mathbf{A} \cdot \left(\mathbf{v}^* \times \left[\frac{1}{\mu}\nabla \times \mathbf{E}\right]\right) = -\mathbf{v}^* \cdot \left(d\mathbf{A}^T \times \left[\frac{1}{\mu}\nabla \times \mathbf{E}\right]\right) \qquad \text{triple product}$$

$$= -\left\langle\mathbf{v}^*, d\mathbf{A}^T \times \left[\frac{1}{\mu}\nabla \times \mathbf{E}\right]\right\rangle \qquad \text{inner product.}$$

We transform the latter expression based on the transform mediated by the complex stretching according to the transformation rules given in Section 8.3.3:

$$\left\langle\mathbf{v}^*, d\mathbf{A}^T \times \left[\frac{1}{\mu}\nabla \times \mathbf{E}\right]\right\rangle = \left\langle J^{-T}\widehat{\mathbf{v}^*}, J^{-T}|J|\,d\hat{\mathbf{A}}^T \times \left[\frac{1}{\mu}\frac{J}{|J|}\nabla \times \hat{\mathbf{E}}\right]\right\rangle$$

$$= \frac{1}{|J|}\left\langle\widehat{\mathbf{v}^*}, |J|\,d\hat{\mathbf{A}}^T \times \left[J^T\frac{1}{\mu}\frac{J}{|J|}\nabla \times \hat{\mathbf{E}}\right]\right\rangle \qquad \text{extract } J^{-T}$$

$$= \left\langle\widehat{\mathbf{v}^*}, d\hat{\mathbf{A}}^T \times \left[J^T\frac{1}{\mu}\frac{J}{|J|}\nabla \times \hat{\mathbf{E}}\right]\right\rangle.$$

Together this means

$$d\mathbf{A} \cdot \left(\mathbf{v}^* \times \left[\frac{1}{\mu}\nabla \times \mathbf{E}\right]\right) = -\left\langle\mathbf{v}^*, d\mathbf{A}^T \times \left[\frac{1}{\mu}\nabla \times \mathbf{E}\right]\right\rangle \qquad (8.77)$$

$$= -\left\langle\widehat{\mathbf{v}^*}, d\hat{\mathbf{A}}^T \times \left[J^T\frac{1}{\mu}\frac{J}{|J|}\nabla \times \hat{\mathbf{E}}\right]\right\rangle. \qquad (8.78)$$

This is the desired result. The boundary term is independent of the complex mapping if we use (8.62) to transform μ.

8.4 FEM FOR HELMHOLTZ SCATTERING IN 2D AND 3D

We want to generalize the FEM approach for the solution of 1D scattering problems discussed in Section 8.1 to 2D and 3D situations. The general electromagnetic scattering problem encounters two additional difficulties, compared to the 1D approach:

1. The higher space dimension
2. The vectorial character of the solution

To keep the presentation of the FEM simple, we start with the Helmholtz equation since this provides a convenient entry to higher-dimensional FEM approaches. After the study of the Helmholtz equation, we will return in Section 8.5 to our original goal, namely the FEM-based solution of time-harmonic electromagnetic vectorial scattering problems.

Typically, FEM approaches are associated with triangular or tetrahedral meshes. However, they can be derived for a whole variety of meshes including rectangular meshes and meshes mixed from quadrilateral and triangular meshes.

8.4.1 RECTANGULAR MESHES

First, we discuss the use of rectangular meshes. This yields a realization of finite elements most closely related to methods like FDTD, which are often based on uniform rectangular structured meshes. Figure 8.15 displays the mesh we want to describe. It is constructed from the tensor product of N_x points (vertices) in the x-direction and N_y points in the y-direction with coordinates listed in the vectors $x = (x_1, \ldots, x_{N_x})$ and $y = (y_1, \ldots, y_{N_y})$. We consider the rectangular mesh, as other

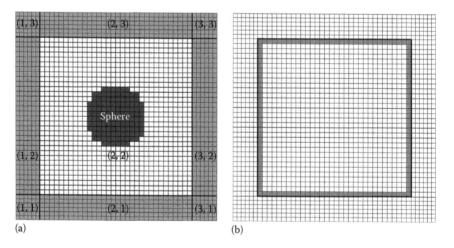

(a)

(b)

FIGURE 8.15   (See colour insert.) (a) Discretized computational domain containing a sphere. Discretization by rectangles. There are nine subdomains with possibly different discretizations $(1, 1), \ldots, (3, 3)$. The outer grey areas are the PML areas. (b) Boundary edges can be identified via the grey-shaded boundary cells.

meshes in 2D, to be built from vertices, edges, and faces (cells). We need to address them, as well as some subsets, in particular:

Vertices: We denote the set of vertices by V. Each vertex gets a unique integer identifier. They are counted globally from the lower-left to the upper-right corner in row-major order, that is if the index of the vertex is (i,j) with $1 \leq i,j \leq N_x, N_j$, its global index is $id = i + N_x (j-1)$.

Boundary vertices: This is the subset of vertices placed on the boundary of the (inner) computational domain; compare Figure 8.15. The boundary vertices are needed to compute the boundary integral in the Helmholtz case.

Edges: We denote the set of edges by E. Each edge gets a unique integer identifier. All edges are counted globally from the lower-left to the upper-right corner, first all horizontal edges, then the vertical edges in row-major order.

Boundary edges: This is the subset of edges placed on the boundary of the computational domain; compare Figure 8.15. The boundary edges are needed to compute the boundary integral in Maxwell's case.

Cells/faces: Each cell gets a unique integer identifier. All cells are counted globally from the lower-left to the upper-right corner in row-major order, that is if the index of the cell is (i,j) with $i,j \geq 1$, its global index is $id = i + (N_x - 1)(j-1)$.

8.4.2 MESH AND ASSEMBLY PROCESS: GENERAL SCHEME

At the end of the assembly procedure, we want to have a linear system $Ax = b$, whose solution solves our discretized scattering problem. We do this as follows:

1. Step through all cells according to their global index. For each cell, do the following:

   a. Compute a local matrix A_{loc}. The matrix A_{loc} depends on the shape of the actual cell and on the chosen field representation (linear, quadratic, higher order, triangular, rectangular, type, etc.). There are many different possibilities; we will discuss some in Section 8.4.3:

$$A_{\text{loc}} = \begin{pmatrix} a_{11} & a_{12} & \cdots & a_{1,\text{DOF}_{\text{cell}}} \\ a_{21} & \ddots & & \\ \vdots & & & \\ & & & a_{\text{DOF}_{\text{cell}},\text{DOF}_{\text{cell}}} \end{pmatrix} \in \mathbb{C}^{\text{DOF}_{\text{cell}} \times \text{DOF}_{\text{cell}}}.$$

   Here, the *degrees of freedom* (DOFs) are the number of unknowns related to the cell.

   b. Compute a local vector b_{loc} which accounts for the field sources:

$$b_{\text{loc}} = \begin{pmatrix} b_1 \\ \vdots \\ \end{pmatrix} \in \mathbb{C}^{\text{DOF}_{\text{cell}}}.$$

As for the local matrices, this depends, apart from the source fields, on the chosen local field representation.

c. Plug the local matrices and vectors into global ones, without considering a mutual interaction. That is in the first step, the local representations are unconnected global arrays:

$$A = \begin{pmatrix} \blacksquare & & & & & & \\ & \blacksquare & & & & & \\ & & \blacksquare & & & & \\ & & & \blacksquare & & & \\ & & & & \ddots & & \\ & & & & & \blacksquare & \\ & & & & & & \blacksquare \end{pmatrix} \in \mathbb{C}^{(N_{\text{cells}} \cdot \text{DOF}_{\text{cell}}) \times (N_{\text{cells}} \cdot \text{DOF}_{\text{cell}})}$$

$$b = \begin{pmatrix} \blacksquare \\ \blacksquare \\ \blacksquare \\ \blacksquare \\ \vdots \\ \blacksquare \\ \blacksquare \end{pmatrix} \in \mathbb{C}^{(N_{\text{cells}} \cdot \text{DOF}_{\text{cell}})},$$

where the $\blacksquare$ represents entire local matrices and vectors, respectively.

2. The global interdependence of the local DOFs is realized via the connectivity matrix C. This matrix connects the local and the global DOFs and must be constructed in a way which is dependent on both the geometry and the local representation:

$$C^T A C x = C^T b.$$

8.4.3 Finite Elements for Rectangular Meshes

In order to realize the FEM, we have to provide three basic ingredients:

1. The geometrical patches K on which the field has to be approximated (here, a rectangle)
2. The polynomial space $\mathcal{P}_K$ consisting of shape functions $N_i(\mathbf{r})$, which we want to use for the approximation
3. The set of functionals $M_j(\cdot)$ we want to use to measure the contribution of the individual basis functions

8.4.3.1 Rectangular Elements

The reference patch $\hat{K}$ is just the unit square. We map the canonical element, the unit square with points $(\hat{x}, \hat{y})$, to an arbitrary rectangular physical domain with points $(x, y) = Q(\hat{x}, \hat{y})$ via

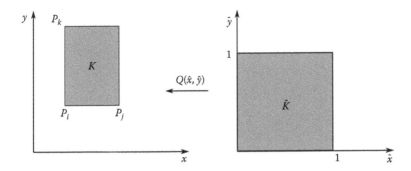

FIGURE 8.16 Mapping from the unit square to a rectangle in the physical domain.

$$\begin{pmatrix} x \\ y \end{pmatrix} = \begin{pmatrix} x_i \\ y_i \end{pmatrix} + \left(\begin{pmatrix} x_j \\ y_j \end{pmatrix} - \begin{pmatrix} x_i \\ y_i \end{pmatrix} \right) \hat{x} + \left(\begin{pmatrix} x_k \\ y_k \end{pmatrix} - \begin{pmatrix} x_i \\ y_i \end{pmatrix} \right) \hat{y}$$

$$= P_i + (P_j - P_i)\,\hat{x} + (P_k - P_i)\,\hat{y}. \tag{8.79}$$

(see Figure 8.16).

8.4.3.2 Polynomial Space

It is natural to extend the linear 1D polynomial space $\mathcal{P}^{(1)} = \{1, \hat{x}\}$ defined on the unit segment to the (tensor) polynomial space $\mathcal{P}^{(1,1)} = \{1, \hat{x}, \hat{x}\,\hat{y}, \hat{y}\}$ consisting of all polynomials which are linear in $\hat{x}$ or $\hat{y}$. The basis functions in the 1D case were $\{\hat{x}, 1 - \hat{x}\}$; hence, we try in the 2D case the tensor product $\{\hat{x}, 1 - \hat{x}\} \times \{\hat{y}, 1 - \hat{y}\} = \{(1 - \hat{x})\,(1 - \hat{y}),\, \hat{x}\,(1 - \hat{y}),\, \hat{x}\,\hat{y},\, (1 - \hat{x})\,\hat{y}\}$. Thus, we have four shape functions defined on each rectangle. We associate the shape functions with the corners of the rectangle on which they exist (see Figure 8.17). Note that the mapping (8.79) is a special case of the so-called bilinear mapping

$$\begin{pmatrix} x \\ y \end{pmatrix} = P_i \left(1 - \hat{x}\right) \left(1 - \hat{y}\right) + P_j \hat{x} \left(1 - \hat{y}\right) + P_k \left(1 - \hat{x}\right) \hat{y} + P_l \hat{x}\,\hat{y} \tag{8.80}$$

which maps the unit square to the convex hull of the points P_i, P_j, P_k, and P_l. Here, in complete analogue to (8.18), we used the shape functions to construct a general domain mapping.

Next, we use these shape functions and the DOFs to construct the finite element basis functions.

8.4.3.3 DOFs on a Rectangle

Let us consider a part of the full mesh, for example the four rectangles shown in Figure 8.18. Each of the shape functions may contribute to the solution. The question is how do we measure the contribution? Meaning how do we normalize the basis functions? In the 1D case, Figure 8.3 and (8.17), the hat functions were normalized such that the weight of the hat function is the value of the function at the interval ends. Of course other choices are possible and useful, in particular, in Maxwell's

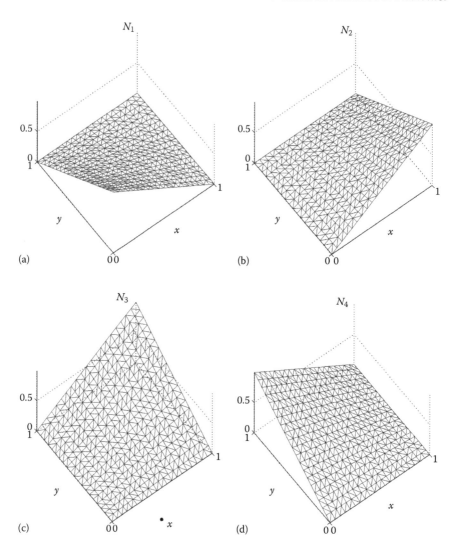

FIGURE 8.17   The four bilinear shape functions $N_1, \ldots, N_4$ on a square. (a) Shape function N_1 associated with vertex (0; 0), (b) shape function N_2 associated with vertex (1; 0), (c) shape function N_3 associated with vertex (1; 1), (a) shape function N_4 associated with vertex (0; 1).

case, where point data are less meaningful. However, in our 2D linear scalar case, we continue to use the nodal values for normalization. Taking the four functionals $M_j(\cdot)$ associated to a unit square with corners (x_j, y_j), $j = 1, \ldots, 4$, as

$$M_j(v(x, y)) = \int_{\hat{K}} \delta(x_j - x)\delta(y_j - y)v(x, y)\, dx\, dy$$

$$= v(x_j, y_j) \tag{8.81}$$

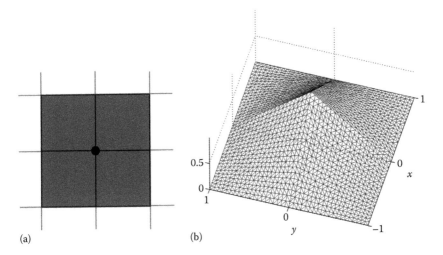

(a)   (b)

FIGURE 8.18 (a) Four neighbouring rectangles with a common vertex (b) building the support of a piecewise bilinear basis function and the corresponding basis function. The basis function is composed of four shape functions.

we find, as desired,

$$M_j(N_k) = \delta_{jk}$$

with Kronecker delta δ_{ij}.

We may use higher-order shape functions as shown in Figure 8.19 to construct other types of basis functions. This time the basis function is identified with an

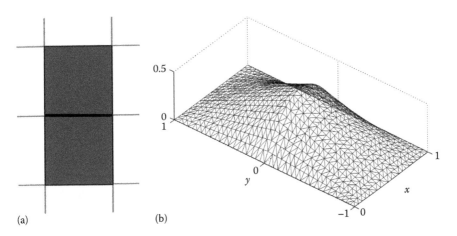

(a)   (b)

FIGURE 8.19 (a) Two neighbouring rectangles with a common edge (b) building the support of a quadratic basis function and the corresponding basis function. The basis function is composed from two shape functions.

edge shared by two neighbouring rectangles. The trace of the basis function over the common edge is a quadratic bubble. How do we construct such kinds of shape functions on each rectangle?

First, we extend the linear polynomial ansatz space $\mathcal{P}^{(1,1)}$ to an ansatz space quadratic in x- and y-directions. This is

$$\mathcal{P}^{(2,2)} = \left\{ 1, x, y, xy, x^2, y^2, x^2 y, xy^2, x^2 y^2 \right\}.$$

It has nine elements. The first four span the linear space $\mathcal{P}^{(1,1)}$ and the remaining five are new functions. We will associate them with the four edges and the face of the rectangle. In classical finite element theory, one assigns the DOFs of higher-order shape functions also with the values of the functions at properly defined positions. Instead, we can also use the following definition [3]:

$$M_j(v(s)) = \int_{\hat{k}} \partial_s p(s) \partial_s v(s)\, ds, \quad j = 5, \ldots, 8 \qquad \text{along the edges} \qquad (8.82)$$

$$M_j(v(x,y)) = \int_{\hat{k}} \nabla p(x,y) \cdot \nabla v(x,y)\, dx\, dy, \quad j = 9 \qquad \text{on the faces,} \qquad (8.83)$$

which gives a more direct connection between the geometric entities *edge* and *face* and the DOFs.

8.4.3.4 Bilinear Finite Element Discretization

We want to use the four shape functions

$$\{ v_i : i = 1, \ldots, 4 \} = \left\{ (1 - \hat{x})(1 - \hat{y}), \hat{x}(1 - \hat{y}), \hat{x}\hat{y}, (1 - \hat{x})\hat{y} \right\}$$

from Figure 8.17 and the domain mapping (8.79) to derive a bilinear finite element discretization. Since

$$Q(\hat{x}, \hat{y}) = P_i + (P_j - P_i)\hat{x} + (P_k - P_i)\hat{y} \qquad (8.84)$$

and $x_i = x_k$ and $y_i = y_j$, we compute

$$J = \begin{pmatrix} x_j - x_i & x_k - x_i \\ y_j - y_i & y_k - y_i \end{pmatrix} = \begin{pmatrix} x_j - x_i & 0 \\ 0 & y_k - y_i \end{pmatrix} \qquad \text{and}$$

$$|J| = (x_j - x_i)(y_k - y_i). \qquad (8.85)$$

Thus, we have four shape functions $\hat{v}_1, \ldots, \hat{v}_4$ defined on the reference domain $\hat{K}$. We map the functions to the original rectangles and combine them to get continuous basis functions which are used to solve the Helmholtz scattering problem (8.73). This amounts to compute

$$(A)_{i,j} = a(v_i, v_j) = \underbrace{\int_{\Omega_{\text{cd}}} \nabla v_i^* \cdot \nabla v_j\, dx\, dy}_{\text{stiffness part } s_{ij}} - \underbrace{\int_{\Omega_{\text{cd}}} v_i^* k^2 v_j\, dx\, dy}_{\text{mass part } m_{ij}}, \qquad (8.86)$$

where Ω_{cd} denotes the entire computational domain (interior domain plus PML domain) and the boundary integral on the boundary $\partial\Omega$

$$(b)_i = \int_{\partial\Omega} v_i^* \nabla u_{\text{source}} \, d\mathbf{f} + a_{\text{ext}}(v_i, g). \tag{8.87}$$

Using our mapping rules, we compute the elements of the local mass and stiffness matrices as

$$S_{ij} = \int_{\hat{K}} |J| \, d\hat{x} d\hat{y} \left(\left(\nabla_{\hat{x}\hat{y}} \hat{v}_i^* \right)^T J^{-1} J^{-T} \nabla_{\hat{x}\hat{y}} \hat{v}_j \right) \tag{8.88}$$

$$m_{ij} = \int_{\hat{K}} |J| \, d\hat{x} d\hat{y} \, k^2 \hat{v}_i^* \, \hat{v}_j. \tag{8.89}$$

With a counterclockwise enumeration of the vertices of the rectangle and the corresponding bilinear shape functions and setting $a = x_j - x_i$ and $b = y_k - y_i$, we get

$$S^{\text{loc}} = \begin{pmatrix} \dfrac{b^2 + a^2}{3ab} & -\dfrac{2b^2 - a^2}{6ab} & \dfrac{-b^2 - a^2}{6ab} & \dfrac{b^2 - 2a^2}{6ab} \\[2mm] -\dfrac{2b^2 - a^2}{6ab} & \dfrac{b^2 + a^2}{3ab} & \dfrac{b^2 - 2a^2}{6ab} & \dfrac{-b^2 - a^2}{6ab} \\[2mm] \dfrac{-b^2 - a^2}{6ab} & \dfrac{b^2 - 2a^2}{6ab} & \dfrac{b^2 + a^2}{3ab} & -\dfrac{2b^2 - a^2}{6ab} \\[2mm] \dfrac{b^2 - 2a^2}{6ab} & \dfrac{-b^2 - a^2}{6ab} & -\dfrac{2b^2 - a^2}{6ab} & \dfrac{b^2 + a^2}{3ab} \end{pmatrix}$$

and

$$M^{\text{loc}} = k^2 \begin{pmatrix} \dfrac{ab}{9} & \dfrac{ab}{18} & \dfrac{ab}{36} & \dfrac{ab}{18} \\[2mm] \dfrac{ab}{18} & \dfrac{ab}{9} & \dfrac{ab}{18} & \dfrac{ab}{36} \\[2mm] \dfrac{ab}{36} & \dfrac{ab}{18} & \dfrac{ab}{9} & \dfrac{ab}{18} \\[2mm] \dfrac{ab}{18} & \dfrac{ab}{36} & \dfrac{ab}{18} & \dfrac{ab}{9} \end{pmatrix}.$$

The discrete version of the right-hand side of (8.87) is computed the same way. First, the sesquilinear term $a_{\text{ext}}(v_i, g)$ is assembled on the PML domain. Second, the boundary integral term $\int_{\partial\Omega} v_i^* \nabla u_{\text{source}} \, d\mathbf{f}$ as part of (8.87) has to be computed.

Here, and in all following examples, we consider only plane waves $E \exp(i\mathbf{kr})$ as source fields. Hence, the boundary term related to the surface of the computational domain becomes

$$(b_{\text{face}}) := \int_{\text{face}} v_i^* \nabla u_{\text{source}} \, d\mathbf{f} = i E \mathbf{k} \int_{\text{face}} v_i^* \exp(i\mathbf{kr}) \, d\mathbf{f}. \tag{8.90}$$

We do it in a general 3D setting to prepare the corresponding computation for triangular meshes and for Maxwell's case.

8.4.3.5 Boundary Integral

We want to carry out the integration of (8.90)

$$(b_{\text{face}})_i = iE\mathbf{k} \int_{\text{face}} v_i^* \exp(i\mathbf{r}\mathbf{k}) \, d\mathbf{f}. \tag{8.91}$$

The situation is illustrated in Figure 8.20 for a general situation. We want to compute the integral in the reference domain. To this end, we need a parametrized description of the face. We describe the plane defined by the two vectors $\mathbf{f}_1$ and $\mathbf{f}_2$ by

$$\begin{pmatrix} x \\ y \\ z \end{pmatrix} = \begin{pmatrix} x_0 \\ y_0 \\ z_0 \end{pmatrix} + \mathbf{f}_1 \tau_1 + \mathbf{f}_2 \tau_2, \quad \tau_1, \tau_2 \in \mathbb{R}.$$

We use this parametrization to define a special mapping Q_{face}, which maps the face from the $\hat{x}\,\hat{y}$-plane of a $\hat{x}\,\hat{y}\,\hat{z}$-reference system to the original space:

$$Q_{\text{face}} : \mathbb{R}^3 \to \mathbb{R}^3$$

$$\begin{pmatrix} x \\ y \\ z \end{pmatrix} = \begin{pmatrix} x_0 \\ y_0 \\ z_0 \end{pmatrix} + \mathbf{f}_1 \hat{x} + \mathbf{f}_2 \hat{y} + \mathbf{f}_1 \times \mathbf{f}_2 \hat{z}.$$

The Jacobian of Q_{face} is

$$J_{\text{face}} = (\mathbf{f}_1, \mathbf{f}_2, \mathbf{f}_1 \times \mathbf{f}_2).$$

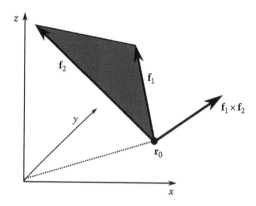

FIGURE 8.20   Boundary face in the original domain.

Hence, we can rewrite the integral (8.91) to

$$iE\mathbf{k} \int_{\text{face}} v_i^* \exp(i\mathbf{rk}) \, d\mathbf{f} \tag{8.92}$$

$$= iE\mathbf{k} \int_{\widehat{\text{face}}} \left(\hat{v}_i^* \exp(i\mathbf{rk}) \, d\hat{\mathbf{f}} \, |J|_{\text{face}} J_{\text{face}}^{-1} \right)$$

$$= iE\mathbf{k} \exp(i\mathbf{r}_0\mathbf{k}) \int_{\widehat{\text{face}}} \left(\hat{v}_i^* \exp\left(i\left(\mathbf{f}_1\hat{x} + \mathbf{f}_2\hat{y}\right)\mathbf{k}\right) d\hat{x}d\hat{y} \right) \mathbf{e}_{\hat{z}} \, |J|_{\text{face}} J_{\text{face}}^{-1}. \tag{8.93}$$

We introduce

$$\hat{I}_i := \int_{\widehat{\text{face}}} \hat{v}_i^* \exp\left(i\left(\mathbf{f}_1\hat{x} + \mathbf{f}_2\hat{y}\right)\mathbf{k}\right) d\hat{x}d\hat{y} \tag{8.94}$$

Further, using the result of Exercise 8.4, we compute

$$\mathbf{e}_{\hat{z}} \, |J|_{\text{face}} J_{\text{face}}^{-1} = (0, 0, 1) \begin{pmatrix} \mathbf{f}_2 \times (\mathbf{f}_1 \times \mathbf{f}_2) \\ (\mathbf{f}_1 \times \mathbf{f}_2) \times \mathbf{f}_1 \\ \mathbf{f}_1 \times \mathbf{f}_2 \end{pmatrix}$$

$$= \mathbf{f}_1 \times \mathbf{f}_2. \tag{8.95}$$

This is the expected result for the face element, computed based on the Jacobian. Alternatively, we could derive $d\mathbf{f} = d\mathbf{f}_1 \times d\mathbf{f}_2 = d\hat{x}\,\mathbf{f}_1 \times d\hat{y}\,\mathbf{f}_2 = d\hat{x}d\hat{y}\,\mathbf{f}_1 \times \mathbf{f}_2$. These simplifications together enable us to rewrite the integral of (8.93) to

$$\int_{\text{face}} v_i^* \exp(i\mathbf{rk}) \, d\mathbf{f} = \exp(i\mathbf{r}_0\mathbf{k}) \, \hat{I}_i \mathbf{f}_1 \times \mathbf{f}_2, \tag{8.96}$$

which yields the complete expression for the boundary integral (8.90)

$$(b_{\text{face}})_i = iE\mathbf{k} \int_{\text{face}} v_i^* \exp(i\mathbf{kr}) \, d\mathbf{f} = iE\mathbf{k} \exp(i\mathbf{r}_0\mathbf{k}) \, \hat{I}_i \mathbf{f}_1 \times \mathbf{f}_2$$

$$= iE \, |\mathbf{k}, \mathbf{f}_1, \mathbf{f}_2| \exp(i\mathbf{r}_0\mathbf{k}) \, \hat{I}_i. \tag{8.97}$$

In Exercise 8.8, it will be shown that in 2D and for linear elements, the factor $\hat{I}_i$ on a single boundary segment pointing along $\mathbf{f}_1$ computes to

$$(b_{\text{face}})_i = iE \, |\mathbf{k}, \mathbf{f}_1| \exp(i\mathbf{r}_0 \cdot \mathbf{k}) \cdot \begin{cases} \dfrac{1}{(i\mathbf{f}_1\mathbf{k})^2} \left(e^{i\mathbf{f}_1\mathbf{k}} - i\mathbf{f}_1\mathbf{k} - 1 \right) & \text{if } v_i = v_0 \\[4mm] \dfrac{1}{(i\mathbf{f}_1\mathbf{k})^2} \left((i\mathbf{f}_1\mathbf{k} - 1)e^{i\mathbf{f}_1\mathbf{k}} + 1 \right) & \text{if } v_i = v_1, \end{cases} \tag{8.98}$$

where v_0 is associated with the first point and v_1 with the last point of $\mathbf{f}_1$ and $\mathbf{k}$ and $\mathbf{f}_1$ are two component vectors defined in the 2D plane describing the geometry.

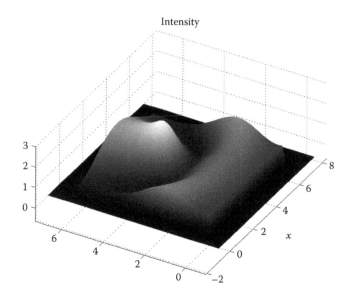

FIGURE 8.21 **(See colour insert.)** Intensity distribution related to the mesh from Figure 8.15 for a lens with $n = 2$ embedded in air.

Finally, we give an example. Using the mesh from Figure 8.15, a plane wave with $k = 1$ entering the computational domain from below, and a lens with refractive index $n = 2$ embedded in air, we compute an intensity pattern as it is depicted in Figure 8.21.

8.4.4 FINITE ELEMENTS FOR TRIANGULAR MESHES

We derive the FEM on triangles for linear shape functions. We follow the same scheme presented in Section 8.4.3 for rectangular meshes. First, we need the mapping from the unit triangle to the physical triangle as shown in Figure 8.22.

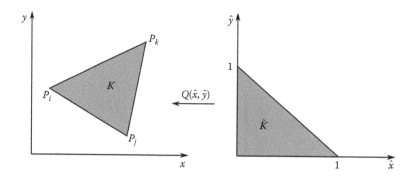

FIGURE 8.22 Map from unit triangle to the physical triangle.

We relate the given nodes P_i, P_j, and P_k to the unit coordinates $(0,0)$, $(1,0)$, and $(0,1)$ and consider the mapping

$$P = P_i + (P_j - P_i)\hat{x} + (P_k - P_i)\hat{y}. \tag{8.99}$$

Each point of K can be addressed if

$$0 \leq \hat{x}$$
$$0 \leq \hat{y}$$
$$\hat{x} + \hat{y} \leq 1.$$

Especially the last equation becomes clear, if we use the equals sign, which gives the diagonal in the right part of Figure 8.22 and in fact the line $P_k P_j^*$ in the physical domain

$$P = P_i + (P_j - P_i)\hat{x} + (P_k - P_i)(1 - \hat{x})$$
$$= P_k + (P_j - P_k)\hat{x}.$$

In components, (8.99) reads

$$x = x_i + (x_j - x_i)\hat{x} + (x_k - x_i)\hat{y}$$
$$y = y_i + (y_j - y_i)\hat{x} + (y_k - y_i)\hat{y}. \tag{8.100}$$

This gives the desired transformation from the unit triangle to the triangle with indices i, j, and k. Obviously, the Jacobian of this mapping J and its determinant $|J|$ are

$$J = \begin{pmatrix} x_j - x_i & x_k - x_i \\ y_j - y_i & y_k - y_i \end{pmatrix} \tag{8.101}$$

$$|J| = 2 \text{ times area of triangle } P_i, P_j, P_k, \tag{8.102}$$

compared to the similar result for the rectangular elements (8.85). What are the shape functions on the unit triangle? The 1D shape functions suggest

$$\hat{v}_i(\hat{x}, \hat{y}) := 1 - \hat{x} - \hat{y}$$
$$\hat{v}_j(\hat{x}, \hat{y}) = \hat{x}$$
$$\hat{v}_k(\hat{x}, \hat{y}) = \hat{y}.$$

Hence, we have everything at hand to compute the local stiffness and mass matrices and the local right-hand sides as for the rectangular cases (8.88), (8.87), and (8.89). With

$$\nabla \hat{v}_i = \begin{pmatrix} -1 \\ -1 \end{pmatrix} \quad \nabla \hat{v}_j = \begin{pmatrix} 1 \\ 0 \end{pmatrix} \quad \nabla \hat{v}_k = \begin{pmatrix} 0 \\ 1 \end{pmatrix}$$

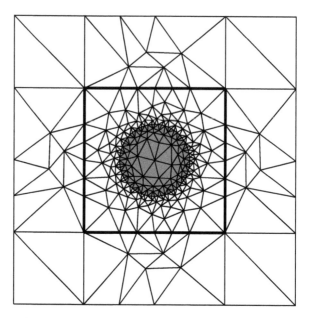

FIGURE 8.23   Triangular mesh. It is much better suited to adopt to curved geometric shapes, then the rectangular mesh (Figure 8.15). The thick line indicates the boundary on which the source field is coupled into the interior domain. The domain exterior to it is the PML domain.

we compute the local stiffness matrix according to (8.88)

$$S^{\text{loc}} = \frac{|J|}{2} \begin{pmatrix} -1 & 1 & 0 \\ -1 & 0 & 1 \end{pmatrix}^T J^{-1} J^{-T} \begin{pmatrix} -1 & 1 & 0 \\ -1 & 0 & 1 \end{pmatrix} \qquad (8.103)$$

and the local mass matrix to

$$M^{\text{loc}} = \frac{|J| \, k^2}{24} \begin{pmatrix} 2 & 1 & 1 \\ 1 & 2 & 1 \\ 1 & 1 & 2 \end{pmatrix}. \qquad (8.104)$$

Since there is no difference in the computation of the boundary term compared to the case of rectangular meshes, we refer to (8.90). Having defined the linear shape functions for triangles we can assemble the matrices for triangular meshes. Figure 8.23 shows such a triangular mesh as a counterpart to the previously discussed rectilinear mesh in Figure 8.15.

8.4.4.1   Global Data Structure and Connectivity Matrix

Once the local computations are known, we need the global data structure to assemble the entire problem. A convenient data structure consists of the three arrays, P, E, and T (the points, the edges, and the triangles, respectively), which are defined as follows:

Vertex Number	x	y		Edge Number	First Vertex	Second Vertex
$P =$ 1	x_1	y_1	$E =$	1	p_i	p_j
2	x_2	y_2		2	p_k	p_l
$\vdots$	$\vdots$	$\vdots$		$\vdots$	$\vdots$	$\vdots$

Triangle Number	First Edge	Second Edge	Third Edge	Domain Index
$T =$ 1	$\pm e_i$	$\pm e_j$	$\pm e_k$	id_1
2	$\pm e_l$	$\pm e_m$	$\pm e_n$	id_2
$\vdots$	$\vdots$	$\vdots$	$\vdots$	$\vdots$

The edges are globally oriented using the vertex number. They are positively oriented from the lower vertex number towards the higher number. The triangles are defined via the edges, where the edges locally, that is with respect to the triangle, are considered as positively oriented in counterclockwise orientation. If this orientation coincides with global orientation, the corresponding edge in the triangle structure T is given by its edge number, otherwise with its edge number multiplied with -1. We need to construct the connectivity matrix C. Suppose we are going to compute the local matrices for triangle t_l which is defined via $T(l, 2 : 4) = \{e_i, e_j, e_k\}$. The first entries of the edges are the desired global vertices, that is $v_1 = e_i(1)$, $v_2 = e_j(1)$, $v_3 = e_k(1)$. Hence, we assign the local vertices $\{1, 2, 3\}$ of the reference domain and the vertices $\{v_1, v_2, v_3\}$.

8.5 FEM FOR MAXWELL'S SCATTERING IN 2D AND 3D

All the main aspects for realizing FEM scattering problem have been discussed using the Helmholtz case. The essential new item in Maxwell's case is the treatment of vectorial fields. The deep understanding of the relation between the scalar Helmholtz case and the vectorial Maxwell's case relies greatly on the work of Whitney and Nédélec [6,7], Bosssavit [2], and Webb [10]. A huge contribution of many other authors has added to the field, which makes the FEM today a very flexible, effective, and well-studied method for electrodynamical simulations. A survey of finite element studies with focus on nano-optical applications has been given in Ref. [9].

8.5.1 Finite Elements for Rectangular Meshes

We derive the FEM for Maxwell's scattering for rectangular meshes in complete analogy to the Helmholtz case discussed in Section 8.4.3.

8.5.1.1 Polynomial Space

The polynomial space used for the construction of the trial functions in the Helmholtz case was $\mathcal{P}^{(1,1)} = \{1, \hat{x}, \hat{x}\,\hat{y}, \hat{y}\}$, that is first order in each direction. For Maxwell's case, we need vectorial functions. If we compute

$$\nabla \mathcal{P}^{(1,1)} = \left\{ \begin{pmatrix} 1 \\ 0 \end{pmatrix}, \begin{pmatrix} y \\ x \end{pmatrix}, \begin{pmatrix} 0 \\ 1 \end{pmatrix} \right\}$$

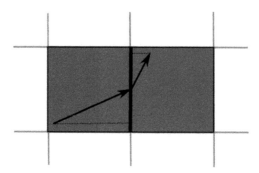

FIGURE 8.24 The projection of the field to the separating edge must be continuous. It defines the DOF attributed this edge. The union of both rectangles gives the support of the corresponding finite element.

we get three vectorial functions. This cannot be enough since the computation of the curl of each of these basis functions vanishes and we must be able to represent fields with nonvanishing curl. The most direct enhancement of this space is

$$\left\{ \begin{pmatrix} 1 \\ 0 \end{pmatrix}, \begin{pmatrix} y \\ 0 \end{pmatrix}, \begin{pmatrix} 0 \\ x \end{pmatrix}, \begin{pmatrix} 0 \\ 1 \end{pmatrix} \right\}. \tag{8.105}$$

This space has four basis functions which we want to attribute to the four edges. The reason for choosing the edges instead of the vertices as in the Helmholtz case lies in the nature of the equation. Whereas in the Helmholtz case it was reasonable to look for functions which are continuous across the element boundaries, continuity conditions for Maxwell's equations require only the tangential components to be continuous. Hence, one has to compute the tangential projection of the field to the boundary of the element, the rectangle, and to ensure that the tangential projection of two neighbouring elements to the common boundary is continuous; compare Figure 8.24. Consequently, a possible definition of the DOFs is the following.

8.5.1.2 DOFs on a Rectangle

In Ref. [4], a definition of the DOFs has been proposed, which takes care of a possible hierarchical construction of high-order elements. We restrict ourselves to the lowest-order edge elements. The DOFs associated with an edge are computed as the integrated projection along an edge:

$$M_j\left(\mathbf{v}_i(\mathbf{r})\right) = \int_{edge_j} \mathbf{v}_i(\mathbf{r}) \, d\mathbf{r}$$

$$= \delta_{ij}. \tag{8.106}$$

We look for shape functions v_j resulting from a superposition of the four functions (8.105) satisfying this relation. Obviously, we obtain this if we combine them as follows:

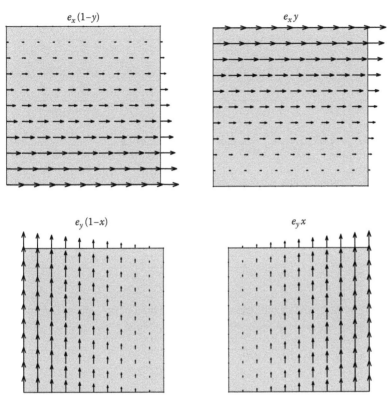

FIGURE 8.25 The four lowest-order edge shape functions on a square. Positive orientation has been chosen along the direction of the coordinate axes.

$$\hat{\mathbf{v}}_1 = \begin{pmatrix} 1 \\ 0 \end{pmatrix} - \begin{pmatrix} y \\ 0 \end{pmatrix} = \mathbf{e}_x(1 - y) \tag{8.107}$$

$$\hat{\mathbf{v}}_2 = \begin{pmatrix} y \\ 0 \end{pmatrix} = \mathbf{e}_x y \tag{8.108}$$

$$\hat{\mathbf{v}}_3 = \begin{pmatrix} 0 \\ 1 \end{pmatrix} - \begin{pmatrix} 0 \\ x \end{pmatrix} = \mathbf{e}_y(1 - x) \tag{8.109}$$

$$\hat{\mathbf{v}}_4 = \begin{pmatrix} 0 \\ x \end{pmatrix} = \mathbf{e}_y x. \tag{8.110}$$

Figure 8.25 shows the corresponding shape functions. This graphical representation makes clear that (8.106) is in fact satisfied.

8.5.1.3 Linear Finite Element Discretization

This is very similar to the procedure for the Helmholtz problem on rectangular meshes in Section 8.4.3. We use the same domain mapping (8.84) and the same Jacobi matrix (8.85). The difference lies in the different mass and stiffness matrices, the different right-hand side, and the different connectivity matrix. According to Maxwell's scattering problem formulation (8.76), we have to evaluate

$$(A)_{i,j} = a\left(\mathbf{v}_i, \mathbf{v}_j\right) = \underbrace{\int_{\Omega_{cd}} dx\,dy\,\left(\nabla \times \mathbf{v}_i^*\right) \cdot \frac{1}{\mu}\left(\nabla \times \mathbf{v_j}\right)}_{\text{stiffness part } s_{ij}} - \underbrace{\omega^2 \int_{\Omega_{cd}} dx\,dy\, \mathbf{v}_i^* \varepsilon \mathbf{v}_j}_{\text{mass part } m_{ij}}$$

$$(b)_i = \int_{\partial\Omega} d\mathbf{f} \cdot \left(\mathbf{v}_i^* \times \left[\frac{1}{\mu}\nabla \times \mathbf{E}_{\text{source}}\right]\right) + a_{\text{ext}}(\mathbf{v}_i, \mathbf{g}). \qquad (8.111)$$

Using our mapping rules, we compute the elements of the local mass and stiffness matrices in the reference domain as

$$s_{ij} = \int_{\hat{K}} d\hat{x}\,d\hat{y}\,\left(\nabla_{\hat{x}\hat{y}} \times \hat{\mathbf{v}}_i^*\right) \cdot (\mu^c)^{-1}\left(\nabla \times \hat{\mathbf{v}}_j\right) \qquad (8.112)$$

$$m_{ij} = \omega^2 \int_{\Omega_{cd}} d\hat{x}\,d\hat{y}\, \hat{\mathbf{v}}_i^* \varepsilon^c \hat{\mathbf{v}}_j. \qquad (8.113)$$

As shown in (8.61) and (8.62), the entire transformation, both the shape transformation *and* the complexification for the PML, is contained in the transformed material quantities

$$\varepsilon^c = |J|\,J^{-1}\varepsilon J^{-T}$$

$$(\mu^c)^{-1} = \frac{1}{|J|}J^T\mu^{-1}J.$$

Using these expressions together with shape functions $\hat{\mathbf{v}}_1, \ldots, \hat{\mathbf{v}}_4$, we obtain

$$M^{\text{loc}} = \omega^2 \begin{pmatrix} \frac{b}{3a} & \frac{b}{6a} & 0 & 0 \\ \frac{b}{6a} & \frac{b}{3a} & 0 & 0 \\ 0 & 0 & \frac{a}{3b} & \frac{a}{6b} \\ 0 & 0 & \frac{a}{6b} & \frac{a}{3b} \end{pmatrix}$$

and

$$S^{\text{loc}} = \frac{1}{ab} \begin{pmatrix} 1 & -1 & -1 & 1 \\ -1 & 1 & 1 & -1 \\ -1 & 1 & 1 & -1 \\ 1 & -1 & -1 & 1 \end{pmatrix}.$$

The expression $a_{\text{ext}}(\mathbf{v}_i, \mathbf{g})$ of the right-hand side of (8.111) is computed exactly the same way, just restricted to the exterior domain. The remaining boundary term, associated to a particular face (an edge in 2D), is

$$(b_{\text{face}})_i = \int_{\widehat{\text{face}}} d\hat{\mathbf{f}} \cdot \left(\hat{\mathbf{v}}_i^* \times \left[(\mu^c)^{-1}\left(\nabla \times \mathbf{E}_{\text{source}}\right)_{\hat{x},\hat{y},\hat{z}}\right]\right). \qquad (8.114)$$

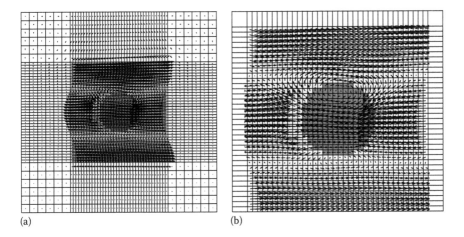

FIGURE 8.26 Maxwell's scattering of a plane wave entering from below from a cylinder computed based on first-order rectangular elements. (a) Full domain, including PML domain. (b) Zoom.

Its evaluation in case that $\mathbf{E}_{source}$ is a plane wave $\mathbf{E}_0 \exp(i\mathbf{kr})$ can be done in a similar way as in the Helmholtz case (8.90). The expression $(\mu^c)^{-1}$ is a constant matrix on the rectangle. Further, it holds $\nabla \times \mathbf{E}_0 \exp(i\mathbf{kr}) = i\mathbf{k} \times \mathbf{E}_0 \exp(i\mathbf{kr})$. Hence, one has to compute

$$(b_{\text{face}})_i = \exp(i\mathbf{kr}_0) \int_{\widehat{\text{face}}} \exp(i\mathbf{kr}) \left(d\hat{\mathbf{f}} \cdot \left(\hat{\mathbf{v}}_i^*(\mathbf{r}) \times \left[(\mu^c)^{-1} (i\mathbf{k} \times \mathbf{E}_0)_{\hat{x},\hat{y},\hat{z}} \right] \right) \right),$$

where $\mathbf{r}_0$ is the lower-left corner at the global position of the rectangle. Cyclic interchange in the triple product yields

$$(b_{\text{face}})_i = \exp(i\mathbf{kr}_0) \left[(\mu^c)^{-1} (i\mathbf{k} \times \mathbf{E}_0)_{\hat{x},\hat{y},\hat{z}} \right] \cdot \int_{\widehat{\text{face}}} \exp(i\mathbf{kr}) \left(d\hat{\mathbf{f}} \times \hat{\mathbf{v}}_i^* \right). \qquad (8.115)$$

This is of a very similar structure as (8.90) and will be computed the same way. In Exercise 8.10, we study reformulations of this expression and its application to plane waves. Figure 8.26 shows the scattering of a plane wave from a cylinder simulated with vectorial linear elements, compare with the scalar version depicted in Figure 8.21.

8.5.2 FINITE ELEMENTS FOR TRIANGULAR MESHES

We follow the presentation in Section 8.4.4.

8.5.2.1 Triangular Elements

We want to find a representation of the lowest-order finite elements for triangles comparable to those for rectangles (see again Figure 8.25). The main difficulty here is to suggest a suitable polynomial space for the approximation of the vectorial fields on a triangle. Whereas the rectangular case could be treated just by extending the 1D case using tensor products, a new idea for triangles is needed.

8.5.2.2 Polynomial Space

The polynomial space should be the lowest-order space where the **curl**-operation does not vanish. The most simple polynomial function of this kind is

$$\mathbf{v} = \begin{pmatrix} -y \\ x \end{pmatrix},$$

whose center is the origin. Therefore, taking the reference triangle as in Figure 8.22, we shift this vector field to the three vertices of the reference triangle and obtain a guess for the vectorial polynomial space:

$$\left\{ \begin{pmatrix} -y \\ x \end{pmatrix}, \begin{pmatrix} -y \\ x-1 \end{pmatrix}, \begin{pmatrix} -(y-1) \\ x \end{pmatrix} \right\}. \tag{8.116}$$

This is the counterpart of the polynomial space (8.105), the unit square with four basis elements. As in the rectangular case, we must distribute the basis functions such that the resulting shape functions can be attributed to the edges of the triangle.

8.5.2.3 DOFs on a Triangle

We use again (8.106) and apply it to the edges of the reference triangle. Let us compute $\mathbf{v}_1$ which should be assigned to the edge $(0,0) - (1,0)$. We set $\mathbf{v}_1$ as an arbitrary superposition of the basis functions (8.116):

$$\hat{\mathbf{v}}_1 = a_1 \begin{pmatrix} -\hat{y} \\ \hat{x} \end{pmatrix} + a_2 \begin{pmatrix} -\hat{y} \\ \hat{x}-1 \end{pmatrix} + a_3 \begin{pmatrix} -(\hat{y}-1) \\ \hat{x} \end{pmatrix}$$

with unknown coefficients a_1, a_2, and a_3. Application of (8.106) to the first edge $(0,0) - (1,0)$ results in

$$1 \overset{!}{=} \int_0^1 d\hat{x}\, \mathbf{e}_{\mathbf{x}} \cdot \left(a_1 \begin{pmatrix} 0 \\ \hat{x} \end{pmatrix} + a_2 \begin{pmatrix} 0 \\ \hat{x}-1 \end{pmatrix} + a_3 \begin{pmatrix} 1 \\ \hat{x} \end{pmatrix} \right)$$

$$= \int_0^1 d\hat{x}\, a_3.$$

Hence, we obtain $a_3 = 1$, and performing the edge integration along the other two edges, we find $a_1 = a_2 = 0$ and therefore

$$\hat{\mathbf{v}}_1 = \begin{pmatrix} -(\hat{y}-1) \\ \hat{x} \end{pmatrix}. \tag{8.117}$$

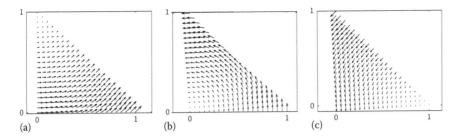

FIGURE 8.27   The lowest-order edge elements for triangles. (a) Linear element associated with edge between vertices $(0; 0)$; $(0; 1)$, (b) linear element associated with edge between vertices $(0; 0)$; $(0; 1)$, (c) linear element associated with edge between vertices $(0; 1)$; $(0; 0)$.

The same way, we get the other two shape functions

$$\hat{\mathbf{v}}_2 = \begin{pmatrix} -\hat{y} \\ \hat{x} \end{pmatrix} \qquad \hat{\mathbf{v}}_3 = \begin{pmatrix} -\hat{y} \\ \hat{x} - 1 \end{pmatrix}. \tag{8.118}$$

Figure 8.27 displays the computed shape functions.

8.5.2.4   Linear Finite Element Discretization

As in the other linear finite element cases, we can compute the stiffness and the mass matrix and the contribution to the right-hand side directly. We refer again to Figure 8.22 and introduce the two vectors from point P_i to the points P_j and P_k

$$\mathbf{l}_2 = P_j - P_i \quad \mathbf{l}_3 = P_k - P_i.$$

The square of the Euclidean length of these vectors is $\|\mathbf{l}_2\|^2 = \mathbf{l}_2 \cdot \mathbf{l}_2$, $\|\mathbf{l}_3\|^2 = \mathbf{l}_2 \cdot \mathbf{l}_2$, and $\mathbf{l}_2 \cdot \mathbf{l}_3$ denotes the scalar product between these vectors. We have to compute the local mass and stiffness matrices (8.112) and (8.112). A direct computation yields

$$M^{\text{loc}} = \frac{\varepsilon}{12\,|J|}$$

$$\times \begin{pmatrix} \|\mathbf{l}_2\|^2 + 3\,\|\mathbf{l}_3\|^2 - \mathbf{l}_2 \cdot \mathbf{l}_3 & \|\mathbf{l}_2\|^2 - \|\mathbf{l}_3\|^2 - \mathbf{l}_2 \cdot \mathbf{l}_3 & -\|\mathbf{l}_2\|^2 - \|\mathbf{l}_3\|^2 - 3\mathbf{l}_2 \cdot \mathbf{l}_3 \\ \|\mathbf{l}_2\|^2 - \|\mathbf{l}_3\|^2 - \mathbf{l}_2 \cdot \mathbf{l}_3 & \|\mathbf{l}_2\|^2 + \|\mathbf{l}_3\|^2 + \mathbf{l}_2 \cdot \mathbf{l}_3 & -\|\mathbf{l}_2\|^2 + \|\mathbf{l}_3\|^2 - \mathbf{l}_2 \cdot \mathbf{l}_3 \\ -\|\mathbf{l}_2\|^2 - \|\mathbf{l}_3\|^2 + 3\mathbf{l}_2 \cdot \mathbf{l}_3 & -\|\mathbf{l}_2\|^2 + \|\mathbf{l}_3\|^2 - \mathbf{l}_2 \cdot \mathbf{l}_3 & 3\,\|\mathbf{l}_2\|^2 + \|\mathbf{l}_3\|^2 - 3\mathbf{l}_2 \cdot \mathbf{l}_3 \end{pmatrix}$$

and

$$S^{\text{loc}} = \frac{2}{\mu\,|J|} \begin{pmatrix} 1 & 1 & 1 \\ 1 & 1 & 1 \\ 1 & 1 & 1 \end{pmatrix}.$$

As an example for the mass matrix computation, the coefficient m_{22} is computed in Exercise 8.7. The contribution of a plane wave to the right-hand side follows directly from (8.115) and the definition of the shape functions (8.117) and (8.118).

EXERCISES

8.1 Segment mapping
Show that the linear mapping (8.18) based on the hat functions is in fact a linear scaling from the unit interval to the physical interval $[x_i, x_j]$.
Answer

$$x = x_i \widehat{v_1}(\hat{x}) + x_j \widehat{v_2}(\hat{x})$$
$$= x_i(1 - \hat{x}) + x_j \hat{x}$$
$$= x_i + \hat{x}(x_j - x_i), \quad \hat{x} \in [0, 1].$$

8.2 Uniform discretization in 1D
Given a uniform discretization of interval $(0, L)$ into N segments with stepsize $h = L/N$.
(a) Show that the $N \times N$ system matrix A from Algorithm 8.1 takes the form

$$A = \frac{1}{h}\begin{pmatrix} 2 & -1 & & & \\ -1 & 2 & -1 & & \\ & & \ddots & & \\ & & -1 & 2 & -1 \\ & & & -1 & 1 \end{pmatrix} - \begin{pmatrix} 0 & & & \\ & \ddots & & \\ & & 0 & \\ & & & ik \end{pmatrix}$$

$$- \frac{hk^2}{6}\begin{pmatrix} 4 & 1 & & & \\ 1 & 4 & 1 & & \\ & & \ddots & & \\ & & 1 & 4 & 1 \\ & & & 1 & 2 \end{pmatrix}.$$

(b) Compare the result to the discretization of the Helmholtz equation when a uniform finite- difference stencil is applied. What is the difference?
Answer
(a) Just add the local matrices.
(b) Only the mass matrix is different. In the finite difference, it is an identity matrix.

8.3 Rectangular domain mapping
Show that the bilinear domain mapping (8.119) in fact maps the unit square to a quadrilateral.
Answer The mapping $(\hat{x}, \hat{y}) \mapsto (x, y)$ is given by

$$\begin{pmatrix} x \\ y \end{pmatrix} = P_i(1 - \hat{x})(1 - \hat{y}) + P_j\hat{x}(1 - \hat{y}) + P_k(1 - \hat{x})\hat{y} + P_l\hat{x}\hat{y} \quad (8.119)$$

1. Obviously, the boundary of the unit square maps to the boundary of the quadrilateral. For example setting $\hat{y} = 0$ and letting $\hat{x}$ move from 0 to 1 moves the point (x, y) from P_i to P_j.

2. If we fix $\hat{y}$, the mapping is linear in $\hat{x}$ and vice versa.
3. If we fix $\hat{y}$ and let $\hat{x}$ go from 0 to 1, the entire line lies in the convex hull of the given four points.

8.4 Face transform

Equation (8.56) shows that the coordinates of a face vector in the reference system transform to the coordinates in the physical system via right multiplication with $|J| J^{-1}$. Suppose the constant Jacobian is given by three column vectors

$$J = (\mathbf{a}, \mathbf{b}, \mathbf{c}). \tag{8.120}$$

Show the following:

$$|J| J^{-1} = \begin{pmatrix} \mathbf{b} \times \mathbf{c} \\ \mathbf{c} \times \mathbf{a} \\ \mathbf{a} \times \mathbf{b} \end{pmatrix}, \tag{8.121}$$

where the vectors resulting from the cross products are row vectors.

Answer

We compute the inverse matrix of the Jacobi matrix:

$$\begin{pmatrix} a_1 & b_1 & c_1 \\ a_2 & b_2 & c_2 \\ a_3 & b_3 & c_3 \end{pmatrix} \cdot J^{-1} = \begin{pmatrix} 1 & & \\ & 1 & \\ & & 1 \end{pmatrix}.$$

Let us compute the first row of J^{-1} by means of Cramer's rule:

$$\left(J^{-1}\right)_{11} = \frac{1}{|J|} \begin{vmatrix} 1 & b_1 & c_1 \\ 0 & b_2 & c_2 \\ 0 & b_3 & c_3 \end{vmatrix} = \frac{1}{|J|} (b_2 c_3 - b_3 c_2)$$

$$\left(J^{-1}\right)_{12} = \frac{1}{|J|} \begin{vmatrix} 0 & b_1 & c_1 \\ 1 & b_2 & c_2 \\ 0 & b_3 & c_3 \end{vmatrix} = -\frac{1}{|J|} (b_1 c_3 - b_3 c_1)$$

$$\left(J^{-1}\right)_{13} = \frac{1}{|J|} \begin{vmatrix} 0 & b_1 & c_1 \\ 0 & b_2 & c_2 \\ 1 & b_3 & c_3 \end{vmatrix} = \frac{1}{|J|} (b_1 c_2 - b_2 c_1).$$

Hence, the components of the first row are computed by $(\mathbf{b} \times \mathbf{c})/|J|$. The other rows follow the same way.

8.5 Transform of μ^{-1}

Let the Jacobian be given as in Exercise 8.4 and let in the physical domain the permeability μ be a scalar. Show that

$$\hat{\mu}^{-1} = \frac{\mu^{-1}}{|J|} \begin{pmatrix} \mathbf{aa} & \mathbf{ab} & \mathbf{ac} \\ \mathbf{ab} & \mathbf{bb} & \mathbf{bc} \\ \mathbf{ac} & \mathbf{bc} & \mathbf{cc} \end{pmatrix}$$

using (8.62). Show further that this simplifies in 2D to

$$\hat{\mu}^{-1} = \frac{\mu^{-1}}{|J|} \begin{pmatrix} aa & ab & 0 \\ ab & bb & 0 \\ 0 & 0 & 1 \end{pmatrix}$$

with **a** and **b** vectors in the *xy*-plane and $z = \hat{z}$.

Answer

Apply

$$J = (\mathbf{a}, \mathbf{b}, \mathbf{c}) \tag{8.122}$$

to the definition (8.61). For the second part, use the Jacobian

$$J = \begin{pmatrix} a_x & b_x & 0 \\ a_y & b_y & 0 \\ 0 & 0 & 1 \end{pmatrix}.$$

8.6 Transform of ε

Let the Jacobian be given as in Exercise 8.4 and let the permittivity ε be a scalar. Show that

$$\hat{\varepsilon} = \frac{\varepsilon}{|J|} \begin{pmatrix} (\mathbf{b} \times \mathbf{c})^2 & -ab & -ac \\ -ab & (\mathbf{c} \times \mathbf{a})^2 & -bc \\ -ac & -bc & (\mathbf{a} \times \mathbf{b})^2 \end{pmatrix}$$

using (8.61). Show further that this simplifies in 2D to

$$\hat{\varepsilon} = \frac{\varepsilon}{|J|} \begin{pmatrix} bb & -ab & 0 \\ -ab & aa & 0 \\ 0 & 0 & (\mathbf{a} \times \mathbf{b})^2 \end{pmatrix}$$

with **a** and **b** vectors in the *xy*-plane and $z = \hat{z}$.

Answer

$$\hat{\varepsilon} = |J| J^{-1} \varepsilon J^{-T}$$

$$= \begin{pmatrix} \mathbf{b} \times \mathbf{c} \\ \mathbf{c} \times \mathbf{a} \\ \mathbf{a} \times \mathbf{b} \end{pmatrix} \varepsilon \frac{|J|}{|J|} J^{-T}$$

$$= \frac{\varepsilon}{|J|} \begin{pmatrix} \mathbf{b} \times \mathbf{c} \\ \mathbf{c} \times \mathbf{a} \\ \mathbf{a} \times \mathbf{b} \end{pmatrix} \cdot (\mathbf{b} \times \mathbf{c}, \quad \mathbf{c} \times \mathbf{a}, \quad \mathbf{a} \times \mathbf{b}),$$

which gives the desired result using $\mathbf{c} = (0, 0, 1)^T$ and, for example

$$(\mathbf{c} \times \mathbf{a}) \cdot (\mathbf{b} \times \mathbf{c}) = -ab.$$

The reduction to 2D follows as in Exercise 8.5.

8.7 Helmholtz equation: Local mass matrix elements for triangles

As an example of the mass matrix computation, determine the coefficient m_{22} of the mass matrix for the linear shape functions on the triangle. Use the shape functions (8.118) and the transformed permittivity $\hat{\epsilon}$ from (8.61).

Answer

$$m_{22} = \int_0^1 \int_0^{1-\hat{y}} \hat{\mathbf{v}}_2^T \hat{\epsilon} \, \hat{\mathbf{v}}_2 \, d\hat{x} d\hat{y}$$

$$= \int_0^1 \int_0^{1-\hat{y}} (-\hat{y}, \hat{x}, 0) \, \hat{\epsilon} \begin{pmatrix} -\hat{y} \\ \hat{x} \\ 0 \end{pmatrix} d\hat{x} d\hat{y}$$

$$= \int_0^1 \int_0^{1-\hat{y}} (\hat{y}^2 \mathbf{b}^2 + 2\mathbf{ab}\hat{x}\hat{y} + \hat{x}^2 \mathbf{a}^2) \, d\hat{x} d\hat{y}$$

$$= \frac{\mathbf{b}^2 + \mathbf{ab} + \mathbf{a}^2}{12}$$

8.8 Helmholtz equation 2D: Boundary integral terms with linear elements

Consider a y-invariant structure whose geometry is completely described by a piecewise linear polygon in the xz-plane. We pick one boundary segment $\mathbf{f}_1 = (x_1 - x_0, z_1 - z_0)$ pointing from (x_0, z_0) to (x_1, z_1) in counterclockwise direction. Let the linear finite element function assigned to point (x_0, z_0) be v_0 and the one assigned to point (x_1, z_1) be v_1. Show that the contribution of this boundary segment to the boundary integral (8.90) becomes

$$(b_{\text{face}})_i = iE \, |\mathbf{k}, \mathbf{f}_1| \exp(i\mathbf{r}_0 \cdot \mathbf{k}) \cdot \begin{cases} \dfrac{1}{(i f_1 \mathbf{k})^2} \left(e^{i f_1 \mathbf{k}} - i f_1 \mathbf{k} - 1 \right) & \text{if } v_i = v_0 \\ \dfrac{1}{i f_1 \mathbf{k})^2} \left((i f_1 \mathbf{k} - 1) e^{i f_1 \mathbf{k}} + 1 \right) & \text{if } v_i = v_1. \end{cases}$$

$$(8.123)$$

Answer

Due to the structure of the problem, we consider only wave vectors $\mathbf{k}$ with components in the xz-plane, $\mathbf{k} = (k_x, 0, k_z)^T$. We start with (8.90)

$$(b_{\text{face}})_i = \left(\int_{\text{face}} d\mathbf{f} \, v_i^* \exp(i\mathbf{k}\mathbf{r}) \right) iE\mathbf{k}$$

and map the integral to the reference coordinates

$$(b_{\text{face}})_i = \left(\int_{\widehat{\text{face}}} d\hat{\mathbf{f}} \, |J| J^{-1} v_i^* \exp(i\mathbf{k}\mathbf{r}) \right) iE\mathbf{k}.$$

Using $d\hat{\mathbf{f}} = d\hat{x}d\hat{y}\,\mathbf{e}_z$ and (8.95), this means

$$(b_{\text{face}})_i = \left(\int_{\widehat{\text{face}}} d\hat{x}d\hat{y}\,\mathbf{e}_y \times \mathbf{f}_1 v_i^* \exp{(i\mathbf{kr})} \right) iE\mathbf{k}.$$

Using $\mathbf{r} = \mathbf{r}_0 + \mathbf{f}_1\hat{x} + \mathbf{e}_y\hat{y}$, we obtain

$$(b_{\text{face}})_i = \left(\int_0^1 \int_0^1 d\hat{x}d\hat{y}\,v_i^* \exp{(i\mathbf{kr})} \right) i\mathbf{k} \cdot \left(\mathbf{e}_y \times \mathbf{f}_1 \right) E$$

$$= \exp(i\mathbf{r}_0 \cdot \mathbf{k}) \left(\int_0^1 \int_0^1 d\hat{x}d\hat{y}\,v_i^* \exp\!\left(i\left(\mathbf{f}_1\hat{x} + \mathbf{e}_y\hat{y}\right) \cdot \mathbf{k}\right) \right) i\mathbf{k} \cdot \left(\mathbf{e}_y \times \mathbf{f}_1 \right) E$$

$$= \exp(i\mathbf{r}_0 \cdot \mathbf{k}) \left(\int_0^1 d\hat{x}\,v_i^* \exp{\left(i\hat{x}\mathbf{f}_1 \cdot \mathbf{k}\right)} \right) i\mathbf{k} \cdot \left(\mathbf{e}_y \times \mathbf{f}_1 \right) E$$

using

$$\mathbf{k} \cdot \left(\mathbf{e}_y \times \mathbf{f}_1 \right) = \left| \mathbf{k}, \mathbf{e}_y, \mathbf{f}_1 \right| = -\begin{vmatrix} 0 & k_x & f_{1x} \\ 1 & 0 & 0 \\ 0 & k_z & f_{1z} \end{vmatrix} = \begin{vmatrix} k_x & f_{1x} \\ k_z & f_{1z} \end{vmatrix} = |\mathbf{k}, \mathbf{f}_1| .$$

Performing the integration, we get the desired result as follows:

$$(b_{\text{face}})_i = iE\,|\mathbf{k}, \mathbf{f}_1|\exp{(i\mathbf{r}_0 \cdot \mathbf{k})} \begin{cases} \dfrac{1}{(i\mathbf{f}_1\mathbf{k})^2}\left(e^{i\mathbf{f}_1\mathbf{k}} - i\mathbf{f}_1\mathbf{k} - 1\right) & \text{if } v_i = v_0 \\[3mm] \dfrac{1}{i\mathbf{f}_1\mathbf{k})^2}\left((i\mathbf{f}_1\mathbf{k} - 1)e^{i\mathbf{f}_1\mathbf{k}} + 1\right) & \text{if } v_i = v_1. \end{cases}$$

$$(8.124)$$

8.9 Plane-wave propagation in 2D: Variational formulation

The plane-wave propagation on a homogeneous domain is the simplest type of a scattering problem, since there is no scattered field at all, just the propagating source field. Let an axis-parallel rectangular domain be given. Let the plane wave be $u_{\text{source}} = \exp(ikx)$, with $k = |\mathbf{k}|$ the modulus of the wave vector. Show that the following is a possible variational formulation:

$$\text{Find } u \in V: \quad a(v, u) = \int_{\partial\Omega} v^* iku_{\text{source}}\,d\mathbf{s} \quad \text{for all } v \in V$$

with

$$a(v, u) = \int_\Omega \nabla v^* \cdot \nabla u\,dx\,dy - \int_\Omega v^* k^2 u\,dx\,dy.$$

Answer

Start from

$$\int_{\partial\Omega} v^*\nabla u\, ds - \int_\Omega \nabla v^* \cdot \nabla u\, dx\, dy + \int_\Omega v^* k^2 u\, dx\, dy = 0.$$

This gives

$$\int_\Omega \nabla v^* \cdot \nabla u\, dx\, dy - \int_\Omega v^* k^2 u\, dx\, dy = \int_{\partial\Omega} v^*\nabla u\, ds$$

$$= \int_{\partial\Omega} v^*\nabla u_{\text{source}}\, ds$$

$$= \int_{\partial\Omega} v^* i k u_{\text{source}}\, ds,$$

since there is no scattered field.

8.10 Maxwell's equations: General boundary integral term

Let $\mathbf{f}_1$ and $\mathbf{f}_2$ span a face segment in 3D.

(a) Show that the contribution of this face to the boundary integral (8.114) becomes

$$(b_{\text{face}})_i = -\int_{\widehat{\text{face}}} d\hat{x}d\hat{y}\,\widehat{\mathbf{v}}^*_i \cdot \frac{1}{\mu}\begin{pmatrix} -\mathbf{f}_2 \cdot (\nabla \times \mathbf{E}_{\text{source}}) \\ \mathbf{f}_1 \cdot (\nabla \times \mathbf{E}_{\text{source}}) \\ 0 \end{pmatrix}. \tag{8.125}$$

(b) Derive the corresponding equation if the source field is a plane wave $\mathbf{E} = \mathbf{E}_0 \exp(i\mathbf{r}\mathbf{k})$

Hint: Proceed as follows:

1. Start from the boundary term formulation (8.78).
2. Transform $\nabla \times \hat{\mathbf{E}}$ back to original coordinates using (8.59).
3. Express $d\hat{\mathbf{A}}$ via unit vectors.

Answer

We start from (8.78)

$$(b_{\text{face}})_i = -\int_{\widehat{\text{face}}} \left\langle \widehat{\mathbf{v}}^*_i, d\hat{\mathbf{A}}^T \times \left[J^T \frac{1}{\mu}J\frac{1}{|J|}\nabla \times \hat{\mathbf{E}} \right] \right\rangle$$

Here, all quantities are given in reference coordinates. Since we mapped the face to the xy-plane, we have $d\hat{\mathbf{A}} = dx\, dy\, \mathbf{e}_x \times \mathbf{e}_y = dx\, dy\, \mathbf{e}_z$. Further, we transform $\nabla \times \hat{\mathbf{E}}$ back to original coordinates to perform a direct computation of $\nabla \times \mathbf{E}$. We get

$$(b_{\text{face}})_i = -\int_{\widehat{\text{face}}} d\hat{x}d\hat{y} \left\langle \widehat{\mathbf{v}}^*_i, \mathbf{e}_z \times \left[J^T \frac{1}{\mu}\nabla \times \mathbf{E} \right] \right\rangle.$$

For the expression in square brackets, we compute

$$\left[J^T \frac{1}{\mu} \nabla \times \mathbf{E} \right] = \frac{1}{\mu} \begin{pmatrix} \mathbf{f}_1^T \\ \mathbf{f}_2^T \\ (\mathbf{f}_1 \times \mathbf{f}_2)^T \end{pmatrix} \nabla \times \mathbf{E}$$

$$= \frac{1}{\mu} \begin{pmatrix} \mathbf{f}_1 \cdot (\nabla \times \mathbf{E}) \\ \mathbf{f}_2 \cdot (\nabla \times \mathbf{E}) \\ (\mathbf{f}_1 \times \mathbf{f}_2) \cdot (\nabla \times \mathbf{E}) \end{pmatrix}.$$

Consequently, we find

$$\mathbf{e}_z \times \left[J^T \frac{1}{\mu} \nabla \times \mathbf{E} \right] = \frac{1}{\mu} \begin{pmatrix} -\mathbf{f}_2 \cdot (\nabla \times \mathbf{E}) \\ \mathbf{f}_1 \cdot (\nabla \times \mathbf{E}) \\ 0 \end{pmatrix}.$$

With respect to the boundary integral, this gives

$$(b_{\text{face}})_i = -\int\limits_{\text{face}} d\hat{x} d\hat{y}\, \widehat{\mathbf{v}}^*_i \cdot \frac{1}{\mu} \begin{pmatrix} -\mathbf{f}_2 \cdot (\nabla \times \mathbf{E}) \\ \mathbf{f}_1 \cdot (\nabla \times \mathbf{E}) \\ 0 \end{pmatrix}.$$

In the case of a plane wave $\mathbf{E} = \mathbf{E}_0 \exp(i\mathbf{r}\mathbf{k})$, this results in

$$(b_{\text{face}})_i = -\int\limits_{\text{face}} d\hat{x} d\hat{y}\, \widehat{\mathbf{v}}^*_i \cdot \frac{1}{\mu} \begin{pmatrix} -\mathbf{f}_2 \cdot (i\mathbf{k} \times \mathbf{E}) \\ \mathbf{f}_1 \cdot (i\mathbf{k} \times \mathbf{E}) \\ 0 \end{pmatrix}.$$

Using $\mathbf{r} = \mathbf{r}_0 + \mathbf{f}_1\hat{x} + \mathbf{f}_2\hat{y}$, this results in

$$(b_{\text{face}})_i = e^{i\mathbf{r}_0 \cdot \mathbf{k}} \int\limits_{\text{face}} d\hat{x} d\hat{y}\, e^{i(\mathbf{f}_1\hat{x} + \mathbf{f}_2\hat{y}) \cdot \mathbf{k}} \widehat{\mathbf{v}}^*_i \cdot \frac{1}{\mu} \begin{pmatrix} \mathbf{f}_2 \cdot (i\mathbf{k} \times \mathbf{E}_0) \\ -\mathbf{f}_1 \cdot (i\mathbf{k} \times \mathbf{E}_0) \\ 0 \end{pmatrix}. \qquad (8.126)$$

8.11 Maxwell's equations: Boundary integral terms in 2D with linear elements

This is Maxwell's analogue to Exercise 8.8. Consider the same 2D problem with a piecewise linear polygon as boundary. We pick one boundary segment $\mathbf{f}_1 = (x_1 - x_0, z_1 - z_0)$ pointing from (x_0, z_0) to (x_1, z_1) in counterclockwise direction.

Let the lowest-order finite element function assigned to this edge and let edge and field have the same orientation. Show that the contribution of this boundary segment to the boundary integral (8.115) becomes

$$(b_{\text{face}})_i = i\,|\mathbf{k}, \mathbf{E}_0|\, e^{i\mathbf{r}_0 \cdot \mathbf{k}} \frac{1}{i\mathbf{f}_1 \cdot \mathbf{k}} \left(e^{i\mathbf{f}_1 \cdot \mathbf{k}} - 1 \right),$$

with $|\mathbf{k}, \mathbf{E}_0|$ the determinant of the column vectors $\mathbf{k}, \mathbf{E}_0$.

Answer
From (8.107), we know the linear edge element assigned to the x-axis of the unit square element

$$\hat{\mathbf{v}}_1 = \begin{pmatrix} 1 - \hat{z} \\ 0 \\ 0 \end{pmatrix}$$

and, from (8.117), the linear edge element assigned to the x-axis of the unit triangular element

$$\hat{\mathbf{v}}_1 = \begin{pmatrix} -(\hat{z} - 1) \\ 0 \\ \hat{x} \end{pmatrix}.$$

Based on (8.126), we get

$$(b_{\text{face}})_i = e^{i\mathbf{r}_0 \cdot \mathbf{k}} \int_0^1 d\hat{x}\, e^{i\hat{x}\mathbf{f}_1 \cdot \mathbf{k}} \hat{\mathbf{v}}^*_i \cdot \frac{1}{\mu} \begin{pmatrix} -\mathbf{e}_y \cdot (i\mathbf{k} \times \mathbf{E}_0) \\ -\mathbf{f}_1 \cdot (i\mathbf{k} \times \mathbf{E}_0) \\ 0 \end{pmatrix}.$$

For both test functions, we get

$$(b_{\text{face}})_i = e^{i\mathbf{r}_0 \cdot \mathbf{k}} \int_0^1 d\hat{x}\, e^{i\hat{x}(-\mathbf{e}_y) \cdot \mathbf{k}} \frac{1}{\mu} (-\mathbf{e}_y) \cdot (i\mathbf{k} \times \mathbf{E}_0)$$

$$= i \left| -\mathbf{e}_y, \mathbf{k}, \mathbf{E}_0 \right| e^{i\mathbf{r}_0 \cdot \mathbf{k}} \int_0^1 d\hat{x}\, e^{i\hat{x}\mathbf{f}_1 \cdot \mathbf{k}}$$

$$= i \left| \mathbf{k}, \mathbf{E}_0 \right| e^{i\mathbf{r}_0 \cdot \mathbf{k}} \frac{1}{i\mathbf{f}_1 \cdot \mathbf{k}} \left(e^{i\mathbf{f}_1 \cdot \mathbf{k}} - 1 \right).$$

REFERENCES

1. R. Becker and R. Rannacher, An optimal control approach to a posteriori error estimation in finite element methods, *Acta Numer.*, 10(1), 1–102 (2001).
2. A. Bossavit, Whitney forms: A class of finite elements for three-dimensional computations in electromagnetism, *IEEE Proc. A Phys. Sci. Meas. Instrum., Manage. Educ. Rev.*, 135(8), 493–500 (1988).
3. L. Demkowicz, *Computing with hp-Adaptive Finite Elements: Volume 1. One and Two Dimensional Elliptic and Maxwell Problems*. Boca Raton, FL: CRC Press, 2006.
4. L. Demkowicz, P. Monk, L. Vardapetyan, and W. Rachowicz, De Rham diagram for hp finite element spaces, *Comput. Math. Appl.*, 39(7), 29–38 (2000).
5. P. Monk, *Finite Element Methods for Maxwell's Equations*. Oxford, U.K.: Oxford University Press, 2003.
6. J.-C. Nédélec, Mixed finite elements in $\mathbb{R}^3$, *Num. Math.*, 35(3), 315–341 (1980).
7. J.-C. Nédélec, A new family of mixed finite elements in $\mathbb{R}^3$. *Num. Math.*, 50, 57–81 (1986).

8. D. Pardo, L. Demkowicz, C. Torres-Verdan, and M. Paszynski, A self-adaptive goal-oriented hp-finite element method with electromagnetic applications. Part II: Electrodynamics, *Comput. Methods Appl. Mech. Eng.*, 196(37–40), 3585–3597 (2007).

9. J. Pomplun, S. Burger, L. Zschiedrich, and F. Schmidt, Adaptive finite element method for simulation of optical nano structures, *Phys. Stat. Solidi (b)*, 244(10), 3419–3434 (2007).

10. J. P. Webb, Hierarchal vector basis functions of arbitrary order for triangular and tetrahedral finite elements, *IEEE Trans. Ant. Propag.*, 47(8), 1244–1253 (1999).

11. L. Zschiedrich, S. Burger, J. Pomplun, and F. Schmidt, Goal oriented adaptive finite element method for precise simulation of optical components, in W. Sidorin (Ed.), *Integrated Optics: Devices, Materials, and Technologies XI*, Proc. SPIE 6475, 6475–16 (2007).

Index

Milton Keynes UK
Ingram Content Group UK Ltd.
UKHW022037141024
449569UK00014B/635